Woven textiles

The Textile Institute and Woodhead Publishing

The Textile Institute is a unique organisation in textiles, clothing and footwear. Incorporated in England by a Royal Charter granted in 1925, the Institute has individual and corporate members in over 90 countries. The aim of the Institute is to facilitate learning, recognise achievement, reward excellence and disseminate information within the global textiles, clothing and footwear industries.

Historically, The Textile Institute has published books of interest to its members and the textile industry. To maintain this policy, the Institute has entered into partnership with Woodhead Publishing Limited to ensure that Institute members and the textile industry continue to have access to high calibre titles on textile science and technology.

Most Woodhead titles on textiles are now published in collaboration with The Textile Institute. Through this arrangement, the Institute provides an Editorial Board which advises Woodhead on appropriate titles for future publication and suggests possible editors and authors for these books. Each book published under this arrangement carries the Institute's logo.

Woodhead books published in collaboration with The Textile Institute are offered to Textile Institute members at a substantial discount. These books, together with those published by The Textile Institute that are still in print, are offered on the Woodhead web site at: www.woodheadpublishing.com. Textile Institute books still in print are also available directly from the Institute's web site at: www.textileinstitutebooks.com.

A list of Woodhead books on textiles science and technology, most of which have been published in collaboration with the Textile Institute, can be found towards the end of the contents pages.

Woodhead Publishing Series in Textiles: Number 125

Woven textiles

Principles, developments and applications

Edited by K. L. Gandhi

The Textile Institute

WOODHEAD PUBLISHING

Oxford Cambridge Philadelphia New Delhi

Published by Woodhead Publishing Limited in association with The Textile Institute
Woodhead Publishing Limited, 80 High Street, Sawston,
Cambridge CB22 3HJ, UK
www.woodheadpublishing.com
www.woodheadpublishingonline.com

Woodhead Publishing, 1518 Walnut Street, Suite 1100, Philadelphia,
PA 19102-3406, USA

Woodhead Publishing India Private Limited, G-2, Vardaan House, 7/28 Ansari Road,
Daryaganj, New Delhi – 110002, India
www.woodheadpublishingindia.com

First published 2012, Woodhead Publishing Limited

British Library Cataloguing in Publication Data
A catalogue record for this book is available from the British Library.

Library of Congress Control Number: 2012942831

ISBN 978-1-84569-930-7 (print)
ISBN 978-0-85709-558-9 (online)
ISSN 2042-0803 Woodhead Publishing Series in Textiles (print)
ISSN 2042-0811 Woodhead Publishing Series in Textiles (online)

The publisher's policy is to use permanent paper from mills that operate a sustainable forestry policy, and which has been manufactured from pulp which is processed using acid-free and elemental chlorine-free practices. Furthermore, the publisher ensures that the text paper and cover board used have met acceptable environmental accreditation standards.

Typeset by Replika Press Pvt Ltd, India
Printed by Lightning Source

Contents

Contributor contact details

(* = main contact)

Editor and Chapters 2, 3, 4 and 5

Dr Kim Gandhi
6 Greenleaf Close
Boothstown
Worsley
Manchester M28 1HR
UK

E-mail: kim.gandhi@virgin.net

Chapter 1

Professor Pramod Kumar Hari
1027 Sector A, Pocket B&C,
Vasant Kunj, New Delhi-110070
India

E-mail: haripramodkumar@gmail.com

Chapter 6

Jacquie Wilson
Manchester
UK

E-mail: jacquie.wilson@live.co.uk

Chapter 7

Dr Pammi Sinha
Department of Art, Design and Architecture
University of Huddersfield
Queensgate
Huddersfield HD1 3DH
UK

E-mail: p.sinha@hud.ac.uk

Visiting Research Fellow
Textiles and Paper
University of Manchester
Sackville Street
Manchester M13 9PL
UK

E-mail: pammi.sinha@manchester.ac.uk

Chapter 8

B. K. Behera
Department of Textile Technology
Indian Institute of Technology Delhi
Hauz Khas
New Delhi - 110016
India

E-mail: behera@textile.iitd.ac.in

Chapter 9

Dr M. Amirul Islam
Bally Ribbon Mills
Bally, PA 19503
USA

E-mail: AmirIslam@ballyribbon.com

Chapter 10

J. M. Hardcastle
Manchester
UK

E-mail: mikehardcastle@talktalk.net

Chapter 11

Nicola A. Redmore
School of Art, Design and Architecture
University of Huddersfield
Queens gate
Huddersfield HD1 3DH
UK

E-mail: n.a.redmore@hud.ac.uk

Chapter 12

Amit Rawal
Department of Textile Technology
Indian Institute of Technology Delhi
Hauz Khas
New Delhi – 110016
India

E-mail: arawal@textile.iitd.ac.in; amitrawal77@hotmail.com

Chapter 13

Dr Xiaogang Chen
School of Materials
University of Manchester
Sackville Street
Manchester M13 9PL
UK

E-mail: xiaogang.chen@manchester.ac.uk

Chapter 14

S. Rajendran* and S. C. Anand
Institute for Materials Research and Innovation
University of Bolton
Bolton BL3 5AB
UK

E-mail: sr2@bolton.ac.uk

Woodhead Publishing Series in Textiles

1 **Watson's textile design and colour Seventh edition**
Edited by Z. Grosicki

2 **Watson's advanced textile design**
Edited by Z. Grosicki

3 **Weaving Second edition**
P. R. Lord and M. H. Mohamed

4 **Handbook of textile fibres Vol 1: Natural fibres**
J. Gordon Cook

5 **Handbook of textile fibres Vol 2: Man-made fibres**
J. Gordon Cook

6 **Recycling textile and plastic waste**
Edited by A. R. Horrocks

7 **New fibers Second edition**
T. Hongu and G. O. Phillips

8 **Atlas of fibre fracture and damage to textiles Second edition**
J. W. S. Hearle, B. Lomas and W. D. Cooke

9 **Ecotextile '98**
Edited by A. R. Horrocks

10 **Physical testing of textiles**
B. P. Saville

11 **Geometric symmetry in patterns and tilings**
C. E. Horne

12 **Handbook of technical textiles**
Edited by A. R. Horrocks and S. C. Anand

13 **Textiles in automotive engineering**
W. Fung and J. M. Hardcastle

14 **Handbook of textile design**
J. Wilson

15 **High-performance fibres**
Edited by J. W. S. Hearle

16 **Knitting technology Third edition**
D. J. Spencer

17 **Medical textiles**
Edited by S. C. Anand

18 **Regenerated cellulose fibres**
Edited by C. Woodings

19 **Silk, mohair, cashmere and other luxury fibres**
Edited by R. R. Franck

20 **Smart fibres, fabrics and clothing**
Edited by X. M. Tao

21 **Yarn texturing technology**
J. W. S. Hearle, L. Hollick and D. K. Wilson

22 **Encyclopedia of textile finishing**
H-K. Rouette

23 **Coated and laminated textiles**
W. Fung

24 **Fancy yarns**
R. H. Gong and R. M. Wright

25 **Wool: Science and technology**
Edited by W. S. Simpson and G. Crawshaw

26 **Dictionary of textile finishing**
H-K. Rouette

27 **Environmental impact of textiles**
K. Slater

28 **Handbook of yarn production**
P. R. Lord

29 **Textile processing with enzymes**
Edited by A. Cavaco-Paulo and G. Gübitz

30 **The China and Hong Kong denim industry**
Y. Li, L. Yao and K. W. Yeung

31 **The World Trade Organization and international denim trading**
Y. Li, Y. Shen, L. Yao and E. Newton

32 **Chemical finishing of textiles**
W. D. Schindler and P. J. Hauser

33 **Clothing appearance and fit**
J. Fan, W. Yu and L. Hunter

34 **Handbook of fibre rope technology**
H. A. McKenna, J. W. S. Hearle and N. O'Hear

35 **Structure and mechanics of woven fabrics**
J. Hu

36 **Synthetic fibres: nylon, polyester, acrylic, polyolefin**
Edited by J. E. McIntyre

37 **Woollen and worsted woven fabric design**
E. G. Gilligan

38 **Analytical electrochemistry in textiles**
P. Westbroek, G. Priniotakis and P. Kiekens

39 **Bast and other plant fibres**
R. R. Franck

40 **Chemical testing of textiles**
Edited by Q. Fan

41 **Design and manufacture of textile composites**
Edited by A. C. Long

42 **Effect of mechanical and physical properties on fabric hand**
Edited by H. M. Behery

43 **New millennium fibers**
T. Hongu, M. Takigami and G. O. Phillips

44 **Textiles for protection**
Edited by R. A. Scott

45 **Textiles in sport**
Edited by R. Shishoo

46 **Wearable electronics and photonics**
Edited by X. M. Tao

47 **Biodegradable and sustainable fibres**
Edited by R. S. Blackburn

48 **Medical textiles and biomaterials for healthcare**
Edited by S. C. Anand, M. Miraftab, S. Rajendran and J. F. Kennedy

49 **Total colour management in textiles**
Edited by J. Xin

50 **Recycling in textiles**
Edited by Y. Wang

51 **Clothing biosensory engineering**
Y. Li and A. S. W. Wong

52 **Biomechanical engineering of textiles and clothing**
Edited by Y. Li and D. X-Q. Dai

53 **Digital printing of textiles**
Edited by H. Ujiie

54 **Intelligent textiles and clothing**
Edited by H. R. Mattila

55 **Innovation and technology of women's intimate apparel**
W. Yu, J. Fan, S. C. Harlock and S. P. Ng

56 **Thermal and moisture transport in fibrous materials**
Edited by N. Pan and P. Gibson

57 **Geosynthetics in civil engineering**
Edited by R. W. Sarsby

58 **Handbook of nonwovens**
Edited by S. Russell

59 **Cotton: Science and technology**
Edited by S. Gordon and Y-L. Hsieh

60 **Ecotextiles**
Edited by M. Miraftab and A. R. Horrocks

61 **Composite forming technologies**
Edited by A. C. Long

62 **Plasma technology for textiles**
Edited by R. Shishoo

63 **Smart textiles for medicine and healthcare**
Edited by L. Van Langenhove

64 **Sizing in clothing**
Edited by S. Ashdown

65 **Shape memory polymers and textiles**
J. Hu

66 **Environmental aspects of textile dyeing**
Edited by R. Christie

67 **Nanofibers and nanotechnology in textiles**
Edited by P. Brown and K. Stevens

68 **Physical properties of textile fibres Fourth edition**
W. E. Morton and J. W. S. Hearle

69 **Advances in apparel production**
Edited by C. Fairhurst

70 **Advances in fire retardant materials**
Edited by A. R. Horrocks and D. Price

71 **Polyesters and polyamides**
Edited by B. L. Deopura, R. Alagirusamy, M. Joshi and B. S. Gupta

72 **Advances in wool technology**
Edited by N. A. G. Johnson and I. Russell

73 **Military textiles**
Edited by E. Wilusz

74 **3D fibrous assemblies: Properties, applications and modelling of three-dimensional textile structures**
J. Hu

75 **Medical and healthcare textiles**
Edited by S. C. Anand, J. F. Kennedy, M. Miraftab and S. Rajendran

76 **Fabric testing**
Edited by J. Hu

77 **Biologically inspired textiles**
Edited by A. Abbott and M. Ellison

78 **Friction in textile materials**
Edited by B. S. Gupta

79 **Textile advances in the automotive industry**
Edited by R. Shishoo

80 **Structure and mechanics of textile fibre assemblies**
Edited by P. Schwartz

81 **Engineering textiles: Integrating the design and manufacture of textile products**
Edited by Y. E. El-Mogahzy

82 **Polyolefin fibres: Industrial and medical applications**
Edited by S. C. O. Ugbolue

83 **Smart clothes and wearable technology**
Edited by J. McCann and D. Bryson

84 **Identification of textile fibres**
Edited by M. Houck

85 **Advanced textiles for wound care**
Edited by S. Rajendran

86 **Fatigue failure of textile fibres**
Edited by M. Miraftab

87 **Advances in carpet technology**
Edited by K. Goswami

88 **Handbook of textile fibre structure Volume 1 and Volume 2**
Edited by S. J. Eichhorn, J. W. S. Hearle, M. Jaffe and T. Kikutani

89 **Advances in knitting technology**
Edited by K-F. Au

90 **Smart textile coatings and laminates**
Edited by W. C. Smith

91 **Handbook of tensile properties of textile and technical fibres**
Edited by A. R. Bunsell

92 **Interior textiles: Design and developments**
Edited by T. Rowe

93 **Textiles for cold weather apparel**
Edited by J. T. Williams

94 **Modelling and predicting textile behaviour**
Edited by X. Chen

95 **Textiles, polymers and composites for buildings**
Edited by G. Pohl

96 **Engineering apparel fabrics and garments**
J. Fan and L. Hunter

97 **Surface modification of textiles**
Edited by Q. Wei

98 **Sustainable textiles**
Edited by R. S. Blackburn

99 **Advances in yarn spinning technology**
Edited by C. A. Lawrence

100 **Handbook of medical textiles**
Edited by V. T. Bartels

101 **Technical textile yarns**
Edited by R. Alagirusamy and A. Das

102 **Applications of nonwovens in technical textiles**
Edited by R. A. Chapman

103 **Colour measurement: Principles, advances and industrial applications**
Edited by M. L. Gulrajani

104 **Fibrous and composite materials for civil engineering applications**
Edited by R. Fangueiro

105 **New product development in textiles: Innovation and production**
Edited by L. Horne

106 **Improving comfort in clothing**
Edited by G. Song

107 **Advances in textile biotechnology**
Edited by V. A. Nierstrasz and A. Cavaco-Paulo

108 **Textiles for hygiene and infection control**
Edited by B. McCarthy

109 **Nanofunctional textiles**
Edited by Y. Li

110 **Joining textiles: Principles and applications**
Edited by I. Jones and G. Stylios

111 **Soft computing in textile engineering**
Edited by A. Majumdar

112 **Textile design**
Edited by A. Briggs-Goode and K. Townsend

113 **Biotextiles as medical implants**
Edited by M. King and B. Gupta

114 **Textile thermal bioengineering**
Edited by Y. Li

115 **Woven textile structure**
B. K. Behera and P. K. Hari

116 **Handbook of textile and industrial dyeing. Volume 1: Principles, processes and types of dyes**
Edited by M. Clark

117 **Handbook of textile and industrial dyeing. Volume 2: Applications of dyes**
Edited by M. Clark

118 **Handbook of natural fibres. Volume 1: Types, properties and factors affecting breeding and cultivation**
Edited by R. Kozlowski

119 **Handbook of natural fibres. Volume 2: Processing and applications**
Edited by R. Kozlowski

120 **Functional textiles for improved performance, protection and health**
Edited by N. Pan and G. Sun

121 **Computer technology for textiles and apparel**
Edited by J. Hu

122 **Advances in military textiles and personal equipment**
Edited by E. Sparks

123 **Specialist yarn and fabric structures**
Edited by R. H. Gong

124 **Handbook of sustainable textile production**
M. I. Tobler-Rohr

125 **Woven textiles: Principles, developments and applications**
Edited by K. Gandhi

126 **Textiles and fashion: Materials design and technology**
Edited by R. Sinclair

127 **Industrial cutting of textile materials**
I. Viļumsone-Nemes

128 **Colour design: Theories and applications**
Edited by J. Best

129 **False twist textured yarns**
C. Atkinson

130 **Modelling, simulation and control of the dyeing process**
R. Shamey and X. Zhao

131 **Process control in textile manufacturing**
Edited by A. Majumdar, A. Das, R. Alagirusamy and V. K. Kothari

132 **Understanding and improving the durability of textiles**
Edited by P. A. Annis

133 **Smart textiles for protection**
Edited by R. Chapman

134 **Functional nanofibres and applications**
Edited by Q. Wei

135 **The global textile and clothing industry: Technological advances and future challenges**
Edited by R. Shishoo

136 **Simulation in textile technology: Theory and applications**
Edited by D. Veit

137 **Pattern cutting for clothing using CAD: How to use Lectra Modaris pattern cutting software**
M. Stott

Part I

Yarns and weaving technology

1
Types and properties of fibres and yarns used in weaving

P. K. HARI, Consultant, India

Abstract: This chapter reviews the types of fibres used in spinning, from natural to regenerated and synthetic fibres. It discusses key fibre properties such as length, linear density/fineness, cross-section and crimp. The chapter also reviews yarn properties such as linear density, diameter, packing and twist as well as fabric properties such as cover, mass, volume and thickness.

Key words: natural fibres, regenerated fibres, synthetic fibres, fibre properties, yarn properties, fabric properties.

1.1 Introduction

Woven fabrics consist of interlacements of yarns in two mutually perpendicular directions. Their structure permits a variation in the length and distribution of interlacements, resulting in a range of fabric deformations. Such control also offers scope for the generation of a diverse range of patterns and properties in fabrics. Triaxial fabrics are a special class of woven fabrics in which the warp threads constitute two sheets inclined at different angles, instead of being vertical. The weft thread is interlaced horizontally with the warp; this configuration thus has yarns in the fabric in three axes, hence the term 'triaxial' fabrics.

The building block of a woven fabric is yarn but the basic unit is fibre. Strength, extension, fineness, length and surface properties are some of the primary attributes of fibres which, using different types or blends of fibres, create a wide spectrum of useful and desirable physical and mechanical properties in the yarn and fabric produced. There are two basic types of fibres or filaments used in yarn:

- discontinuous fibres, i.e. short/staple fibres (including most natural fibres); and
- continuous fibres (such as silk and synthetic fibres).

Natural fibres have traditionally offered a variety of physical and mechanical properties. Using a natural, regenerated and synthetic filament yarn increases the range of physical and mechanical properties available and, indeed, allows functional properties to be tailor-made for a particular end use. Man-made filaments can be modified by texturing to alter bulk or stretch, or cut to a desired length to produce staple fibres which, either alone or blended with other fibres, are then blended as spun yarns. A knowledge of fibre properties enables a design engineer to select the proper materials to create fabrics for specific end uses.

1.2 Types of natural and regenerated fibres

Cotton, wool, flax and silk are the most important natural textile fibres, followed by fibres such as jute and kenaf. Natural fibres vary in length and in their other properties. They are not consistent and contain unwanted impurities from their original source. Such fibres need to be processed to remove these impurities and reduce the variation in their length. Such fibres can be made coherent, continuous and load bearing by providing sufficient twist to the strand.

The mechanical properties of silk fibres are a combination of high strength, extensibility and compressibility. Jute and kenaf are strong fibres, exhibiting brittle fracture and with a small extension at break; they have high initial modulus but show very little recoverable elasticity. Cotton is the most important natural cellulosic textile fibre. Length and fineness are important measures of fibre quality. Cotton fibres range in dimensions from a length of 5 cm to 1.5 cm and a linear density of 1 dtex to 3 dtex. Amongst its many important properties, cotton fibre is hydrophilic and porous; on immersion in water, it swells and its internal pores fill with the water. Chemical modification can add many desired properties to a cotton fabric, including colour, permanent press, flame resistance, soil release and antimicrobial properties.

Regenerated fibres are manufactured from natural polymers. Two of the most widely used regenerated fibres are rayon and acetate. There is a wide variety of different rayon fibres, including regular rayons, high-tenacity rayons, low wet modulus rayons, high wet modulus and modal rayons, high-strength/high-elongation rayons, polynosic rayons, flame-retardant rayons, high absorbency (alloy rayons), hollow rayons and cuprammonium rayons. High-tenacity rayons are used where strength, toughness and durability are required. These fibres are dimensionally stable when used as reinforcement in tyres, conveyor belts, drive belts and hoses. Other applications for this type of rayon include industrial sewing threads, tent fabrics and tarpaulins.

A range of cross-sectional shapes such as round, flat, Y-shaped, E-shaped, U-shaped, T-shaped and irregular (crenulated) shaped fibres are produced in rayon. It is generally accepted that the crenulated structure is formed by greater shrinkage of the fibre skin than of the core. Given its aesthetic properties, particularly its silk-like sheen, finer denier fibres are used for shawls, scarves, blouses and coat linings. Blends with cotton, polyester and flax help overcome inherent defects such as poor dimensional stability and mushy hand. Polynosic rayon fibres have high wet and dry strength, low elongation (8–11%), relatively low water retention and a very high wet modulus of 1 g/den. They have the firm and crisp hand of cotton in fabrics.

Cellulose acetate and triacetate textile fibres produce fabrics with an excellent hand, good dyeability, softness (comfort) and draping quality. The traditional market for cellulose acetate and triacetate fabrics is women's apparel. Acetate yarns are used for both tricot knit and woven constructions. Principal products within women's apparel include dresses, blouses and lingerie. Acetate has replaced rayon as a liner in men's suits and in non-apparel applications such as curtains. Cellulose triacetate offers the unique combination of 'ease-of-care' and aesthetic properties. A particularly important application of triacetate is in surface-finished fabrics such as fleece, velour and suede-like fabrics for dresses. Triacetate is also desirable for print fabrics as it produces bright sharp colours. Cellulose acetate and triacetate fibres are blended with nylon and polyester for numerous end-uses to compensate for the lower strength and durability of the weaker acetate and triacetate fibres. Yarns of this type, which are textured by air entanglement and bulking processes, have unique characteristics and aesthetics which permit their use in upholstery fabrics. Triacetate–nylon blended filament yarns are lightweight with good strength and bulk, and are useful in producing silk-type woven fabrics, such as plain taffetas, fine crepes and delicate jacquard figured styles for blouse and lingerie wear.

1.3 Types of synthetic fibres

An understanding of the chemistry and physics of natural textile fibres led to the creation of a range of man-made fibres such as viscose, rayon, nylon, polyester, acrylic and polypropylene. These synthetic fibres provide a range of specific properties, depending on purpose. Because they are manufactured, synthetic fibres can be produced consistently and are free of natural impurities. Synthetics have much to offer in terms of ease of care and durability but the problems of garment comfort, fabric cover, lustre, drape and garment tailorability have been major obstacles limiting the types of woven apparel fabrics made from 100% synthetic filament yarn. Instead, man-made/synthetic filaments are often cut into regular short lengths (staples) so that they are compatible with natural fibres and can be blended and spun into yarns with the right mix of properties for apparel applications.

1.3.1 Polyester

Polyester fibre, specifically polyethylene terephthalate (PET), is the most important synthetic fibre worldwide in terms of production volume[1] due to its low cost, the ease with which it can be processed and excellent performance. Amongst its properties, PET is a strong fibre (5 g/decitex). Polyester fibres need to be treated with surface finishes or lubricants to allow high-speed processing. The various processing steps such as drawing, bulking and spinning would be impossible without spin finishes which alter the frictional properties of the fibre.

Polyester technology is responsible for a large number of products that range from cotton-blended staple to high-performance tyre cord. Drawn filament yarn can be treated in a number of ways. It can simply be twisted on a ring frame, or bulked by a process such as false-twist texturing. Many apparel filament yarns need to be textured to give desirable aesthetic properties, particularly for women's wear markets. The principle of false-twist texturing is to create minor side-to-side variations across a filament yarn, causing the yarn to bend during controlled thermal shrinkage to create a 3D structure with a bulky feel. For industrial use, such as for tyre cord, high-tenacity yarns are drawn under conditions where low heat shrinkage, low extension and high modulus products are produced. Polyester staple fibre can also be crimped to blend with cotton, wool or other natural fibres at the carding stage to prepare appropriate yarn blends. PET staple blends with wool and cotton are highly successful but, in a low twist yarn, the fibre has many loose ends and pill formation can take place.

Synthetic fibres like PET and nylon are normally round in cross-section (whilst natural fibres have different cross-sections). These fibre shapes affect yarn and fabric aesthetics in terms of feel, drape, handle and appearance. Trilobal polyester yarns glitter because incident light reflects off the fibre surface. By contrast, octalobal yarns produce fabrics with an opaque matte effect, as light is effectively absorbed by multiple reflections off the many acute angles in the yarn. Sharp-edged filaments have the same prized rustle and high frictional characteristics of pure silk. Flat rectangular filaments give fabrics an unpleasant slimy handle. Synthetic fibres in general, and PET in particular, are hydrophobic materials. PET has a moisture regain of 0.4% at 65% RH and rapidly builds up static electrical charges by friction to give a very unpleasant electric shock. Static charges can also lead to the attraction of dust and dirt, and, to avoid these problems, polyester is blended with wash-fast hydrophilic materials.

1.3.2 Polyamides (including nylon)

Polyamide textile fibres include nylon 6,6 and nylon-6. Excellent mechanical properties, strength, fatigue resistance and good adhesion to other materials such as rubber are some of the reasons for the predominance of polyamide fibres in a wide range of applications. Nylon fibres, for example, offer a number of attributes such as high strength, abrasion resistance, lustre, chemical and oil resistance, washability, elasticity and dyeability in a wide range of colours. In general, most nylon fibre products require high strength and round cross-sections. However, for textile and carpet applications, fibres with modified cross-sections have been developed to achieve different aesthetics (such as lustre, opacity and better insulation).

Polyamides are used for women's hosiery, intimate apparel and stretch fabrics such as blouses, dresses, lingerie, ski apparel and swimwear. Polyamides have also been used for house furnishings such as bedspreads, carpets, curtains and upholstery. Other technical applications include truck and airplane tyres, seat belts, parachutes, ropes, fishing lines, nets, sleeping bags, tarpaulins, tents, sewing thread, ropes, monofilament fishing line and dental floss. High strength, toughness and abrasion resistance are also the main factors for selecting polyamide fibres for a wide range of military applications. In general, most nylon fibre products require high strength and round cross-sections. However, for textile and carpet applications, fibres with modified cross-sections have been developed to achieve different aesthetics (such as lustre, opacity and better insulation). In the area of general apparel, polyester has gained considerable significant market share at the expense of polyamides because of its easy-care characteristics.

1.3.3 Acrylic

Acrylics exist in a range of cross-sections, from almost perfectly round to dog-bone-shaped. Dry-spun acrylic fibres have a lower bending modulus due to their dog-bone shaped cross-section; as a result, a softer yarn with better fabric aesthetics can be obtained at a comparable fibre denier. The dog-bone-shaped cross-section prevents the fibres from packing closely to give a bulky yarn. Acrylics have a high resistance to sunlight, making continuous filament acrylics valuable in outdoor applications such as for tents and awnings.

In the general apparel markets, continuous filament acrylics and modacrylics find use as artificial fur; the fibres in these applications are made in a coarse denier with a special cross-section and surface modification to simulate natural animal hairs. Continuous filament yarn in very fine deniers is valued as a silk replacement for high-fashion dress fabrics, satins and poplins or to produce a cloth suitable for surface raising to give a suede or fine velour effect. Fine-denier filaments may also be used to produce fabrics with very high filament density and low air permeability, useful as insulating material, for example, in quilted interliners for parkas and outerwear for skiing. Other prominent uses for acrylics include rugs, curtains and upholstery.

1.3.4 Polypropylene

Some properties, such as toughness, low density and chemical stability are inherent to all polypropylene products. Others, such as colour stability,

UV and thermal stability result from the use of additives. The quality of polypropylene fibre has been improved to enhance properties such as light weight, mouldability, light stability, coloration, flame resistance and antistatic properties. Polypropylene is increasingly used for automotive interior furnishings whilst, for furniture, the current trend is to produce upholstery with a polypropylene tufted fabric of medium-denier using both air-textured and draw-textured yarn.

Woven fabrics from low denier polypropylene fibre (<1.5 dpf) are effective for thermal protection. They provide good aeration during perspiration and are waterproof. In apparel, their use can prevent skin dryness and body odour from perspiration, thus providing comfort and personal hygiene. Its major usage lies in winter wear, sportswear for hiking and skiing, work clothes and underwear. A smaller amount of this fibre is used in blends with nylon or acetate fibres for fashion wear. Polypropylene fibre itself does not ignite but shrinks, melts and pulls away on exposure to a flame, thus making it a more fire-retardant fibre than most synthetic fibres. A high-shrinkage fibre fabric processed at high temperature gives greater fibre density and bulk. There are two types of high-shrinkage polypropylene staple: cotton and wool types. Cotton staple is typically a 0.13 tex filament and 38 mm long. Wool staple is typically a 0.39 tex filament and 60 mm or 90 mm long. High-shrinkage polypropylene fibre is mixed with conventional fibres in spinning to yield a fluffy yarn.

1.4 Key fibre properties and how they are measured

To be suitable as a textile yarn, a fibre must have some essential and other desirable properties. The most important fibre characteristics[2] affecting fabric appearance, properties and performance include:

- length/linear density (fineness)
- cross-sectional shape
- surface characteristics
- crimp
- moisture content/regain
- shrinkage
- lustre/colour
- strength
- stiffness (flexibility)
- elastic recovery
- tenacity/elongation
- abrasion/chemical/flame resistance.

The moisture regain and tenacity/elongation of fibres are given in Tables 1.1 and 1.2. The following sections focus on fibre length/linear density, cross-section and crimp.

1.4.1 Fibre length and fineness/linear density

Length is a very important parameter for staple fibres affecting processability, as well as governing the physical and mechanical properties of the spun yarn itself. The lengths of staple fibres used in the production of spun yarns vary. Examples are:

- for polyester/viscose, 31–51 mm (for ring spinning) and 60–90 mm (for a worsted system)
- for cotton, 20–45 mm (for ring spinning).

Table 1.1 Moisture regain for fibres

Fibre	Moisture regain (%)
Wool	13.6
Silk	11.0
Rayon	11.0
Linen	8.75
Cotton yarn	7.0
Cellulose acetate	6.5
Nylon	4.5
Cellulose triacetate	3.5
Acrylic	1.5
Polyester	0.4

Table 1.2 Tenacity/elongation for fibres

	Tenacity (g/den)	Elongation (%)
Nylon (regular)	4.5–5.0	25–18
Nylon (high tenacity)	6.5–7.7	20–14
Nylon (staple)	3.8–4.5	37–25
Dacron polyester	4.4–5.0	22–18
Dacron polyester	3.0–3.9	40–25
Orlon acrylic	4.7–5.2	17–15
Orlon acrylic	2.0–2.5	45–20
Viscose (regular)	1.5–2.4	30–15
Viscose (high tenacity)	2.4–4.6	20–9
Acetate rayon	1.3–1.5	30–23
Wool	1.0–1.7	35–25
Cotton	2.0–5.0	7–3
Silk	2.2–4.6	25–10

Length variability is generally high for natural fibres (40% for cotton), but it should be as low as possible to improve yarn regularity. Longer fibres are advantageous since they require less twist for a desired yarn strength than shorter fibres. Low twist makes it possible to produce bulkier yarns.

The linear density, that is, the fineness of fibres, is expressed in decitex for natural fibres or denier for synthetic fibres. Examples include: cotton (1.4 decitex); silk (1 decitex); wool (1.7 decitex); and synthetic fibres (0.7–1.6 dpf). Many important properties of a yarn, such as packing and flexural rigidity, are influenced by fibre linear density. Finer fibres have a greater specific surface area which helps to improve the cohesion between fibres and gives better capillary action within the yarn structure. The total potential area of contact between fibres is governed by the specific surface area (surface area/unit volume) of fibres. The cohesion between fibres depends on the degree of inter-fibre contact generated by lateral pressure as a result of twist or wrapping. Torsional resistance of fibres also increases rapidly with increase in linear density. A coarser fibre is more difficult to twist and, once twisted, stores a higher level of torsional strain energy than finer fibres. This may be manifested in a yarn in the form of kinks and snarls in a relaxed state.

The uniformity of a yarn improves with a greater number of fibres in the yarn cross-section. For a given linear density of yarn, there will be greater number of finer fibres in the yarn cross-section. A uniform yarn automatically reduces the variation in other properties too. To spin a yarn successfully there should be a minimum number of fibres in the yarn cross-section (40–90 for ring spinning; 100 in rotor spinning).

Rigidity and cohesion contribute significantly to yarn strength. Strength is needed to withstand the spinning and weaving processes and to give strength to the fabric. Flexibility and fineness enable the fibres to be spun and woven to give the fabric its unique draping characteristics. Bending rigidity is proportional to the square of linear density of fibres. An increase in the linear density increases resistance to bending which makes the fibres stiffer. Fibres having a linear density of less than 0.33 tex are flexible enough to be used in apparel products. For industrial applications, the linear density ranges between 0.66 and 1.66 tex or more. Higher linear density is also used to improve the abrasion resistance of yarn and fabric.

1.4.2 Fibre cross-section

A fibre's cross-section affects fabric appearance, hand, drape, flexibility and moisture transport. It also affects cohesion, bending rigidity and the packing of fibres in a yarn. Cotton fibre has a U-shaped cross-section with a thin cuticle and a primary wall on the outside, and a hollow collapsed lumen inside. Jute fibre has a polygonal shaped cross-section. Synthetic fibres can be designed with any cross-sectional shape such as a circular, trilobal/multilobal, hollow shape, etc. The cohesion of circular cross-section fibres is generally higher than those with non-circular cross-section due to the greater surface area of contact between adjacent fibres. Cohesion depends on cross-sectional shape, fineness, and the flexural and torsional rigidity of fibres.

1.4.3 Fibre crimp

Natural fibres like cotton have crimp which facilitates processing and improves yarn bulk. Crimp (3.5–5.5 per cm) needs to be introduced in synthetic fibres to aid processing. Crimp geometry in staple fibre prevents fibres getting too close to each other and enhances yarn bulk. Fibre crimp also improves fibre cohesion.

1.5 Comparing fibre properties

The different properties of some common fibres are compared in Table 1.3. Polyester fibres have the best performance in the range of highly desired properties, followed by nylon, acrylics, cellulosics and wool. Acrylic fibres score more poorly than polyester or nylon but are superior to wool and the

Table 1.3 Comparative properties of common fibres, measured on a scale of 1–5, with 5 being the best

	Polyester	Nylon	Acrylic	Wool	Cellulosics
Category 1: Highly desired properties					
Abrasion resistance (durability)	4	5	3	3	3
Strength (durability)	4	5	3	2	3
Wash–wear performance	5	3	3	1	1
Wrinkle resistance	5	3	3	4	1
Pill resistance	3	1	3	3	5
Category 2: Useful properties					
Bulk (cover)	3	3	5	4	1
Water transport (HS comfort)	3	2	5	3	4
Static resistance	2	1	4	5	5
Speed of drying	5	5	5	1	1
Category 3: Properties for specific applications					
Resistance to burning	5	5	3	5	1
Resistance to degradation by dry heat	5	3	5	5	3
Resistance to degradation by wet heat	3	4	5	2	4
Resistance to degradation by sunlight	3	3	5	2	3
Resistance to insects	5	5	5	1	1
Resistance to micro-organisms	5	5	5	1	1

cellulosics. Acrylic fibres are poorer than polyester in terms of strength, abrasion resistance, wash–wear performance and wrinkle resistance. Acrylic fibres rank highest in the categories of useful and specific properties. Some of the properties in these two categories are considered more important for non-apparel markets, such as good resistance to sunlight degradation for drapery. In summary, acrylic fibres exhibit a well-rounded combination of performance properties with no major deficiencies. However, for woven apparel, the key properties of good wash–wear performance, abrasion and wrinkle resistance favour the use of polyester-cotton blends or cross-linked resin-finished cotton fibres.

1.6 New types of fibre

There are a number of emerging types of fibre, some of the most important of which are discussed below.

1.6.1 Polyester fibres based on terephthalic acid

Polybutylene terephlate (PBT) differs only slightly from PET in chemical composition but more substantially in its properties such as its higher crystallization rate and lower melting point. It can be melt-spun at lower temperatures and is inherently whiter than PET. As a fibre, it is much more elastic and has excellent resilience and recovery from small deformations. It dyes easily with disperse dyes at the boil, unlike PET, and resists photo-oxidative yellowing. Resilience and ease of dyeing are assets for carpet fibre.

Another polyester, polytrimethylene terephthalate (PTT), is a fibre with excellent recovery properties. Carpets made from PTT show excellent resilience due to the unique spring-like crystal structure of the fibre. It has a lower static charge and is resistant to stains. It is also used in intimate apparel and inner linings for clothing, and in the automotive and upholstery industries.

1.6.2 Microfibres

One of the most important developments in recent years has been in the technology to extrude extremely fine filaments of less than 1.0 denier, while maintaining the strength, uniformity and processing characteristics expected by textile manufacturers and consumers. Microfibres are fibres finer than 1.2 dtex for polyester and finer than 1.0 dtex for polyamide. Normal polyester filament yarn is around 3–5 dpf, whilst a typical polyester microfibre is only 0.5 dpf. Such fine fibres in the form of yarn give excellent textile properties, being very flexible, and giving a soft hand and excellent drape to fabrics.

Microfibres also lend themselves very well to fabric aesthetics. Dyed fabrics have solid bright colours due to the fine size of individual filaments. The vulnerability to damage from ironing is one disadvantage, the thermal capacity of the tiny filaments being so low that it is easy to overheat them. They can snag easily and need to be handled with care in processing.

The high density of filaments in typical microfibre fabrics makes them inherently windproof and waterproof. There are only tiny gaps for air to move through and the fabrics are largely unwettable since surface tension prevents water from penetrating the interstices in the fabric. Such fabrics are comfortable to wear as water vapour from perspiration evaporates easily. Microfibre fabrics also have heat-insulating properties. A 0.5 dpf polyester filament (density 1.4 g/cm^3) has a diameter of about 7 micrometers, right in the middle of the infrared (IR) wavelength (2–20 mm). Radiation is efficiently scattered by the microfibres and the radiation loss of body heat is reduced. As a result, microfibres are used in outdoor clothing and sportswear (e.g. skiing or mountaineering).

1.6.3 Bicomponent fibres

Bicomponent fibres are filaments made up of different polymers. There are many geometrical arrangements. The three main configurations are:

- side-by-side
- core–sheath
- multiple core (or 'islands in a sea').

So-called 'splittable pie' configurations are also used in the production of microfibres. Side-by-side bicomponent fibres can be used to produce self-bulking yarns. After drawing and relaxing, two PET polymers of different molecular weights (spun as a side-by-side heterofil) curl up like a bimetallic strip and give a spiral crimp. The core–sheath configuration is adaptable because many different polymers may be applied as a sheath over a solid polyester core, thus giving a variety of modified surface properties while maintaining all the major fibre and textile properties of PET. The multiple core or 'islands in a sea' type of heterofil is of interest in connection with microfibres. Hollow fibres are a type of core–sheath heterofil in which the core is composed of air. For end-uses like fabric interlinings, it is desirable to stabilize a filled hollow fibre structure by incorporating an additional thermally bondable bicomponent fibre.

1.6.4 Technical/high performance fibres

There are many kinds of technical textile products produced according to particular applications.[3,4] The variety of textile structures in technical textiles is thus much larger than it is in apparel textiles. There is a also a wider range of materials from which fibres can be manufactured. Examples are glass, ceramic and basalt filaments. Glass filament yarn can be textured, folded or cabled and finds application in aerospace, automotive, marine, civil, sporting goods and in the electrical/electronics industries. Ceramic filament is resistant to very high temperature (2000–3000°C) and finds application in thermocouples, furnace curtains/lining, seals and gaskets. Basalt filament finds application in high service temperature and high performance machinery, automobiles, ship building, aviation, cryogenic technology and on sea platforms.

Aramid and high performance polyethylene filaments are organic polymeric materials. Aramid filaments are used in bulletproof jackets, helmets, automotive, electrical/electronic and medical applications. These filaments are also used in ropes and cables, sail cloth and tennis racket strings. High performance polyethylene filaments are also used in ropes, nets and ballistic protection. Carbon is another useful quasi organic-based polymer. Its strength is its thermal stability and it has application in rockets, satellites, aircraft, fishing equipment, golf clubs, automobiles/bicycles, machinery parts and corrosion resistant equipment.

High-modulus, high-tenacity (HM-HT) fibres with a fully-extended well-oriented structure are produced in two contrasting ways. The first involves manipulation of rigid rod molecules into fibres that are already highly oriented in the as-spun state, as in the case of poly(phenylene terephthalamide) or Kevlar. The other method for manufacturing HM-HT fibres results from the ease with which flexible, inert chain molecules can be pulled out into a fully-extended, oriented form by imparting very high draw ratios. Ultra high molecular weight polyethylene (UHMWPE) fibres such as Spectra and Dyneema fall into this category.

When processed into fibres, oriented thermotropic liquid crystalline polymers (TLCP) such as Vectran (a wholly aromatic polyester) are characterized by very high specific tensile properties and intrinsically low density compared with metals, ceramics and carbon. However, they have inherent weaknesses in shear and compression. The key application areas for LCP fibres include hard armour (vehicles, helmets), soft ballistic protection (vests), cut protection (gloves) and a variety of composite uses that include honeycomb structures, pressure vessels and rubber reinforcement.

1.6.5 Biodegradable/medical fibres

Biodegradable polyesters comprise a diverse field, but the best developed fibre (monofil) market is in resorbable surgical sutures, which slowly disappear *in vivo* and do not need subsequent surgical removal. Biodegradable yarn/fibres are also used for disposable items such as wipes. These yarns are environmentally friendly and hydrolysable at temperatures up to 50°C.

Polylactic acid fibre (PLA) silk is used as an implantable suture and as scaffold for ligament tissue engineering, and is produced by electrospinning to 50–500 nm in diameter. Polycaprolactone (PCL) alginate is used to fabricate scaffolds for tissue engineering.

1.6.6 Electro-conductive fibres

Electro-conductive yarns are used as sensors, for data transfer, wireless communication and for electro-magnetic shielding in intelligent and smart textiles. These are core spun yarns in which the core is a conductive filament. Technical textile yarns such as reflective, luminescent, phosphorescent, photoluminescent and electroluminescent yarns find application in areas such as reflective clothing, road signs and markers. Such yarns are produced by lamination of a silver transfer film onto polyester.

1.6.7 Smart fibres

Fibres are flexible, mechanically strong one-dimensional structures that can be converted into a variety of two- and three-dimensional flexible structures such as ligaments (artificial muscles), woven fabrics, non-woven webs and membranes. Smart fibres, which can change shape in response to various stimuli from the environment, open up new applications in areas such as apparel which allows heat and moisture release in warm weather but retains heat at lower temperatures.[5]

1.7 Yarns and their properties

A textile yarn is a continuous linear assembly of fibres or filaments which exhibits suitable characteristics such as good specific tensile strength and high flexibility.[6] The various parameters used to characterize a yarn are given in Table 1.4. Some of these properties are discussed in more detail below.

1.7.1 Linear density

Linear density, yarn count, yarn number and yarn size are all expressions of the 'fineness' of a yarn. A yarn or fibre's fineness cannot be expressed in terms of diameter because its diameter is not stable and uniform along its length, and its cross-sectional shape may not be circular. Consequently, it is expressed either by measuring the mass of a known length of yarn or by measuring the length of a known mass of yarn. These two basic methods for expression of linear density of any textile strand are known as:

- the direct system (mass/length)
- the indirect system (length/mass)

Table 1.4 Yarn characterization

Parameter	
Dimensional parameters	Linear density Diameter No. of filaments in cross-section
Structural parameters	Twist per cm Wrap density Core content Core–sheath ratio
Special treatment	
Fibre parameters	Length Linear density Crimp Cross-section shape

These systems are discussed in more detail below.

In the direct system, the linear density of plied yarn is the simple summation of linear densities of the individual components, i.e. linear density $(R) = T_1 + T_2 + T_3 + \ldots + T_N$, where $T_1, T_2 \ldots T_N$ are the linear densities of n individual components expressed in tex. The tex system has units measured in grams (g) per 1,000 metres (m). As an example, 20 tex means that the mass of 1,000 m of yarn is 20 g. The higher the value, the coarser the yarn. Prefixes such as kilo and deci are used when the numbers become too large or too small to express fineness:

- 1 Ktex = 1,000 tex
- 1 decitex = 1/10 tex
- 1 militex = 1/1,000 tex

Another expression of fineness for continuous filament yarn is the denier, which is expressed as the mass in grams (g) per 9,000 metres (m). As an example, a 3 denier yarn means that the mass of 9,000 m of yarn is 3 g. In the jute system, mass is expressed in pounds (lbs) per 14,400 yards (yds); a 12-pound yarn count indicates that 14,400 yds of yarn has a mass of 12 lbs.

In the indirect system, fineness is expressed as the ratio of the number of standard lengths per unit mass. A higher designated number indicates a finer yarn. Thus, in the English system, an 840 yard length (Hank) per pound refers to the fineness of the yarn. A '30's' yarn indicates that a 1 pound mass will be 30 times 840 yards in length, that is 25,200 yard of yarn. In the metric system, the number of units of 1,000 m is determined in 1 kg of yarn. In the indirect system the relationship is: $1/R = 1/T_1 + 1/T_2 + \ldots + 1/T_k$, where $T_1, T_2 \ldots T_k$ are the linear densities of k individual components. Units of linear density are given in Table 1.5.

The number of filaments in a yarn is generally specified in the case of a

Table 1.5 Units of linear density

	System	Mass	Length	Unit	Symbol
Direct	tex	g	1,000 m	g/1,000 m	tex
	denier	g	9,000 m	g/9,000 m	den
	jute	lb	14,400 yd	lb/14,400 yd	lb
Indirect	English	lb	840 yd	No. of 840 yd/lb	N_e
	metric	kg	km	No. of km/kg	N_m

filament yarn. It can also be found by taking the ratio of the yarn to filament linear density:

$$n = n_t/(n_f/9) \qquad [1.1]$$

where n = number of filaments/fibres, n_t = yarn linear density (tex), and n_f = fibre linear density (den). A filament yarn designated as 130/24, refers to 130 denier with 24 filaments in the cross-section.

1.7.2 Yarn diameter

It is sometimes necessary to know the approximate diameter of a yarn, even though it is well known that yarn diameter cannot be estimated accurately due to the highly compressible nature of the material. Diameter, however, gives an idea of the extent of cover achievable, or the closeness of yarn packing in a fabric. Fabric thickness also depends on yarn diameter. Yarn diameter in terms of linear density in tex for a general case is given as:

$$d = \frac{1}{280.2} = \sqrt{\frac{T}{\phi\rho_f}} \qquad [1.2]$$

where d = yarn diameter (cm), T = yarn linear density (tex), ρ_f = fibre density (g/cm^3), and ϕ = yarn packing coefficient.

1.7.3 Yarn packing

Packing of fibres in a yarn is an important structural parameter which affects the porosity and bulk (specific volume) of a yarn. It influences properties such as compressibility and water/air transport, etc. Compressibility influences flattening of yarn in a woven or knitted structure. Porosity affects the flow of liquid due to capillary action, whilst yarn bulk affects thickness. Packing depends on the physical/mechanical properties of the fibres, their arrangement in the yarn structure and the level of twist. The extent of packing of fibres in a yarn's cross-section is expressed by a packing coefficient. Some typical values for different yarn types are given in Table 1.6. The packing coefficient is defined as:

Table 1.6 Packing coefficient values for various yarn

Type of yarn	Packing coefficient
Monofilament	1.00
Tape	1.00
Multifilament	
Untwisted	0.25
Regularly twisted	0.60
Hard twisted	0.90
Staple yarn	
Soft twisted	0.33
Hard twisted	0.60
Ring	0.50–0.60
Rotor	0.35–0.55
Air jet	0.50–0.70
Friction	0.30–0.55
Wrap	0.40–0.70

$$\text{Packing coefficient} = \frac{\text{specific volume of fibre}}{\text{specific volume of yarn}} = \frac{\text{density of yarn}}{\text{density of fibre}}$$

For blended yarns, average fibre density is given by the following:

$$\frac{1}{\bar{\rho}} = \sum_{1}^{n} \frac{p_i}{p_{fi}} \qquad [1.3]$$

where $\bar{\rho}$ = average fibre density, p_i = mass fraction of the *i*th component, p_{fi} = fibre density of the *i*th component, and n = number of components of the blend.

1.7.4 Yarn twist

Twist is quantified by the number of turns present in a unit length of yarn. Hence, a 20 tpi means 20 turns per inch length of yarn. The unit length can be inches, centimetres or metres. The basic purpose of twist is to induce structural integrity or coherence in a staple spun yarn so that it makes the yarn strong enough to withstand various kinds of stress. In the case of multifilament yarns, the twist is primarily introduced to bind the filaments together so that it facilitates processing by preventing a tendency for filament fraying. As well as processability and strength, twist also influences many other properties of a yarn. Hence, depending upon the end use, an optimum twist level is selected. The two most important aspects of twist are its direction and level (that is, intensity).

The direction of surface fibres/filaments looking up the yarn determines the direction of twist. If the inclination of the surface fibres matches the centre portion of the letter 'S', it is termed an 'S' twist; the reverse is true for 'Z' twist. Most spun yarn has a Z twist. In a single yarn, twist direction does not influence the property of the yarn. If a plied structure is made from single yarns, the ply twist direction is generally opposite to the direction of the single yarn twist. This gives a balanced (torque free) structure when an optimum level of ply twist is chosen with respect to the single twist. However, ply twist in the same direction as a single yarn is introduced for special end uses. Plied yarns are usually designated by a number, such as 30/4, meaning that four groups of 30 single yarns have been plied together to form a single twisted structure. The specific volume of a yarn is mainly a function of twist, though fineness, crimp, flexural rigidity and length of staple fibres do have some influence. Some typical values of specific volume of different types of yarn are given in Table 1.7.

1.8 Types of yarn for spinning

There are two main types of yarn: filament and spun yarns. Filament yarns in the medium to fine count range are used in fabrics for apparel. Filament yarns in the very fine count range are more appropriate for hosiery and also for carpets, sportswear and as industrial yarns for technical textiles. Spun yarns are suitable for shirts, trousers, suits, home textiles and bed linen.

1.8.1 Filament yarns

With the exception of natural silk, continuous filament yarns are man-made. They can be sourced from natural polymers such as paper pulp to produce viscose, rayon, etc., or from synthetic polymers such as polyester, polypropylene, polyethylene, polyamide, etc. A filament yarn can be either monofilament or multifilament. There are also flat yarns.

Monofilament yarns can be solid or hollow. Monofilament yarns have a single filament with a cross-section which can vary depending on the end

Table 1.7 Specific volume of different types of yarn

Yarn	Specific volume (cm^3/g)
Spun	
Cotton	1.10
Nylon	1.45
Polyester	1.30
Filament	
Nylon	1.35
Polyester	1.16

use. It is normally circular but can be altered to a range of other shapes (e.g., triangular, multilobal, serrated, oval or dogbone-shaped). A non-circular cross-section prevents close packing and helps wicking. Monofilament diameters are typically in the range of 0.1–2 mm. The filament yarn is chopped to a suitable length to produce staple fibres, and these staple fibres are spun alone or blended with natural fibres in to yarn.

Continuous filaments are invariably used as a multi-strand and termed a multifilament yarn. A multifilament yarn consists of a bunch of monofilaments with a nominal twist to provide coherence to the structure. Alternatively, fibres in multifilaments can be held together by intermingling or entanglement. A non-circular cross-section prevents the close packing of multifilament yarns.

Flat filament yarns can be textured to increase bulk or extension by false twisting, intermingled by an air-jet or crimped by a stuffer box. Flat yarns can be twisted or interlaced to produce different effects. Tape yarn is produced by splitting up a thin ribbon-like narrow width of polymer film. Tape yarns are used, for example, for making fabric for packaging applications. A tape yarn can be further split or fibrillated mechanically to produce a regular network of interconnected fibres, a texture similar to multifilament yarn.

1.8.2 Spun yarns

Yarns made by using short fibres (natural fibres such as cotton, wool, jute, coir, etc.), staple fibres (rayon, polyester/nylon/acrylic/polypropylene) or a blend of different fibres are designated as spun yarns. Short staple spun yarns are a linear assembly of short discontinuous fibres. Such an array of fibres cannot sustain any load unless coherence is provided by generating transverse forces. This is achieved by twist or fibre/filament wrapping. Twisted yarns are hairy due to the projecting fibre ends but the yarn surface typically shows fibres in a helical configuration. Twisting in a spun yarn is achieved by conventional ring spinning, or by more unconventional methods using rotor spinning, friction spinning, air-jet spinning and hollow spindle wrap spinning.[7] These yarns have different structures and properties.

A yarn made using a mixture of two or more different fibres is called a blended spun yarn. The proportion of different components in the yarn depends on the desired physical and mechanical properties. The most common blends are polyester-cotton, polyester-viscose and cotton-viscose, usually in proportions of 67:33 or 50:50. A yarn designated as PC (67:33) is a 67% polyester and 33% cotton fibre blended spun yarn.

A spun yarn can either be single, or two or more of such yarns can be twisted to achieve certain desirable effects, i.e. to create a plied or folded yarn. These effects can be to increase strength or reduce hairiness, or to create colour effects by twisting single yarns of different colours. Plied/

folded yarns are an assembly of several individual yarns helically wrapped around each other to form a coherent structure. Several plied yarns are twisted (generally in the opposite direction to the twist in the plied yarn) to form a cable yarn.

1.9 Short staple spinning yarns

Short staple spinning uses 25–50 mm fibres during yarn formation by employing the following techniques.

1.9.1 Ring spun yarns

The structure of ring spun yarns gives them visual, tactile, aesthetic and elastic fabric properties suitable for a wide range of end uses. The fibres which form the body of yarn lie parallel along the Z/S twist helix (usually 'Z' twisted). The yarn can be carded or combed, with the latter having lower hairiness, a higher uniformity and strength due to longer and more parallel fibres. In compact ring spinning, the single ribbon of fibres is compacted before twisting to further reduce hairiness and increase yarn strength. These yarns can be spun from most natural or man-made fibres, or from a blend of short staple fibres in the 25–50 mm range, and to a wide count range (5–300 tex). Two ribbons of fibres separated by a small distance can be ply twisted to give a siro spun yarn, usually in the 20–100 tex (2×10 to 2×50) count range. These yarns are less hairy and more uniform and are equivalent to ply yarns.

1.9.2 Rotor spun yarns

Rotor spun yarns are bulky two-layered yarn structures. The core comprises a bulk of parallel fibres aligned with the Z twist helix, with an outer layer of wrapper fibres along the core length. The fibres in a rotor yarn are less packed than in a ring yarn. Rotor yarn is known to be 5–10% bulkier than a ring yarn. They can be spun to a moderate range from coarse to medium (25–100 tex) yarn counts.

1.9.3 Friction spun yarns

Friction spun yarns have a core sheath element. The surface of this yarn is similar in appearance to a ring spun yarn. Outside fibres in these yarns have a uniform surface appearance along a Z twist helix. The core fibres are disposed in parallel and do not contain any twist. Friction spun yarn has a high extensibility due to the large proportion of hooked and disoriented fibres in the sheath and the low level of tension at which the yarn is spun. Friction spun yarns are the weakest yarns due to the presence of highly disoriented and folded sheath fibres with a low degree of packing. The yarns are spun in a coarse count range (over 40 tex).

1.9.4 Air-jet spun yarns

Air-jet spun yarns have a two-layered yarn structure. The core fibres have no twist and are wrapped with fibres as in a rotor spun yarn. Air-jet yarn is less extensible due to the untwisted, straight and parallel arrangement of core fibres constituting 90% of the yarn. The bending rigidity of an air-jet yarn is high due to the clustering effect of core fibres tightly wound by binding fibres which offers very little freedom of movement. Air-jet spun yarn's abrasion resistance is small, causing disintegration of the yarn once the wrapper fibres are broken. These yarns are spun in the fine count range (7–15 tex).

1.10 Long staple spinning yarns

Long staple spinning uses 51–215 mm fibres during yarn formation using the following techniques.

1.10.1 Wrap spun yarns

Wrap spun yarns consist of two component yarns, with a core of parallel, untwisted staple fibres and a binder wrap of either a filament yarn or staple fibres on the outside. The wrap provides cohesion to the staple fibre core by radial pressures along the wrap helix. The yarn has low hairiness. Surface fibre wrapping is carried out by hollow spindle wrap spinning. The yarn has a bulky and crimped appearance though under tension it is lean and smooth. These yarns are spun in the coarse to medium count range (higher than 16 tex).

1.10.2 Self-twist spun yarns

In self-twist spun yarns, two fibrous strands are separately false twisted by alternating an S and Z twist along their length. They are then plied together by an untwisting torque of single to alternating Z and S twists to create a two-fold yarn. These yarns are spun in the medium to coarse count range (20–100 tex).

1.10.3 Twistless spun yarns

Twistless spun yarn is formed by felting or adhesive bonding of fibres. An entanglement process consolidates an assembly of fibres in the form of a felted yarn. This type of yarn is well suited for carpets. For technical textile applications, staple fibres are bonded to a core filament (glass, polypropylene, nylon or polyester) with a polypropylene resin binder.

1.10.4 Core spun yarns

Core spun yarn is a sheath–core yarn formed by twisting fibres as a sheath around an existing core of either a filament or staple spun yarn. These are two-layered structures, the core filament yarn normally providing strength and durability. Elasticated core yarn is used for stretch comfort. These yarns can be produced by ring spinning, friction spinning, wrap spinning and self-twist spinning.

1.10.5 Ply yarns

Ply yarns are produced by twisting two or more yarns. Doubling or plying helps to achieve special physical characteristics by exploiting the direction and level of the twist in plying. Twisting in a direction opposite to the single yarn enables fibres to orient along the plied yarn axis. This makes the plied yarn smoother, more uniform, lustrous, stronger, more extensible and less hairy. Twist lively plied yarn is produced by twisting in the direction of the single yarn.

1.10.6 Fancy yarns

Fancy yarns are made with a distinctive irregular profile which is different from basic single and folded yarns. This profile enhances the visual and textural properties of the fabric. Fancy yarns are also termed novelty yarns, specialty yarns, effect yarns and fashion-oriented yarns, and can incorporate deliberate 'defects' or variations distributed randomly or in a pre-defined order, along with colour to give visually and texturally attractive difference to the fabric. Colour effects along the yarn length are introduced by twisting different coloured yarns, spinning blended dyed fibres, or by irregularly printing/dyeing spun yarns.

Fancy yarns are invariably used to create fashion fabrics. They are used in ladies outerwear, mens jackets, knitwear, ties, furnishings, curtains and upholstery. Typical fancy yarns include snarl, spiral, loop, chenille, slub and nep. Fancy effects can also be created by changing fibre specifications such as type, length, denier, cross-sectional shape, crimp, initial modulus, colour, lustre, affinity for dyes and shrinkage, etc. A deliberate difference in length, denier, cross-section and modulus of component fibres can be exploited to position them at predetermined zones of the yarn cross-section, with short and coarse fibres occupying the yarn surface. For increased lustre, fibres with a flat cross-section, such as trilobal and ribbon-shaped yarns, are often employed. Slubs of varying dimensions can be produced by using differential fibre length in the core and effect components.

Blending of regular polyester fibres with cationic dyeable polyester can be used to get attractive colours, contrast and fancy effects. Similarly, a blend of fibres having varying shrinkage when combined with colour difference can also give fancy effects. Manufacture of fancy yarns often involves plying and cabling together different yarns of varying counts and twists. The density (tpi) and direction of twist (S/Z) are often manipulated during plying and cabling so as to create particular effects. Differences in individual yarn fineness is employed to make corkscrew and spiral yarns. If the direction of twist in the component yarns is kept different, the plying operation will produce an unbalanced twisted spiral yarn.

One of the most popular and well-known fancy yarns is slub yarn which has been used in various types of apparel such as suits and shirts. Corkscrew, spiral, chenille, gimp, loop, snarl and boucle have all found acceptance in apparels and for the production of furnishings and upholstery. Cotton melange yarn has been highly successful in enhancing the appeal of knitted garments such as T-shirts, etc.

1.11 Physical properties of woven fabrics

Fabric cover, mass, specific volume and thickness are important physical properties and can be determined based on yarn count, thread spacing and yarn crimp.[8]

1.11.1 Fabric cover

In fabric, 'cover' is measured as the fraction of total fabric area covered by the component yarns. For a circular cross-section yarn, the cover factor is given as:

$$\frac{d}{p} = \frac{E\sqrt{T}}{280.2\sqrt{\phi\rho_f}} = \frac{K}{28.02\sqrt{\phi\rho_f}} \qquad [1.4]$$

$$K = E\sqrt{T} \times 10^{-1} \qquad [1.5]$$

where T is yarn tex, d is yarn diameter, E is the number of threads per cm $= 1/p$, K is the cover factor, and ϕ is the packing coefficient.

For $d/p = 1$, cover factor is maximum and given by:

$$K_{max} = 28.02\sqrt{\phi\rho_f} \qquad [1.6]$$

Fractional fabric cover is given by:

$$\frac{d_1}{p_1} + \frac{d_2}{p_2} - \frac{d_1 d_2}{p_1 p_2} = \frac{1}{28.02}\left[K_1 + K_2 - \frac{K_1 K_2}{28.02}\right]$$

Multiplying by 28.02 we get a fabric cover factor:

$$\text{Fabric cover factor} = K_1 + K_2 - \frac{K_1 K_2}{28} \qquad [1.7]$$

1.11.2 Fabric mass

Fabric mass (expressed as grams per square meter (gsm)) is a very commonly used parameter, and is given[9] by:

$$\text{gsm} = [T_1 E_1 (1 + c_1) + T_2 E_2 (1 + c_2)] \times 10^{-1} \qquad [1.8]$$

$$\text{gsm} = \sqrt{T_1}\ [(1 + c_1)\ K_1 + (1 + c_2)\ K_2\ \beta] \qquad [1.9]$$

where E_1, E_2 are ends and picks per cm; T_1, T_2 are warp and weft yarn tex; K_1 and K_2 are the warp and weft cover factors, c is the fractional crimp and $d_2/d_1 = \beta$. In practice, comparison between different fabrics is usually made in terms of gsm. A fabric engineer tries to optimize the fabric parameters for a given gsm.

1.11.3 Specific volume

The apparent specific volume of fabric, v_F, is calculated using the following formula:

$$v_F = \frac{\text{fabric thickness (cm)}}{\text{fabric mass (g/cm}^2\text{)}} \qquad [1.10]$$

where fabric mass (g/cm^2) = 10^{-4} × gsm
and

$$\text{Fabric packing factor, } \phi = \frac{v_f}{v_F} \qquad [1.11]$$

where v_f and v_F are, respectively, fibre and fabric specific volume. A knowledge of fabric specific volume helps in calculating the packing of fibres in the fabric. Such studies are useful in evaluating fabric properties such as warmth, permeability to air or liquid.

1.11.4 Thickness

Fabric thickness for a circular yarn cross-section is given by: $h_1 + d_1$ or $h_2 + d_2$, whichever is greater:

$$h_1 = \frac{4}{3} p_2 \sqrt{c_1} \qquad [1.12]$$

$$h_2 = \frac{4}{3} p_1 \sqrt{c_2} \qquad [1.13]$$

When the two threads project equally, then $h_1 + d_1 = h_2 + d_2$ and, in that case, the fabric's *minimum thickness* is given by:

$$\text{Minimum thickness} = 1/2(h_1 + d_1 + h_2 + d_2) = D;\ h_1 = D - d_1$$

Such a fabric produces a smooth surface and ensures uniform abrasive wear. In a fabric with a coarse thread in one direction and a fine thread in the other, maximum crimp is obtained for the coarse thread by stretching the fine thread straight. In this case the fabric's *maximum thickness* is given by:

$$\text{Maximum thickness} = D + d_{\text{coarse}},\ \text{since } h_{\text{coarse}} = D$$

1.11.5 Packing of yarns in fabric

The maximum number[9] of warp and weft yarns in a fabric depends on the weave and yarn count. For plain weave, the maximum number of threads which can be packed in a fabric depends on the following equation:

$$\sqrt{1 - \left(\frac{p_1}{D}\right)^2} + \sqrt{1 - \left(\frac{p_2}{D}\right)^2} = 1 \qquad [1.14]$$

where p_1 and p_2 are the spacing between the neighbouring warp and weft yarns, respectively, and D is the sum of the yarn diameters.

1.12 Mechanical properties of woven fabrics

Woven fabrics are subjected to a combination of in-plane extension/tension, shear, compression, bending in a perpendicular plane and out-of-plane buckling deformations.[10] The tensile properties of woven fabrics depend on the yarn and the fabric structure, that is the yarn twist, yarn strength, weave, crimp and cover. The degree of yarn twist affects fabric elongation. A reduction in twist improves cover and reduces air permeability.

An increase in picks/cm and twist density within a normal limit significantly improves both tensile strength and surface durability, but a reduction in pick density improves the tear strength in the filling direction. The breaking strength of a textile fabric is enhanced by an increase in the strength of its component yarns and the number of yarns per unit width. Fabrics made from ring spun yarns give a higher breaking load than those made from open-end yarns, though the effect of scouring and bleaching in increasing the breaking load is more pronounced with the former.

Cloth strength is affected by the degree of binding of cross threads which aid inter-fibre frictional force and contribute to tensile strength. The degree of binding is dependent on the density of weave and cloth setting. The elongation at break increases with the increase in these variables. The tensile strength of a fabric decreases with an increase in fabric cover. Crimp decreases the strength of the yarn and fabric. The initial modulus of the fabric decreases with an increase in crimp. The straightening of longitudinal threads during tension applies compressive forces at the points of contact with the cross-yarn. If the cross-yarn jams before crimp exchange is complete, then a reduction in fabric strength and extension occurs.

The buckling, bending and drape characteristics of woven fabric influence its performance during the making-up process and in the actual use of the garment. These properties are important, particularly when the fabric is limp, resulting in large-scale deformation even when a small force is applied. The engineering concept of buckling helps to explain the behaviour of fabric during garment manufacturing as well as during its use as a garment. The folding of a garment, bending of sleeves, bending of trouser legs, cutting and sewing operations, etc., all involve buckling phenomena. In bending, a woven fabric exhibits initial non-linearity due to frictional restraint, followed by the linear deformation-recovery behaviour of the material. The puckering tendency of a cloth during stitching shows similar behaviour as does formability, a measure of the ability of a fabric to absorb compression in its own plane without buckling. The formability value is used as a direct indicator of the likelihood of seam pucker occurring during the sewing of a garment.

Fabrics are subjected to bending deformations in most uses. Bending is a very common phenomenon during the use of garment fabrics. Additionally, it has some influence on the drape of upholstery, furnishings and curtain fabrics. Creasing and wrinkling of fabrics and garments is also associated with bending deformation. Bending resistance is made up of the bending resistance of threads lying in the direction of bending, frictional restraint and interaction between the threads. The four main factors directly influencing minimum bending rigidity G_{min} are fibre tensile modulus, fibre linear density, yarn linear density, and the number of threads per unit cloth width.[11]

Inter-fibre adhesion can increase elastic fabric stiffness considerably. Textile yarns, being visco-elastic, exhibit time-dependent effects such as stress relaxation. This causes decay in the inter-yarn force at the cross-over point and results in setting. In addition, fabrics can be set by finishing treatments to relax internal stresses in the fabric and lower its internal energy level.

1.13 Effects of fibre and yarn properties on the use and application of woven fabrics

The physical and mechanical properties of a fibre, as the unit cell of a textile material, clearly contribute to a fabric's mechanical properties. However, according to the spinning system used, the yarn structure and properties can be varied considerably to alter packing, surface and mechanical properties. The fabric itself imposes another set of variables, such as sett and weave, to further diversify the overall list of fabric properties.

The use of textiles for apparel, home furnishing and industrial/technical applications depends on a unique combination of required properties. Fabrics for apparel, for example, require comfort, hand, drape, breathability, easy care, durability/abrasion, aesthetics, dyeability and, of course, tailorability. The requirements of fabrics for home textiles (such as curtains and upholstery, for example) are drape, aesthetics/colour/design and resistance to stain/soiling and degradation from light. Fabrics for industrial and technical applications require specific characteristics which usually include strength and durability. This section reviews some of these properties and how they are influenced by fibre and yarn properties.

In general, spun yarns of either single or blended fibres are the major building materials in woven fabrics. Continuous filament yarns are used either as textured or as core spun yarns. There needs to be maximum packing of yarn in fabric for several applications. The properties of a textile which govern its conformation to a technical specification are the function of the yarn network (interlacement, diameter, spacing), thickness and mass. A large group of fabric properties are more dependent on pore structure than on yarn network, for example, resistance to the flow of liquids, gas and light.

Woven fabrics are made from interlaced yarns which are flexible. The yarns are made by twisting together long, thin, flexible and strong fibres. The properties of yarn and fabric depend largely on the properties of the constituent fibres. Textiles are low-density materials and contain a large proportion of air within their total structure and thus the characteristic textile properties arise from two basic fabric elements: fibre and air. Another set of attributes rely on interchange in yarn and interstices geometry: tear resistance, fabric drape and sewability. Fabric strength and extension is a function of crimp and yarn twist. It is very important to have an in-depth knowledge of the principles involved in the selection of raw materials and their conversion to desired yarns and fabric structures, followed by dyeing and finishing. Additionally, knowledge of polymers and coating helps when designing technical fabrics for different applications.

1.13.1 Creasing and wrinkling properties

In practical use, textile fabrics are subjected to a wide range of complex deformations such as bending, folding, creasing and wrinkling which may be added deliberately during manufacturing or produced by body movement during use. Of these, wrinkling and creasing can affect the aesthetic characteristic of a fabric. The ability of a fabric to resist creasing or wrinkling and to recover from such deformations is determined by many factors, the most important being the mechanical properties of its fibres, such as their tensile strength, bending and torsion, the geometrical and constructional parameters of the yarn and fabric, and the presence or absence of finishing agents on or within the fibre.

Resistance and recovery from creasing is largely influenced by the type of fibre, yarn and fabric structure. Bending elasticity plays an important role in the creasing phenomenon of a fabric. Among the commonly used textile fibres, linen exhibits creases easily due to a very high degree of orientation of cellulose in the fibre. Cellulosic materials are highly susceptible to creasing. The materials which resist creasing also resist deformation and therefore they behave as a rigid material. Ideally a fabric which is deformed should recover from its deformation rapidly. Wool is well known for its very good elastic recovery and therefore has the ability to resist and recover from creasing.

1.13.2 Shear properties

Shear behaviour is one of the most important characteristics contributing to the performance and appearance of woven fabrics. Shear properties influence drapeability, pliability and hand behaviour of apparel fabrics. The study of shear behaviour of fabrics has special significance, particularly for mass-produced apparel. This is mainly because cloth is deformed in all directions during handling on machines, as well as in actual use. In fact, the ability of a fabric to be deformed by shearing distinguishes it from other flat flexible materials such as paper and polymer thin films. Shear angle is one of the main criteria for characterizing the formability of fabrics. This property enables a 2D fabric to undergo complex deformations and helps a 3D garment to conform to body contours, imparting the desired mechanical comfort.

1.13.3 Compression properties

Compression is defined as a decrease in intrinsic thickness with an increase in pressure – intrinsic thickness being the thickness of the space occupied by a fabric subjected to barely perceptible pressure. Compression of fabrics is an inevitable phenomenon during fabric handling and processing. Compressive properties are concerned with the surface smoothness, softness and fullness of a fabric. The compressive properties of textile materials have direct relevance to bulk and handle of fabrics. In a woven construction, warp and weft yarns are compressed due to inter-yarn pressure. The softness of a fabric is determined by its compressibility which depends on the compressibility of the yarn and the fabric construction. Compressibility of fabric is an integral component of fabric hand.

Tailorability of cloth significantly depends on fabric compressional behaviour. Compressibility of fabrics significantly influences drapeability.

When a fabric is draped on the edges of a contour, there is compressional deformation at the point of bending. Fabrics with higher compressional energy give a higher drape coefficient. A highly compressible fabric has high compressional energy and can absorb or withstand compressive forces to a greater extent at the deforming points. This prevents folding at the deforming points and a higher drape coefficient results.

1.13.4 Abrasion properties

Abrasion is a kind of wear in which rubbing away of component fibres and yarns of the fabric takes place. During abrasion, a series of repeated applications of stress takes place. Therefore abrasion resistance is the capacity to absorb energy. The abrasion resistance of a fabric is dependent upon many factors including fibre properties, yarn structure and size, fabric mass and geometry as well as fabric finish (including lubricating and sizing agents used). For hydrophilic fibres, the amount of moisture present also affects abrasion resistance.

1.13.5 Fabric comfort

The subjective feeling of fabric comfort is a complex result of psychological and physiological responses of the human body and the physical properties of the fabric. Comfort[12] has two attributes: mechanical (evaluated by visual and tactile means) and thermal (assessed by the permeability of a fabric to air, water and heat). Comfort is evaluated in part by fabric hand – a traditionally subjective evaluation made by judging the feel of a fabric between the fingers. The fabric hand is expressed by an index – total hand value (THV) – based on a combination of mechanical properties derived from tensile strength, shear, bending, compression, surface friction and the surface deformation recovery of fabric at low stress. These fabric properties are also useful for tailorability and formability, that is, the forming and sewing of flat fabric pieces into 3D-shaped garments. Thermal comfort is evaluated in terms of air permeability, thermal insulation capacity, water repellency and water vapour transmission.

Fabric characteristics such as fibre type, yarn linear density, thread (yarn) density, weave type and yarn twist affect fabric hand, since these constructional parameters determine fabric surface characteristics, which significantly contributes to the perceived touch of a fabric. Surface properties also help to judge the smoothness and crispness of a fabric. Geometrical roughness is another important factor affecting perceived softness of a fabric. For a given fabric construction, the denier, cross-sectional shape and crimp of the fibres influence the softness of the final product. Very fine filaments are effective for good draping, anti-crease properties and softness of handle, but these yarns are more susceptible to abrasion than coarse-filament yarns. The desired softness of handle can also be achieved through modified fibre cross-section.

1.14 Effects of fibre and yarn properties on woven textiles: apparel and sports textiles

In the apparel market, polyester in staple or filament form has a clear advantage because of its easy-care qualities and low cost. Although 100% polyester has limited appeal because of its less desirable handle, it is extremely popular in blends with cotton. Filament yarns are used for outdoor clothing applications or where a fur-like or silky appearance is desired. Fibres with enhanced or modified properties are in great demand. Yarn bulk is enhanced by using bicomponent or biconstituent fibres. Pilling is reduced by producing fibres that are more brittle, whilst a yarn with exceptionally soft handle is produced by using fine-denier fibres or fibres treated with special friction-reducing finishes. Abrasion resistance is enhanced by modifying the spinning conditions to produce a denser filament microstructure.

Sports textiles are textiles used in sports goods and in sportswear. Sportswear is clothing, including footwear, which is worn for sport or exercise. It includes shorts, tracksuits, T-shirts, polo shirts and trainers. Sportswear is also often worn as casual fashion clothing. For most sports, athletes wear a combination of different items of clothing. These should have protective properties against variable atmospheric conditions during use as well as for protection against physical damage. In addition, they require a high tear strength, resistance to abrasion, shape stability and colour fastness. Almost every piece of sports clothing is designed to be lightweight so the athlete is not encumbered by its weight. The best athletic wear for some forms of exercise, for example cycling, should not create drag or be too bulky. On the other hand, it should be loose enough so as not to restrict movement. Various physically dangerous sports require protective gear, e.g. for fencing, American football or ice hockey.

Garments should allow the wearer to stay cool in hot weather and warm in cold weather. In cold climates the best athletic wear should not only provide warmth but also transfer sweat away from the skin. For activities such as skiing and mountain climbing, this is achieved by using layering. Moisture transferring materials must be worn next to the skin, followed by an insulating layer and wind and water resistant shell garments. Waterproof fabric prevents the penetration and absorption of liquid water. The term 'breathable' implies that the fabric is actively ventilated. Breathable fabrics passively allow water vapour to diffuse through them yet prevent the penetration of liquid water. Water-resistant and moisture-permeable materials may be divided into three main categories: high density fabrics, resin-coated materials and film-laminated materials. These are selected according to the finished garment requirements in casual, athletics, ski or outdoor apparel.

Densely woven waterproof breathable fabrics consist of cotton or synthetic micro-filament yarns with compacted weave structure. One of the most famous waterproof breathable fabrics, known as Ventile, was manufactured by using long staple cotton with minimal space between the fibres. Usually combed yarns are woven parallel to each other with no pores for water to penetrate, and Oxford weave is preferred. When the fabric surface is wetted, the cotton fibres swell transversely, reducing the size of pores in the fabric and requiring very high pressure to cause penetration. Therefore, waterproofing is provided without the application of any water repellent finishing treatment.

Densely woven fabrics can also be produced from microdenier synthetic filament yarns. The individual filaments in these yarns are less than 10 micrometers in diameter, so that fabrics with very small pores can be engineered. The use of superfine or microfibre yarns makes the production of dense fabrics possible enabling capillary action which gives the best wicking properties. The wicking behaviour of a fabric is dependent mainly on the base fibre's moisture properties. While most fabrics, both natural and synthetic, have the ability to wick moisture away from the skin, not all of them are fast-drying and air permeable (two factors which have a direct influence on cooling and perceived comfort). High-tech synthetic fabrics are lightweight and capable of transporting moisture efficiently and dry relatively quickly. It is generally agreed that fabrics with moisture wicking properties can regulate body temperature, improve muscle performance and delay exhaustion. While natural fibres such as cotton may be suitable for clothing worn for low levels of activity, synthetic fabrics made of nylon or polyester are better suited for high levels of activity. They absorb much less water than cotton, but can still wick moisture rapidly through the fabric. No single fibre or blend of different fibres provides an ideal sportswear fabric: the right type of fibre should be in the right place. A two-layer structure is better, with a synthetic fibre layer close to the skin with high wicking properties (e.g. a microdenier polyester, which is ideal for wicking perspiration away from the skin) and an outer layer usually of cotton or rayon which absorbs and evaporates.

1.15 Future trends

Electronic-textile (e-textile) technology holds out the promise of truly wearable computers as well as inexpensive large-scale computational devices. New fibres are being created for inclusion in e-textiles, including battery, conductive and mechanically active fibres. Methods are being developed to attach discrete components to e-textiles, including processors, microphones and speakers. In the category of wearable computing, a garment providing the user with precise location information within a building is being developed.

Virtually all woven e-textiles are expected to use the new fibre batteries and solar cells that have been developed.

Conductive woven fabrics are being made by utilizing metallic conducting wires as flexible wearable electrical circuits; the properties of these nanotube-based yarns can be leveraged to produce more robust systems far lighter and more flexible than metallic wires. As a step in the direction towards the creation of nanotube-based electronic components for fabric applications, super-capacitors have also been woven into a textile fabric. The capacitors have capacitances and life cycles comparable to commercially available super-capacitors. Since nanotubes are also known to have effects similar to piezoelectric materials, such materials can be integrated into structures to produce sensors that monitor stresses. Chemo-, piezo- and thermo-resistive fibres are now available in textile compatible form and their integration into fabrics and garments is being actively pursued.

1.16 Sources of further information and advice

The best source of information is the internet. The following list of links to fibre producers, machinery manufacturers and technical textiles producers gives plenty of innovative information, of interest to both academics and industry:

- http://www.ibena.de/techtex/english/fabric/fabric.html
- http://www.innovationintextiles.com/
- http://ec.europa.eu/enterprise/sectors/textiles/research-innovation/technical-textiles/
- http://www.technicaltextilesguide.com/
- http://www.huntsman.com/textile_effects/eng/Technical_Textiles/Technical_textiles/index.cfm?PageID=629
- http://www.technical-textiles.net/htm/j20050119.557671.htm
- http://www.aramid.com/
- http://www.highbeam.com/doc/1P3-618128741.html
- http://www.ascendnif.com/
- http://www.universalfibres.net/industrialFibres.php
- http://www.ifai.com/
- http://www.sagezander.com/
- http://www.torayusa.com/tomac/products/synthetic.htm

1.17 References

1. McIntyre J E (2004) *Synthetic Fibres: Nylon, polyester, acrylic, polyolefin*, Woodhead Publishing, Cambridge.
2. Gordon C (1984) *Handbook of Textile Fibres*, Woodhead Publishing, Cambridge.

3. Alagirusamy R and Das A (2010) *Technical Textile Yarns: Industrial and medical applications*, Woodhead Publishing, Cambridge.
4. Horrocks A R and Anand S C (2000) *Handbook of Technical Textiles*, Woodhead Publishing, Cambridge.
5. Jassal M and Agrawal A K (2011) 'Functional smart textiles using stimuli-sensitive polymers' in *Functional Textiles for Improved Performance, Protection and Health* (Pan N and Sun G) Woodhead Publishing, Cambridge.
6. Lawrence C A (2002) *Fundamentals of Spun Yarn Technology*, CRC Press, Boca Raton, FL.
7. Lünenschloss J and Brockmanns K J (1985) *Textil Praxis International*, February.
8. Peirce F T (1937) *J. Text. Inst.*, 28, T45–112.
9. Weiner L (1971) *Textile Fabric Design Tables*, Technomic, Lancaster, PA.
10. Behera B K and Hari P K (2010) *Woven Textile Structure: Theory and applications*, Woodhead Publishing, Cambridge.
11. Owen J D (1968) *J. Text. Inst.*, 59, 313.
12. Kawabata S, Niva M and Yamashita Y (1999) *Int. J Clothing Sci & Tech.*, 11, 2/3, 134.

2
Yarn preparation for weaving: winding

K. L. GANDHI, Chartered Consultant, UK

Abstract: Yarn preparation has always been key to the efficient running of the subsequent processes. This fact even becomes even more obvious with the ever increasing production rates in weaving and knitting. With this in view, the chapter begins with the objects and economic importance of the first preparatory process – winding. Methods of clearing the yarns faults are illustrated. Different types of winding machine for spun and filament yarns and the structure of the packages produced have been critically reviewed. The chapter includes characteristics of the manual, semi-automatic and fully automatic winding machines. It finally discusses in detail the hazards posed by knots during weaving and then describes the different methods of producing knot-free yarns (splicing techniques), including their application to variety of yarns.

Key words: yarn clearers, drum winding, precision winding, splicing, angle of wind, wind ratio.

2.1 Introduction to yarn preparation

When a fabric is to be woven on any loom, whether it is a hand loom or a power loom, two sets of yarns are required. One set of yarns, called warp yarn, is used in the lengthwise direction and the other set of yarn, called weft yarn, is used in the horizontal direction. The warp yarns are put on a package called a weaver's beam which is put at the back of the loom as shown in Fig. 2.1(a). The number and length of warp yarn to be put on the weaver's beam depend on the quality of the cloth, i.e. number of warp yarns/cm, also called ends/cm, width of the fabric and the length of fabric to be produced from the weaver's beam. Figure 2.1(b) shows the front of the looms as the fabric is produced.[1]

In order to produce a weaver's beam with a few hundred to a few thousand number of warp yarns and their length (m) on the beam, the spun yarns on small packages produced during the spinning process undergo the following yarn preparation processes for weaving purposes:

- winding
- warping
- sizing.

During the spinning process, as the yarn is produced, it is generally wound on a package called a spinner's package. The length of yarn put on this package can be from a few hundred to a few thousand metres depending

(a)

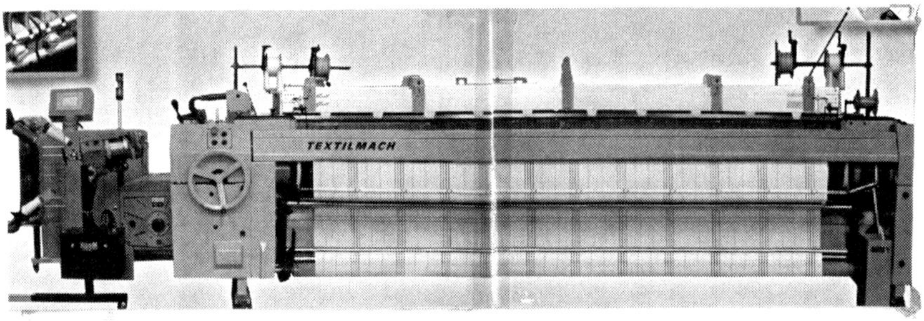

(b)

2.1 (a) Back and (b) front of loom.

on the length and diameter of the spinning package as well as the count, i.e. linear density of the yarn. These small packages produced on the ring-spinning machine are mostly unsuitable for the efficient running of the next processes, i.e. warping, twisting and weft winding (pirn winding), making a winding process necessary.

The sizing process is optional and is carried out mostly for single spun cotton yarns. Woollen and worsted warp yarns are normally produced as two-fold yarns (two yarns twisted together). As such, they do not require a sizing process, in which case only winding and warping preparatory processes are carried out.

2.2 The winding process

Winding is a process of transferring yarn from one package to another or from a number of small packages (spinner's packages, spools or hanks) to a big package (Fig 2.2),[2] i.e. a cone or cheese that is more suitable for the next processes such as warping or knitting or for any other suitable use such as on shuttle-less looms.

Winding is also important in removing defects in the yarn. During the spinning process of traditionally spun staple yarns, for a variety of factors, some defects or imperfections, such as a thick place, thin place, slub or nep, are induced into the yarn. The length and the thickness of these defects can vary significantly (Fig. 2.3).[3] If not removed from the yarn which is going to be used as warp or weft during weaving or knitting, these defects will get woven/knitted and produce a defective fabric. These defects or irregularities from the yarn are therefore removed during the winding process (also called 'yarn clearing'). Not every defect is removed from the yarn. The degree of

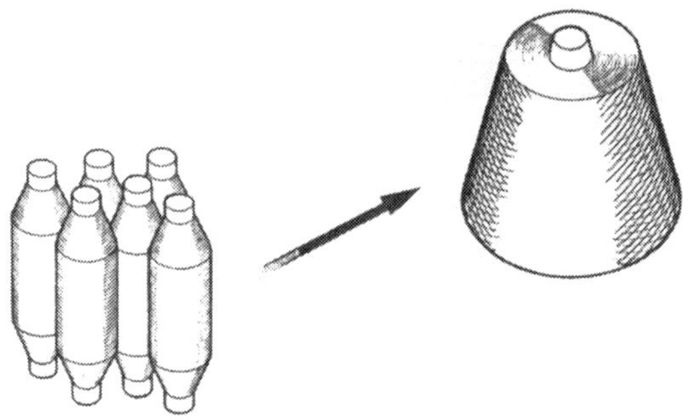

2.2 Yarn from spinner's packages onto cones.

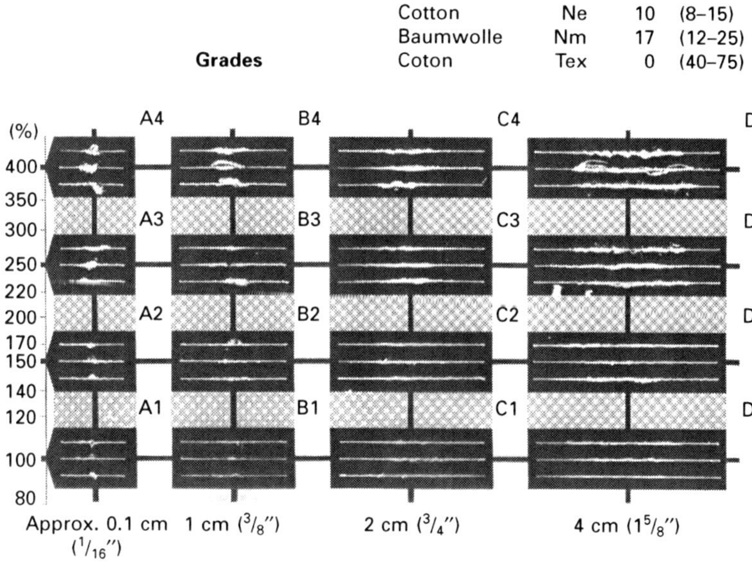

2.3 Defects in yarns of different length and thickness.

clearing of such defects is dependent on what the yarn is going to be used for.

When spun, filament yarns are not produced by twisting of fibres during their production. As a result no yarn irregularities, i.e. thick or thin places, are induced during the spinning process. The winding process for such yarns is carried out to produce bigger packages, if these are required, by transferring yarn from a number of small packages onto big packages suitable for the next process.

2.2.1 Clearing yarn defects during winding

Thick places/slubs in the yarn having a diameter bigger than the normal diameter of yarn are removed either mechanically by a device called a slub catcher or electronically using optical or capacitance sensors.

Mechanical clearers

When cleared mechanically, the yarn is made to pass through a small slit with a definite opening depending on the degree of clearing. Thick places[4,5] and slubs larger than the opening of the slit are arrested, cut and replaced by a knot or a splice (Fig. 2.4(a),(b)). Thin places are usually removed by applying tension to the yarn during the winding process. On a traditional manual winding machine, the yarn is made to pass between the two discs (Fig. 2.4(b) and (c)).[5,6] Suitable weights are added on the top disc. As the yarn passes between the discs it is subjected to tension and the thin places, having a breaking strength lower than the tension applied, are broken and replaced by either a knot or splice.

All modern automatic winding machines are equipped with very efficient tension devices. Oerlikon Schlafhorst offer on their Autoconer 5 a very precise and unique tension control: Autotense *FX* (Fig. 2.5 (a)).[7] The yarn tension in **cN** and its variations are permanently recorded by a sensor and adjusted to the desired level by the electromagnetic tensioner. The yarn tension is measured online and adjusted to the required level in a closed control loop.

The Savio Machine offers a unique tension device called TensorFlex (Fig. 2.5(b))[8] on their Automatic winder Orion E/I. The tension sensor (Tensor) continuously detects actual winding tension since it is positioned immediately before the drum. The Tensor interacts, through the head computer, with the yarn tension to vary the pressure exerted on the yarn as required. The operating range of the yarn tensioner can be set on the machine's computer. When the yarn tensioner reaches the minimum value of this range, it can, if so required, activate the preset speed reduction curve. The Tensor has no moving parts that could influence operation and therefore provides correct reading and correction of tension.

Another electronic yarn tension device (Fig. 2.5(c)),[6] called TENS CONTROL™, with an 'on-line' tension sensor is available in the market. With this device it is possible to constantly control and regulate the yarn tension from the beginning to the end of the package formation, independently from the tension variations coming from the feeder bobbin. On the computerized digital terminal in the drive end, it is possible to set the desired yarn winding tension, which is in turn controlled by a special on-line electronic tension sensor located on each single winding spindle, which again operates a motorized tension device. With this continuously regulating system, the tension device sets in at the slightest variation, keeping yarn tension within the set parameters.

(a)

Key
1 – Slub catcher
2 – Waxing unit
3 – Disc tensioner

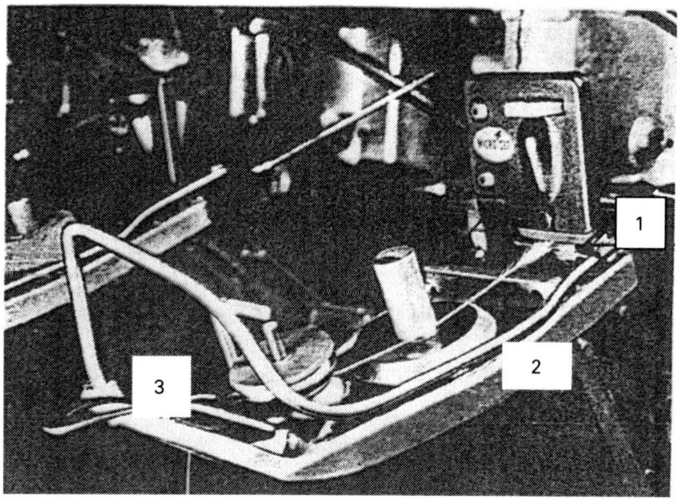

(b)

2.4 (a) Mechanical slub catcher, (b) slub catcher and (c) disc tensioning and waxing.

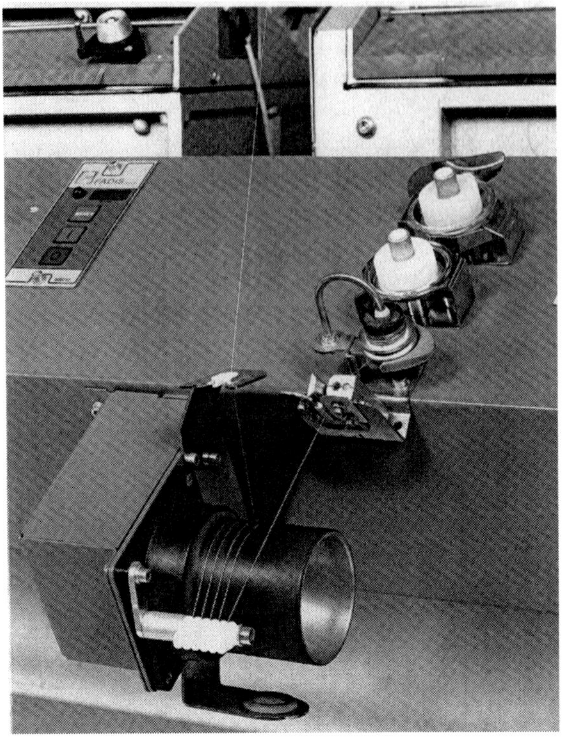

(c)

2.4 Continued

Electronic clearers

In the capacitance clearer, the yarn passes through an air-spaced condenser. The change in the capacitance is related to the mass of the material between the plates. If the mass changes to values beyond pre-set limits, the cutter operates, thus cutting the yarn. It is the mass of the material that is measured and not the physical dimensions, i.e. width or diameter of the defect.[9]

In the photo-electric or optical electronic clearer, the device measures the diameter or profile of the yarn. The yarn is scanned by a light source. The scanned portion interrupts the beam, which thus changes the signal from the photo-electric cell. If limits are exceeded, the cutter is brought into operation. Almost all modern winding machines are equipped with electronic clearers. Fig. 2.6 shows yarn clearer units on Savio winding machines.[8]

Knitting yarns are subjected to lot of friction during the knitting process and, to reduce it, yarn is usually waxed during the winding process. Figure 2.4(c) shows a modern tension and waxing device[6] equipped with large diameter feed roller to eliminate the tension variations and two waxing units.

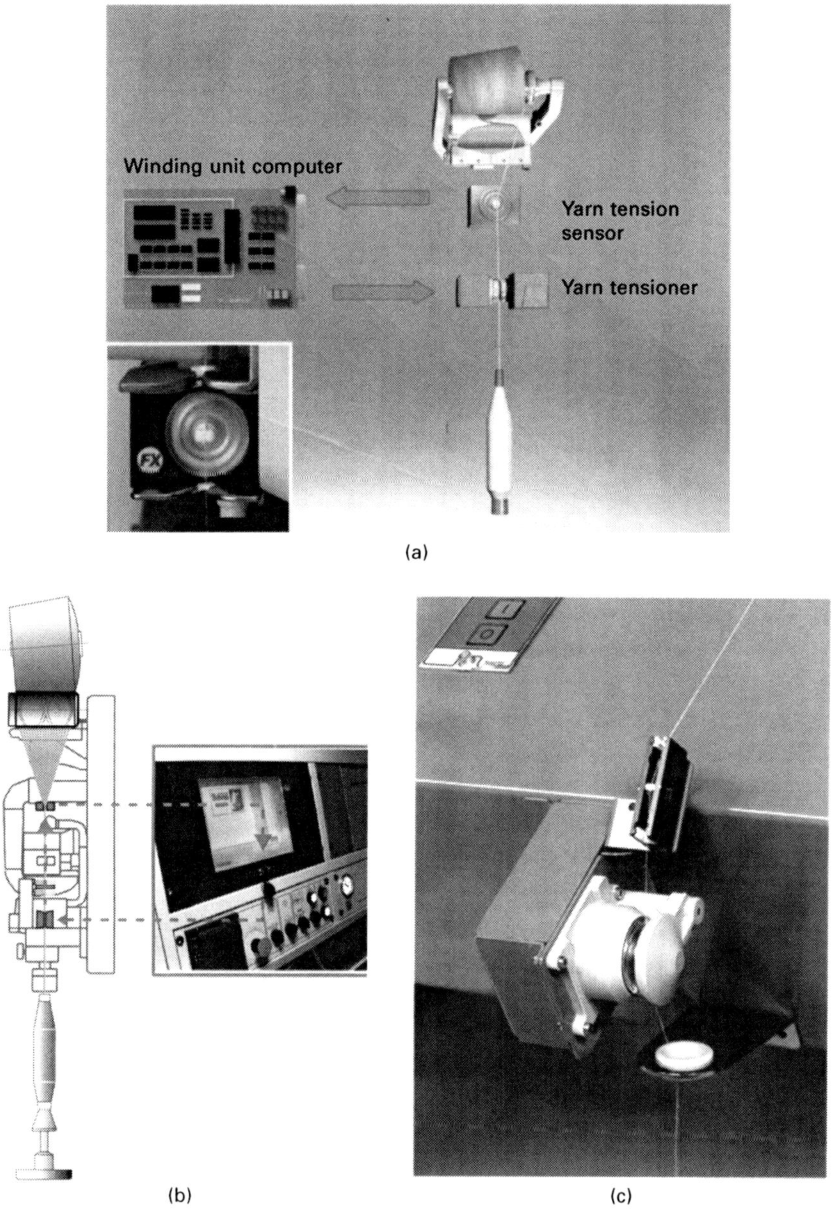

2.5 (a) Autotenser, (b) TensorFlex and (c) tension device TENS CONTROL™.

The winding process is carried out not only for transferring the yarn from a number of small spinning packages onto bigger packages such as cones and cheeses. It is also carried out, when required, for transferring yarn from bigger packages such as cones onto small ones, i.e. pirns or weft bobbins for

2.6 Electronic clearers.

use during weaving as weft yarn. In this case the defects from the yarn have generally already been removed when preparing the cones on the winding machine. In some cases, the yarn is wound from bigger packages produced on an open end spinning machine onto weft packages (pirns) without clearing the defects. Equally, it is common to produce weft bobbins on the spinning frame which are taken straight for weaving.

2.3 Types of winding machines

The following types of winding machines are commonly available:

- Drum winding machines: these produce packages called cones, cheeses, etc.
- Hank winding machines: these unwind hanks, skeins, etc., produced on reeling machines. Figure 2.7(a) shows a hank to cone winding machine SINCRO with electronic yarn guide, a precision crossing angle, a positively driven bobbin and unwinding speed of up to 1000m/min.[6]
- Assembly winding machines: these assemble yarns from two, three or more packages onto one larger package. Figure 2.7(b) shows an assembly winder model SINCRO B from Fadis, Italy.[6] The machine is equipped with a three-ply creel with an individual yarn sensor, TENS CONTROL™, an on-line tension device, yarn cutter and automatic

doffing. Figure 2.7(c) shows an Econoflex residual package winder[10] for winding expensive residual yarn left on packages such as cones and cheeses in the warping section. The model is completely automatic and

(a)

(b)

2.7 (a) Hank to cone winding, (b) assembly winder and (c) Econoflex residual package.

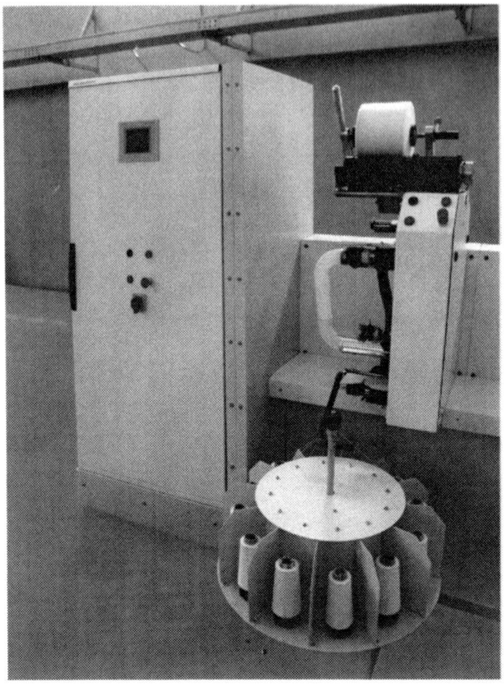

(c)

2.7 Continued

changes the feeding package automatically each time the yarn breaks or ends, reducing the amount of hard waste.

2.3.1 Winding mechanisms

Irrespective of the type of winding machine, two mechanisms essentially are involved in the transferring of yarn from one package onto the desired one. These are:

- rotational movement to the package onto which the yarn is wound
- guiding of the yarn on the package being rotated.

This mechanism is called traversing of the yarn from one side of the package to the other and back. This process continues all the time. Every type of winding machine named above has a rotary and a traversing mechanism.

2.3.2 Cone and cheese winding machines

There are two types of winding machines to wind the yarn in cone or cheese form. These are:

- drum winding machines
- precision winding machines.

2.3.3 Drum winding machines

On a drum winding machine, the rotary movement to the package, whether it is a cone or a cheese, is provided by the frictional contact between the package and the revolving drum surface. The traversing mechanism is provided by the grooves cut into the drum (Fig. 2.8(a)).[5] When the package is brought in contact with the drum, the yarn enters the grooves (Fig. 2.8(b))[8]

(a)

(b)

2.8 (a) and (b) grooved drums and (c) drum winder with cam traverse.

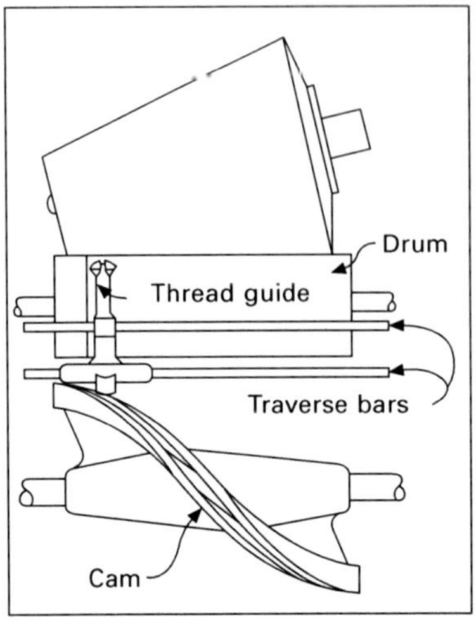

(c)

2.8 Continued

and is guided from one side to the other of the package and back again to continue the process. As the yarn lies in and travels with the groove, moving at approximately the same speed, it is not subject to abrasion likely to cause injury and its properties are unimpaired.

Since the surface of the package is driven with the surface of the drum which has a constant speed, the yarn speed on such machines is always constant. However, the revolutions of the package decrease as the package diameter increases. This kind of winding, normally classed as random winding, is carried out for most staple-spun yarns and is not suitable for continuous filament yarns as the friction between the package and the drum tends to damage the yarn surface. When the yarn breaks as a result of removal of the defect or any other reason, i.e. when the supply package runs out, the package is lifted off the drum and the winding process stops.

The characteristics of random winding are as follows.

- As the package diameter increases, the angle of wind remains the same but wind per double traverse decreases.
- Yarn speed throughout the build of the package remains constant.
- Minimum tension variation occurs during unwinding of the package in subsequent processes.
- The process is capable of producing large diameter packages.
- If dyeing takes place, dye liquid penetration is not very ideal.

2.3.4 Drum winding machines with cam traversing mechanism

Figure 2.8(c) shows the traverse to the yarn achieved with the help of a grooved cam[5] which is mounted on the cam shaft rotating at a definite speed. The grooved cam has a small pin positioned into it and to which is attached a thread guide. When the cam shaft rotates, the pin is guided in the grooved path of the cam, which transfers the motions to the thread guide from one side to the other and back again, thus traversing the yarns on the package.

There exist important relationships between the traverse and the winding speed.[11] The greater the number of coils of yarn in the width of the traverse, the higher the winding speed possible. It is usual, therefore, to operate the traverse mechanism at the maximum speed, to keep this speed constant, and to alter the number of coils or winds by varying the speed of the driving drum, thus altering the ratio between the traverse and the drum speed. The characteristics of the yarn, the type of yarn package, and the subsequent purposes for which the package is required are factors which determine the most satisfactory ratio and, in conjunction with the maximum traverse speed, determine the production of which the machine is capable.

2.3.5 Precision winding machines

On these machines the package is inserted onto a spindle which is rotated at a uniform speed measured in revolutions per minute. When the spindle rotates, the package rotates and yarn is wound on it. The traverse to the yarn is provided by an independent mechanism and usually involves a special designed cam with a groove mounted on a shaft (Fig. 2.9).[5] A small pin with

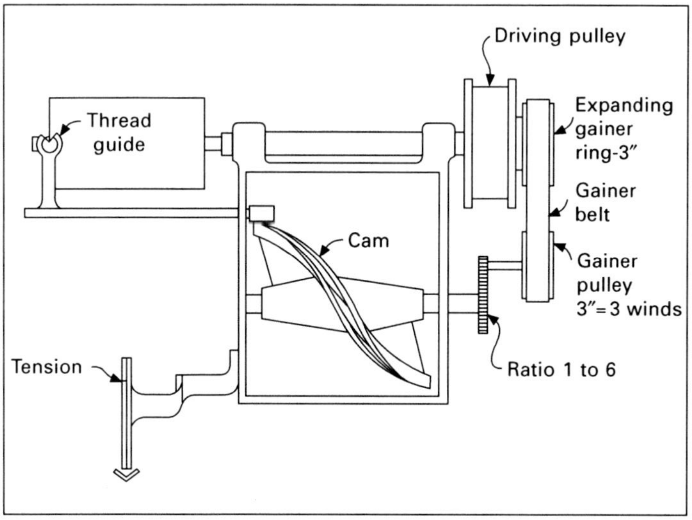

2.9 Cam traverse mechanism.

a thread guide lies in the groove. When the shaft rotates, the cams rotate and the small pin (thread guide) is guided in the groove to transfer motion to the thread guide, thus traversing the yarn on the package.

On these machines, since the package is fixed on the spindle which has a uniform speed in revolutions per minute, the yarn speed increases as the package diameter increases. Precision winding is mostly suitable for continuous filament yarns where the package is rotated independently and there is no frictional contact with any surface. There are some precision winding machines on which the spindle speed decreases as the package diameter increases, in which case the yarn speed is kept constant. This indirectly helps to keep the yarn tension constant through the build of the package. Figure 2.10 shows a different mechanism of traversing the yarn on the Schaerer Precision Cross Winder[12] using counter-rotating propeller blades. It is easy to change the traverse length.

Precision winding offers some of the following advantages:

- higher density of packages
- no pattern zones

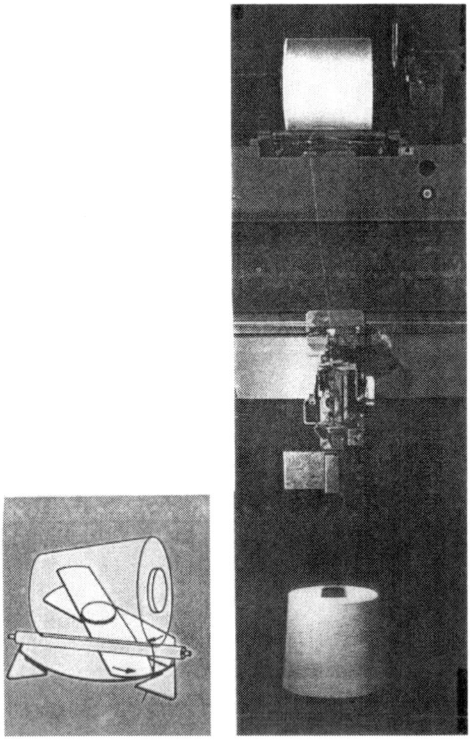

2.10 Counter rotating propeller blades traverse mechanism on precision winder.

- reduced end breakage rate
- higher density of the package gives advantages in twisting operation, i.e. cost of twisting, dyeing, etc.
- easy over-end unwinding
- regular dye liquid penetration
- excellent for winding continuous filament yarn.

2.4 Terminology commonly used in the winding process

2.4.1 Wind

During the winding process, the package rotates, whether by friction of the drum or when mounted onto a spindle, and the yarn is traversed on the package from one side to the other and back. In one stroke of the single traverse, a certain number of coils are wound on the package and this number is defined as the wind. Wind is therefore defined as the number of coils wound around the package in a single traverse.

2.4.2 Wind ratio

This is defined as the number of coils wound on the package per double traverse. This can be found by counting the number of times the package is revolved to traverse and return the yarn to the same point.

2.4.3 Random wound package

This kind of package is mostly produced on a drum winding machine. The wind ratio, i.e. the number of coils per double traverse, decreases as the package diameter increases but the angle of wind remains the same (Fig. 2.11(a)).[6]

2.4.4 Precision wound package

This is mostly wound on a precision winding machine where the package is driven independently. The number of coils per double traverse (wind ratio) remains constant from the start to the finish of the package, i.e. as the package diameter increases, the angle of wind decreases (Fig. 2.11(b)).[6]

2.4.5 Angle of wind

This is the angle between the direction of yarn lay on the package and a perpendicular drawn to package axis (Fig. 2.11). It is formed by the ratio

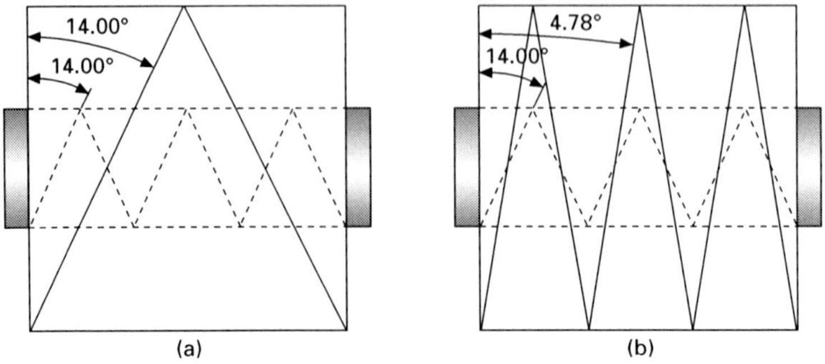

2.11 (a) Random wound and (b) precision wound package.

of two components: speed of yarn guide (traverse velocity)/surface velocity of the package, and is defined as tan Q. The quality of the package during winding is determined by several variables, the most important being the angle at which the yarn is wound onto the package. As a rule, the higher the wind angle, the better the package. The maximum wind angle is determined quantitatively by the package configuration, the properties of the yarn material and the process conditions connected with the winding operation.[13]

The normal range of winding angle is 30–50°. Minimum package density is reached when the winding is 90° or the two sets of threads are perpendicular. Compact packages for warping, shuttleless weaving are normally wound with angle of wind of about 30°. A change in winding angle from 30° to 55° reduces the package density by about 20–25%. Too low an angle of wind will produce a deformed package of 'diabolo' shape when coils will tend to run towards the middle of the package. On the other hand, too large an angle of wind will produce a stitching (webbing) effect when the coils tend to slip over the edges of the package thus causing end breaks during unwinding of the yarn from the cone.

2.4.6 Over-end unwinding

Over-end unwinding occurs when the yarn is unwound, but the package remains stationary (Fig. 2.12).[2] It can be compared with side unwinding when, during unwinding of the yarn, the package rotates (Fig. 2.12).

2.4.7 Open wind

During winding, when each succeeding coil of yarn lies nearest to the preceding coil, the package is said to have an open wind. This will occur when the wind per double traverse happens to be a whole number.

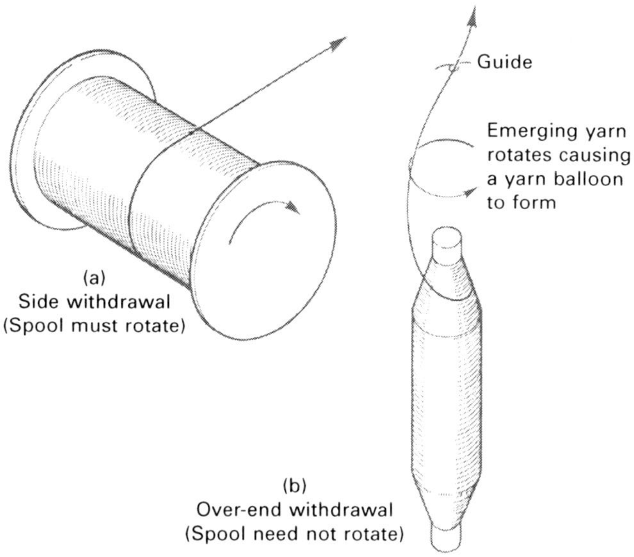

2.12 Over and side unwinding.

2.4.8 Close wind

If the adjacent coil of yarn lies adjacent to the preceding one, then the package is said to have a close wind. This will occur when wind per double traverse happens to be a whole number, plus or minus a small quantity called gain.

2.4.9 Patterning (ribboning)

This is a sort of defect which occurs during winding on a random winding machine when the wind per double traverse becomes exactly an integer. Some coils of yarns tend to superimpose on each other thereby creating a sort of short band or a ribbon. If a package with such a defect is used in subsequent processes such as knitting, warping or weaving, they adversely affect the unwinding properties of the yarn. During unwinding of the yarn at high speed, the coils of yarn from the ribbon tend to slough off in a bunch and cause a machine stoppage.

Almost all random winding machines are equipped with anti-patterning devices of one form or another. The method most commonly employed is that of periodically varying the speed of the driving motors through a control switch which interrupts and restores the current at suitable intervals. The variation in speed of the driving rolls which this action brings about breaks up the synchronism between the rolls and the cones due to the time lag which occurs before the cones respond to the changed speed of the driving rolls.[11]

2.5 Cone types and build

Cones produced by modern high-speed winding machines differ in build and taper according to the type of material and the purpose for which they are used. There are generally three main types[11]:

- cones produced with a close wind on spindle machines with individual cam and reciprocating traverse
- cones with convex base and increasing taper
- cones with little taper, built up with a uniform or slightly accelerated traverse.

The last two types are of open wind and are produced on machines of the drum or driving roll type.

In order to facilitate the overend withdrawal of fine and delicate yarns used, for example, in hosiery, cones of about 9° 15′ taper, built up with a convex base and increasing taper, are used. For warping, cones of smaller taper around 5°, which contain a greater length of material, are generally used. They are built with a uniform traverse and there is tendency for the degree of taper to decrease as they increase in size. There is evidence that cones with an accelerated taper of 4° 20′/9° 15′ give good results when used on shuttleless weaving machines.[14]

2.6 Manual, semi-automatic and fully automatic winding machines

During the ring spinning process, when the ring bobbins have reached their full size on the spinning frame, they are ready to be taken out from the ring spindles. The process is called doffing of the packages and involves running the ring rail down and then running on some wraps of yarn around the base of the bobbins so as to be sure that the end will not be broken in removing the full bobbin from the spindle. During this process the yarn wraps around the bobbin as shown in the Fig. 2.13(a). This is referred to as back wind. However, all these extra wraps of yarn must be removed by the winder before this is put on the supply spindle of the winding machine. When hundreds of bobbins are handled by the operator during the working shift, much of his/her time is wasted in removing the back wind, thereby affecting productivity significantly. The choice of buying a manual winding machine, or using a semi-automatic or fully automatic winding machine to replace some or all of these tasks, depends on the type of activities the particular company is engaged in, i.e. in the production of coarse, medium or fine fabrics, etc. and other requirements.

2.6.1 Manual winding machines

One of greatest disadvantages of a manual winding machine is the number of operations which have to be performed by a winder by hand per day. This depends on factors such as yarn quality, size of the supply package, winding speed, etc., which determine the work load, i.e. number of spindles that can be allotted to a winder. On a manual winding machine almost all the operations are carried out by the operator. These include:[5]

- removal of the back wind from the spinning package (Fig. 2.13(a))
- positioning of the package on the spindle and passing of the yarn through the tension device and slub catcher which eliminate thin and thick places from the yarn (Fig. 2.13(b))
- positioning of the yarn on the cone to commence winding (Fig. 2.13(c))
- when a yarn breaks as a result of the removal of the thick or thin place, or when the supply package runs out, taking the yarn from the running supply package, passing it through the tension device and slub catcher, locating the free end of yarn from the cone and knotting it with the end of yarn from the supply package to commence winding (Fig. 2.13(c))
- when the cone or cheese package becomes full size, removing it from the cone or cheese holder, taking a new empty cone or cheese package and starting the whole process of guiding the yarn from the supply package onto the cone/cheese (Fig. 2.13(d))
- cleaning the machine.

2.6.2 Semi-automatic winding machines

On these machines, some of the above operations are carried out automatically and some manually. The automatic operations might include:

- feed of the supply package onto the winding spindle
- automatic knotting of the yarn from the supply package with the yarn from the cone in case of the yarn breaking as a result of removal of thick or thin places
- automatic knotting when the supply package runs out and a new end from the supply package is tied to the end from the cone
- doffing of the cone, i.e. removal of the cone when it becomes full.

2.6.3 Fully automatic winding machines

On a fully automatic winding machine all of the following functions are performed automatically:

- removal of the back wind from the spinner's package

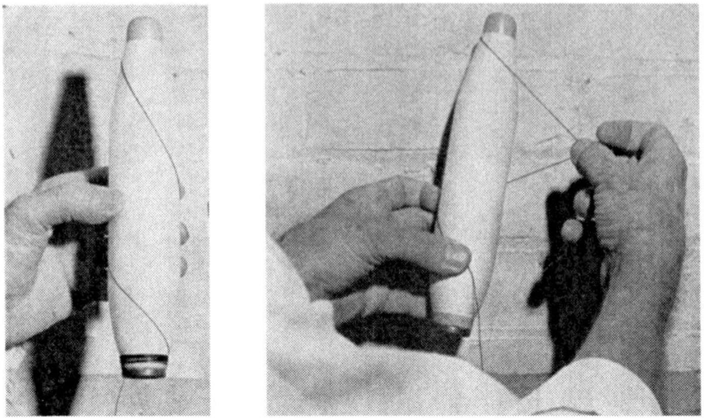

(a)

(b)

2.13 (a) Removal of back wind (wraps of yarn have been unwound from the base of the bobbin on the left), (b) operator feeding the spindle with new package after removal of the back wind, (c) operator trying to join the end from the cone with the end from the package and (d) operator doffing.

- feeding of the supply package to its appropriate spindle
- joining of the two ends, i.e. from the package and the cone or cheese either by putting a knot or a splice when a defect is removed or when the supply package finishes
- doffing, i.e. replacing the full cone (when it has reached its desired diameter) with an empty cone to commence winding.

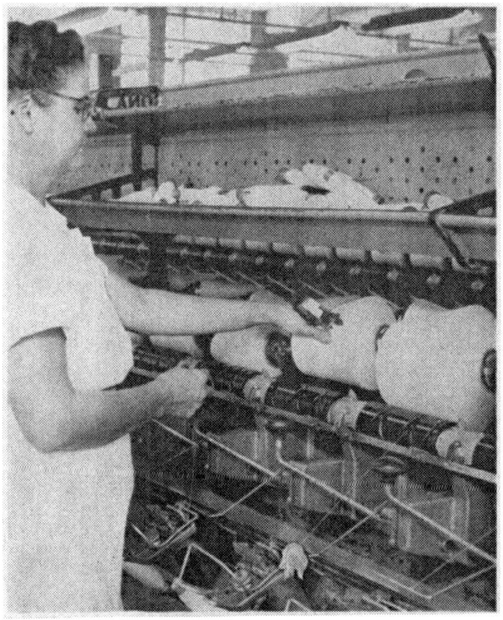

(c)

(d)

2.13 Continued

Figure 2.14(a) shows a Murata fully automatic winding machine equipped to carry out all of the above actions. The automatic doffer as shown in the figure keeps on patrolling the spindles and doffs the cone when required.[15] On modern winding machines, yarn speeds are adjustable from 300 to 2,000 m/min, depending on yarn type, bobbin build and machine specification. Figure 2.14(b) shows a typical automatic back wind removal unit equipped on a fully automatic winding machine. Spinner's packages are dumped into the unit and, after removal of the back wind, the mechanism automatically feeds the packages onto the spindles.

2.7 Hazards from knots during weaving and knitting processes

During the winding process, each time a defect such as a thin, thick place or a slub is removed, or when the spinner's supply package is finished and replaced with a new one, the yarn end from the cone and the supply package is commonly joined either by a fisherman's knot or a weaver's knot, depending on the circumstances. The fisherman's knot is considered to be more reliable and acceptable in most cases other than for coarse yarns used on fine gauze applications. Its knot diameter is three to four times that of the yarn. The weaver's knot, on the other hand, is considered less stable but has the advantage of a diameter of two to three times the yarn diameter and is thinner than the fisherman's knot. This makes it easier to process subsequently through confined spaces. If not knotted properly, knots do tend to untie during the weaving process. Technically speaking, a knot itself can be considered a defect and hazard in subsequent processing of the yarn during knitting or weaving. The only point in its defence is that a knot is much smaller in diameter and length than many of the yarn defects which can extend to over a centimetre in length as well as having a larger diameter.

2.7.1 Frequency and behaviour of knots during weaving

The frequency of knots introduced during winding usually depends on the yarn count, weight of yarn on the spinner's package and the degree of clearing. It is estimated that on average about 15–35 knots/kg of the yarn are introduced during winding. When transferred to weaving, this represents 10–20 knots/metre of warp. It is believed that while weaving cotton warp yarns, knots cause about 20–50% of all the warp breaks which include:

- knot slippage
- jamming of knots between reed wire
- entanglement with other warp threads.

With woollen warps it is believed that up to 40% of all breaks are attributed to knots.

(a)

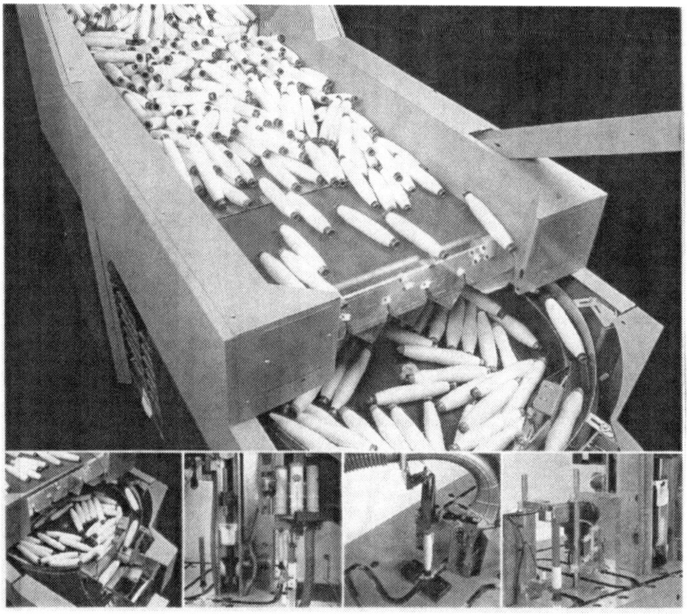

(b)

2.14 (a) Murata fully automatic winding machine with automatic doffer patrolling the spindles and (b) Automatic back wind removal unit equipped on a fully automatic winding machine.

2.7.2 Behaviour of knots during knitting

During knitting, knots either catch in the knitting elements or jam between knitting elements as well as causing temporary interruption of yarn feed thus breaking the yarn, and sometimes even knitting needles. Assuming a knitting machine with 24 feeders each using a package weighing 2 kg, of 50 metric count, which normally might contain between 40 and 50 knots, the total number of knots will be 960–1200. If the yarn usage per loop was 0.2789 cm, with 1800 needles and machine speed of 654 rpm, it is anticipated that a knot will run through the needles of the knitting machine every 19 seconds with the probability of a defect being caused by the knot. During finishing, knots cause holes if the fabric is to be sheared. Figure 2.15 shows the appearance of knotted and spliced yarns of the same quality and their effect when knitted or woven into the fabric.

2.8 Yarn splicing for knot-free yarns

In order to get rid of the problems caused by knots, practically all modern winding machines are equipped with a splicing technique which joins the two ends of yarn without knotting. Figure 2.16 shows the different techniques of splicing the yarns.[16–18]

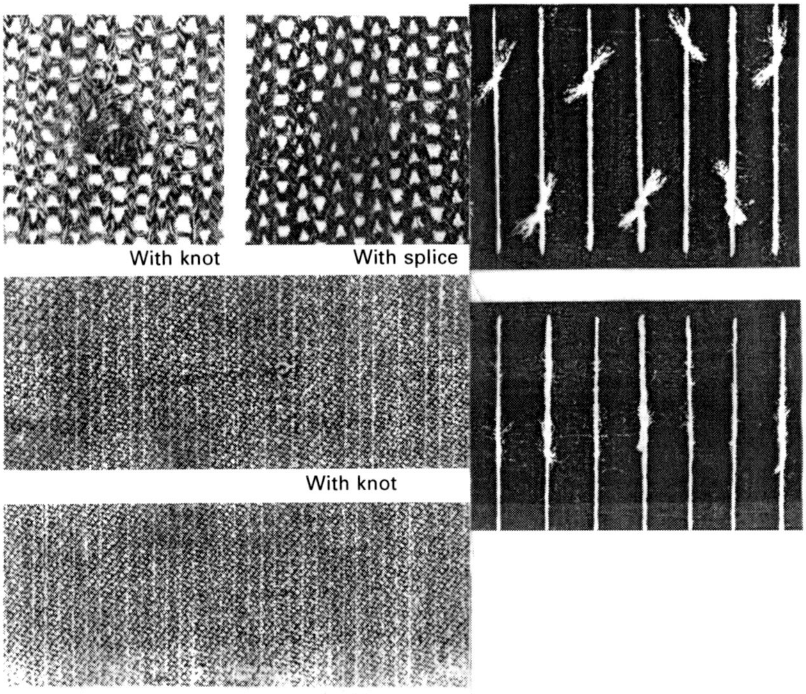

2.15 Knotted and spliced yarns when woven and knitted into a fabirc.

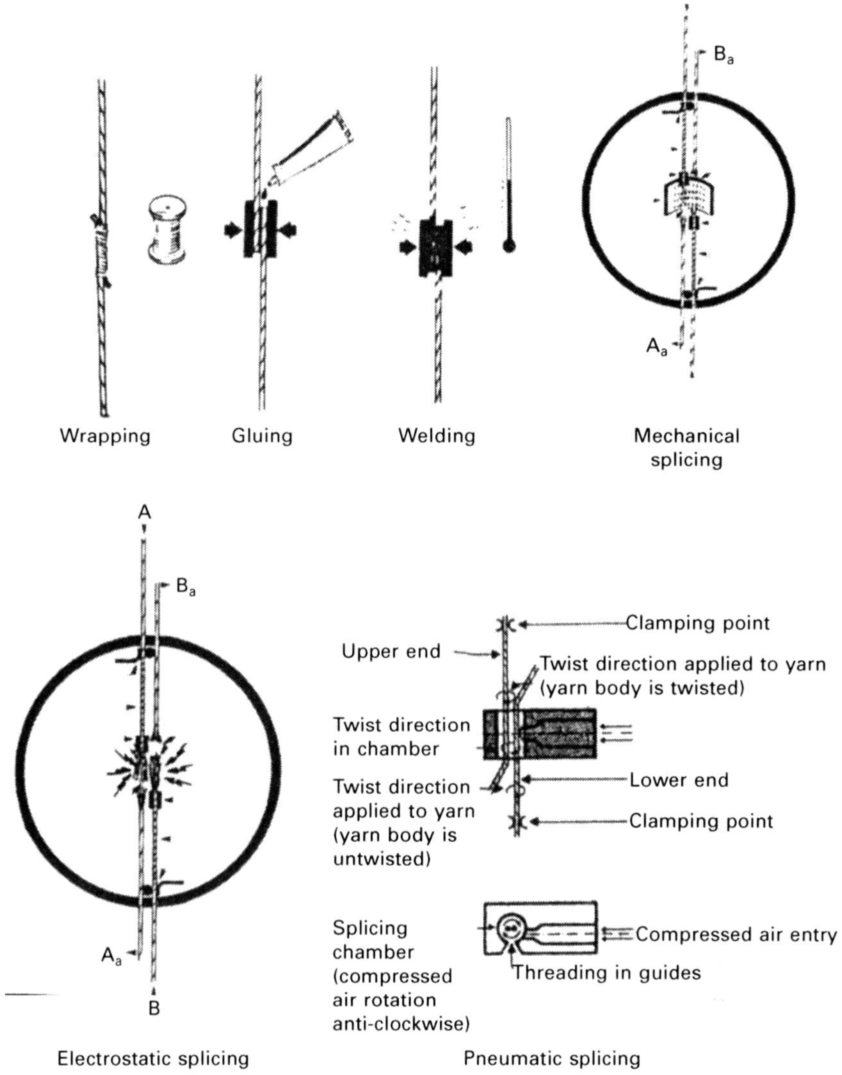

2.16 Different splicing techniques.

2.8.1 Pneumatic splicing

In this method the two ends from the supply package and the cone to be spliced are inserted into a chamber and then overlapped to join them together by means of a strong current of compressed air. The splicing time and air pressure are determined according to fibre type and yarn characteristics. The splicing operation consists of the vertical application of air to the fibres and simultaneous rotational movement of the air to twist/untwist the yarn.

Pneumatic splicing can be applied to a wide range of fibre and yarn types without precise adjustments or settings, thus facilitating efficient winding.

2.8.2 Mechanical splicing

In this process the yarn ends are untwisted to open the fibres. The two ends are overlapped and then twisted together again to essentially the same twist level as in the basic yarn. The technique is ideal for short staple fibres. For long staple yarns it is somewhat difficult to separate fibres at the yarn ends, which has a negative effect on binding of fibres. Different twisting wheels are required for opening and twisting of yarns of different twist levels and made from different staple length fibres. It is also not possible to splice plied yarns as opening by untwisting is not possible.

2.8.3 Electrostatic splicing

In this method of splicing the yarn ends are separated and untwisted in the opening zone. The opened yarn ends are then spread out evenly in an electrostatic field, after which they are intermingled and bound again by a pole change, while twist is simultaneously inserted corresponding to the twist in the basic yarn. The method has some limitations. For yarn containing a large number of fibres in the cross section, the intermingling of fibres at the yarn ends is a problem as fibres obstruct one another. Secondly, plied yarns cannot be spliced due to difficulty in opening fibres at the yarn ends by mechanically untwisting. Thirdly, the time taken for splicing is quite long, thus decreasing winding machine efficiency.

2.8.4 Gluing

In this technique yarns are overlapped and glued together. The process takes a long time and produces a rigid joint.

2.8.5 Welding

In this process the two ends are welded together by heat and are suitable for polyamide yarns.

2.8.6 Wrapping

The two ends of the yarn are overlapped and an auxiliary thread is wrapped around them. This process is normally suitable for joining carpet yarns used on tufting machines. It takes more time and produces a bit thicker, more rigid joint that is stronger than a knot.

2.9 Applications of splicing techniques

Splicing can be applied to a variety of yarns such as:

- long and short staple yarns (wool, cotton, synthetics)
- plied yarns of cotton, wool, synthetics and blends
- siro spun yarns
- woollen spun yarns
- core yarns
- open-end yarns.

It has been conclusively proved that spliced yarns play a significant role in reducing the cost of production and number of fabric defects caused as a result of yarn breakages during the weaving and knitting process. Knot-free yarns have also helped to reduce the processing costs during warping and sizing processes.

2.10 References

1. Joint Stock Company, Textilmach, Russia from brochure.
2. P.R. Lord, M.H. Mohamed, *Weaving: Conversion of Yarn to Fabric*, Merrow Publishing, Watford, 1973.
3. Uster Classimat grading of yarn faults.
4. Schlafhorst Autoconer leaflet.
5. Jesse W. Stribling, *Work load calculations for winding machines*, Universal Winding Company, Leesona Bulletin 139-A, 1952.
6. Fadis Textile Machinery, Italy, General catalogue.
7. Oerlikon Schlafhorst, Autoconer 5 bulletin.
8. Savio Macchine Tessili, Automatic Winder Orion E/I, Italy.
9. J.E. Booth, *Textile Mathematics*, The Textile Institute, Manchester, 1977.
10. GRG Industries Leaflet, www.grgindustries.com.
11. G.A. Bennett, *Introduction to Automatic Weaving*, Harlequin Press Co. Ltd, Manchester, 1948.
12. Schaerer Precision winder leaflet.
13. F. Hebberling, *Textile Research Journal*, September, 1969, vol 39, no 9, p 799.
14. J.B. Aitken, *Automatic Weaving*, Columbine Press, London, 1964.
15. Murüta winding machine, bulletin.
16. J. Lünenschloss, J. Thierhoff, Knottless yarn-joining, *Int Text Bull*, 1/1985, 30–31.
17. Schlafhorst-Dokumentation, *The Theory and Practice of Splicing*, S. Schlafhorst & Co, Germany, 1/1985.
18. B.C. Goswami, R.D. Anandjiwala, D.M. Hall, *Textile Sizing*, Marcel Dekker, New York, 2004.

3
Yarn preparation for weaving: warping

K. L. GANDHI, Chartered Consultant, UK

Abstract: The chapter starts with an introduction to the second stage of yarn preparation – warping – and covers its objectives. It then discusses in detail the methods of warping employing different technologies and the modern creels that are used in the production of warper's or weaver's beam. The chapter also reviews in detail the characteristics of different modern types of tension devices that are used on warping creels. Finally, the chapter describes a single-end warping machine emphasizing its advantages.

Key words: direct beaming, sectional warping, beaming off, warping creel, tensioning units of creels.

3.1 Introduction

Warping is the second stage of the processes used after winding. The process involves transferring yarn from a predetermined number of tubes, cones or cheeses positioned on the creel onto a warper's beam or a weaver's beam. Winding involves either preparing a package called a weaver's beam which is set up on the back of the weaving machine, or preparing a number of packages called warper's beams which are then sent to the next process known as sizing. There are normally two methods of warping employing different technologies:

- direct warping or direct beaming (also sometimes known as beaming)
- indirect warping/sectional warping.

Both methods can be used for spun or continuous filament yarns, depending on the circumstances and the quality of fabric to be produced.

3.2 Direct warping

This method of producing cotton warps is used mainly when the yarn to be woven requires application of starch which is applied in the next yarn preparation process known as sizing, after which it is sent to the loom for weaving. However, as mentioned above, the method can also be employed in the production of beams from continuous filament yarns. A warping machine consists of two units (see Figs 3.1 and 3.5[1,2]):

- warping creel
- head stock.

The number of cones that should ideally be put on the creel to produce a number of warper's beams can be worked out by a simple calculation. For example, a fabric with 20 ends/cm and 100 cm in width is to be produced. This means that the total number of warp yarns in the fabric width will be 2,000 and a weaver's beam with 2,000 ends would be required and positioned at the back of the loom. If the length of fabric to be produced is 20,000 m, this length of yarn would have to be put on the weaver's beam. In actual practice, a small percentage of contraction takes place during the weaving process and is added to the above warp length.

If the yarn for the above fabric was a single cotton yarn, it will have to be sized after warping. However, the preparation for sizing would be done in the warping department. A number of warper's beams are produced and sent to the sizing department to produce a weaver's beam ready to be positioned at the back of the loom. How many warper's beam should be produced? There is no written formula to obtain the answer. It depends on the skill and experience of the person handling the job. However, it also depends on factors such as the creel capacity, i.e. how many cones it can take, as well as how many cones are available. If the creel capacity was, say, 400 cones and

3.1 Direct warping machine (A – creel; B – head stock).

we were to process the above fabric with 2,000 ends, it would be reasonable to take 400 cones and fill the whole creel (see Fig. 3.5).

3.2.1 The direct warping process

In the direct warping process, an end from each of the packages positioned on the creel is drawn from the warp stop motion and tensioning unit located on the creel. It is brought forward to the headstock of the machine located in front of the creel, passed through an expandable V shape comb (Fig. 3.2)[1] and over the measuring roller. The process is complete when all the packages in a sheet form are wound on to a warper's beam (Fig. 3.1). Typically a 20,000 m length of warp containing 400 ends is wound onto the first warper's beam. After this, the warp is cut, the full beam taken out and a new beam put in (Fig. 3.3).[1] Since 2,000 ends are required on the weaver's beam, five warper's beams, each with 400 ends and 20,000 m length of warp, will have to be produced. The process is repeated after each beam is produced.

In the above case, the five warper's beams will be sent to the sizing department and positioned on the creel at the back of the sizing machine (Fig. 3.4).[1] Warp ends are collected from all five beams and passed through the sizing machine. Two thousand ends in sheet form are wound onto the weaver's beam. A number of weaver's beams can be made from the given length of warp on the warper's beams.

3.2.2 Drive to the warper's beam

Two methods of driving the warping beam are available. The majority of warping machines are drum-driven, where the rotational motion to the

3.2 V-shaped expandable comb.

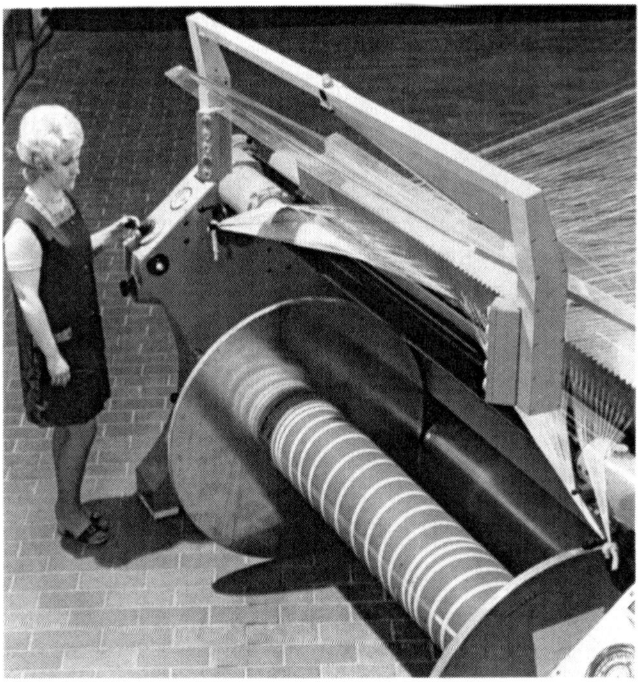

3.3 Full beam being taken out (doffed) and new one put in.

warper's beams is imparted by the frictional contact of the drum (Fig. 3.5).[2] As the beam diameter increases, its rotational speed decreases. This means the warp's linear speed remains constant which maintains constant tension on the warp sheet from the beginning to the end of the warping process.

3.4 Warper's beams at the back of sizing machine.

On some warping machines, the beam is driven by a spindle or with gears (Fig. 3.1). In order to keep the tension uniform throughout the warping process, as the beam diameter increases, its rotational speed decreases and the yarn speed remains constant.

3.2.3 Beam warping for yarns not requiring size application

If the yarn to be processed is a two-fold yarn or another type which generally does not require any application of starch, then beam warping can be carried out in two stages. In the first stage, a number of warper's beams with the desired number of warp ends and warp length on each beam are produced on the direct warping machine. In the second stage, all the warper's beams are positioned on a frame and the warp ends from all the beams are simultaneously unwound and transferred onto either a single weaver's beam or a number of beams which are then ready to be positioned at the back of the loom for weaving.

3.5 Drive to warper's beam with frictional contact of drum (A – creel; B – head stock).

3.3 Indirect/sectional warping

This method of warping is generally carried out for yarns under the following conditions:

- where there is no need for any application of starch, such as two-fold cotton or woollen/worsted warps
- for shorter runs of high class goods
- for coloured yarns, i.e. patterned warps for wool and worsted fabrics, etc.
- where a small number of cones is to be used because of the quality of the fabric to be produced, mostly coloured warps.

Compared to direct warping, the number of cones used is generally much smaller. A sectional warping machine consists of three basic units:

- creel
- winding drum
- the beaming unit.

Warper's beams are not produced in this process and the warp is directly transferred from the drum onto the weaver's beam(s).

The process involves two stages:

1. Warping of a number of sections with the required number of yarns onto a conical drum.
2. Beaming off, i.e. transferring the yarns from all the sections in a sheet form onto a weaver's beam.

An example is the need to produce a fabric 5,000 m in length, 100 cm wide with 20 ends/cm and 100 available cones or cheeses. To produce this fabric requires a weaver's beam with 2,000 ends and 5,000 m length of warp. To produce this weaver's beam, 100 packages will be positioned on the creel (Fig. 3.6(b)). Yarn from each package is drawn through the tension and thread stop motion, brought forward to the front of the machine, and condensed into sections with a small width by passing through one or two different reeds (Fig. 3.6(a),(b)). A section containing 100 ends and 5,000 m in length will be wound onto the conical drum (Fig. 3.6(a)[3],(b)).[4] When the desired length of section has been wound, it is cut and another section adjacent to the previous section is wound for the same length (Fig. 3.6(a)). Since the total number of ends required on the weaver's beam is 2,000, the process is repeated until 2,000 ends of the same length have been wound onto the drum. In total there will be ten sections wound onto the drum.

3.3.1 Beaming off

The next stage, known as beaming off, consists of transferring the yarn from all the sections in a sheet form from the conical drum onto a weaver's beam (Fig. 3.6(c)[3],(d)).[4] Whether the entire length of the warp yarn is transferred from the drum onto a single weaver's beam or a number of beams, depends on the production planning, i.e. how many looms are to run from the same quality of warp.

3.3.2 Waxing of warp

Sometimes a coating of wax is applied to certain types of warp yarns, for example, woollen yarn, during the beaming off process to improve their weavability. The warp sheet is passed over a revolving roller loaded with liquid wax as shown in Fig. 3.6(e)[3] and is then wound onto the weaver's beam.

3.4 Warping creels

A creel is basically a metal frame of varied design to accommodate a few hundred tubes, cone or cheese packages. The frame is fitted with package holders. The number of package holders on the creel frame can vary from 100 to 800 and more in exceptional cases, their number representing the capacity of the creel. The creel is equipped with a tensioning device and thread stop motion through which each running yarn is passed. In modern machines, these devices are equipped with automatic control and centralized tension variation. Depending on the area of application, package type and size of

(a)

(b)

3.6 (a) Sectional warping starting a new section, (b) Sectional warping with section being wound on to drum, (c) beaming off from drum onto weaver's beam, (d) beaming off of solid warp from drum to weaver's beam and (e) waxing of warp.

(c)

(d)

(e)

3.6 Continued

the processed material and the space conditions, a creel must satisfy various requirements. The following types of modern creel design are available from various manufacturers:

- single end creel
- fixed frame creel
- magazine creel
- duplicate creel
- creel with removable bobbin trucks
- swivelling frame creel.

3.4.1 Single end creel

This type of creel is generally used when the factory requires the running of short lengths of warp or where it is necessary to make frequent changes in yarn counts or quality. Only one cone holder is provided for each package. When the packages are exhausted, the empty cones are removed and replaced with fresh ones. During this period, the warping machine remains idle.

3.4.2 Fixed frame creel

Fixed frame creels are often used in pairs enabling one creel to be running while the other is being preloaded. Figure 3.7 shows the design of a fixed

3.7 Fixed creel.

creel (FC) with a fixed double-sided package frame.[5] By opening the tensioned frames which are mounted on ball bearings, the distance between the package top and the yarn tension units can be adjusted, depending on the ballooning of the yarn. The creel can be hydraulically opened and closed by means of a push button. Another widely used new fixed frame creel is available from Huys & Vanhevel, Belgium. The model FC[4] has two designs:

- FCO (fixed creel outside)
- FCI (fixed creel inside).

These feature a fixed steel construction with suspended tensioned frames (Fig. 3.8). By moving the tensioned frames laterally, a loading gangway is created between the packages and the tensioners. Similarly, the distance from the package to the tensioner can be adjusted for optimal yarn balloon formation.

FCO type (exterior yarn sheet)

The FCO type comprises a fixed central package frame with movable exterior tensioner frames. This is the most widely used construction with easy yarn sheet access which facilitates rapid yarn break repairs.

FCI type (inside yarn sheet)

The FCI type comprises a fixed exterior package frame with central movable tensioner frames. This solution provides minimum deviation in angles and is therefore particularly suitable for fine and delicate yarns.

3.4.3 Magazine creel

The magazine creel is most effective when long runs of a single count or quality are required. Two cones/holders are provided for each yarn, one for the running yarn and the other for the reserve cone. The tail end of the running cone is tied to the leading end of the reserve cone, so that when the

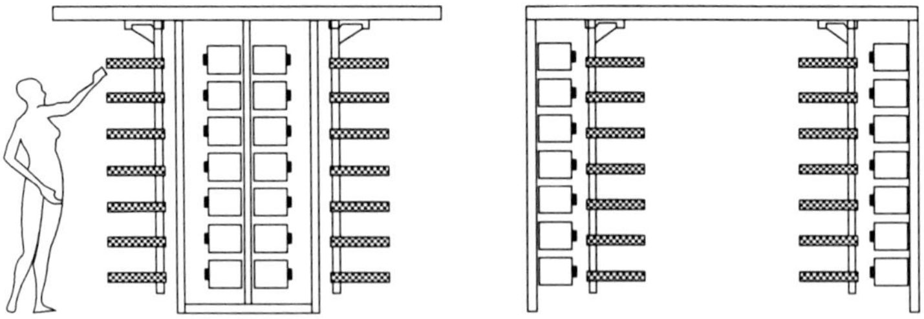

3.8 Fixed creel with outside (left) or inside (right) yarn sheet.

running cone finishes unwinding, the yarn from the reserve package starts and continuity is thus provided without stopping the machine. There will be few leftover packages. This also allows replenishment of the cone holder on the creel with the reserve cones as required. Depending on the application, the creel may be configured for exterior or inside draw off. Figure 3.9 shows model GP-M[6] of a magazine creel made by Benninger AG, Switzerland.

3.4.4 Duplicate creel

The use of this type of creel is the same as that of the single end creel, i.e. where it is necessary to change the quality of the material or yarn counts either at relatively short intervals or when frequent changes are necessary, The design places two creels side by side. The headstock of the warping machine can be moved from one creel to the other. While the running creel is being used, the duplicate creel with the desired yarn quality is prepared ready for use. The headstock is moved to the made-up creel so the warping process may be started without any waste of time. The spare creel then becomes available for the positioning of new packages.

3.4.5 Creel with removable bobbin trucks (RT)

The design of this creel is almost identical with that of the fixed creel. However, the fixed bobbin frame is replaced by removable package trucks[5] or trolleys in the centre of the creel (Fig. 3.10). Individual trucks can be taken out of the creel. The design of this creel also allows the use of a set of spare trucks. These can be loaded with packages outside the creel while warping

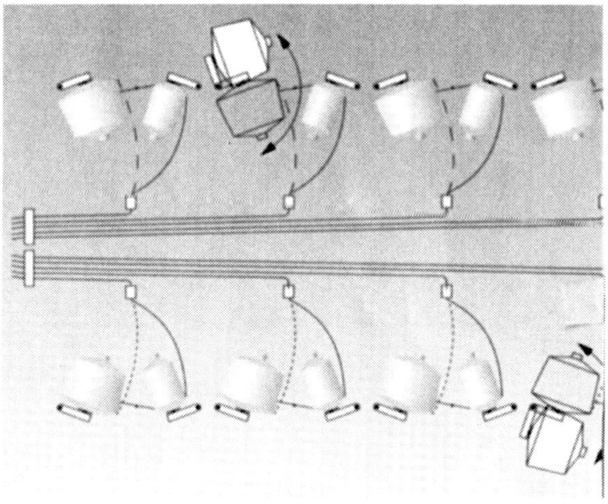

3.9 Magazine creel GP-M for continuous production.

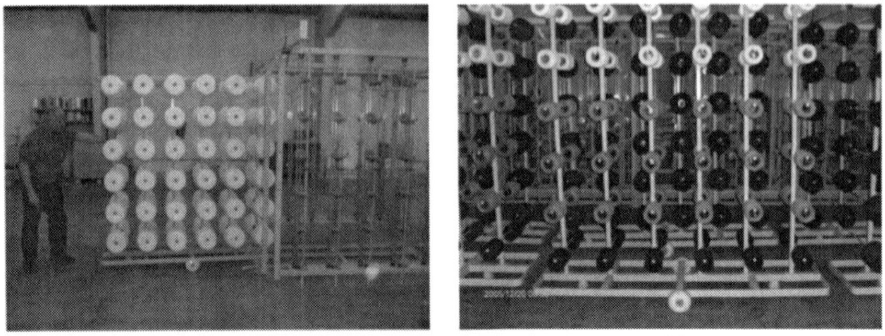

(a) (b)

3.10 Creel with removable package trucks: (a) creel with removable bobbin trucks (b) spare trucks to increase flexibility and speed for warping.

is carried with the other set. This increases the flexibility and productivity of the warping process. The downtime is reduced to that required to move the trucks in and out and to knot the yarns with the finished set. This creel design is particularly useful when there is insufficient room to permit the use of two standard creels. The creel can be hydraulically opened and closed by means of push buttons.

3.4.6 Swivelling frame creel

Figure 3.11(a) shows the design of a swivelling frame creel type CS[4] which is marketed by the Huys Matthys Group from Belgium. This design permits the setting up of other packages when the machine is running. The next set of packages, running for a different quality of fabric, can be loaded on to the creel while the warping machine is still running the previous set. When the running packages are empty, the package frame can be rotated to bring the new set of packages into the active position. The rotating frames have a locating catch for ease of positioning. The creel is particularly suited for large packages of filament yarn. This creel is available in two versions:

- with an exterior yarn sheet: this type is used when the yarn run needs to be accessible for the repair of broken ends
- with an inside yarn sheet: used with heavy packages to facilitate handling and accessibility.

Figure 3.11(b) shows another model of swivelling frame creel GP-D[6] in two versions from Benninger. This creel is ideal for package sizes between 5 and 25 kg. The segments are turned manually but an automatic option is also available.

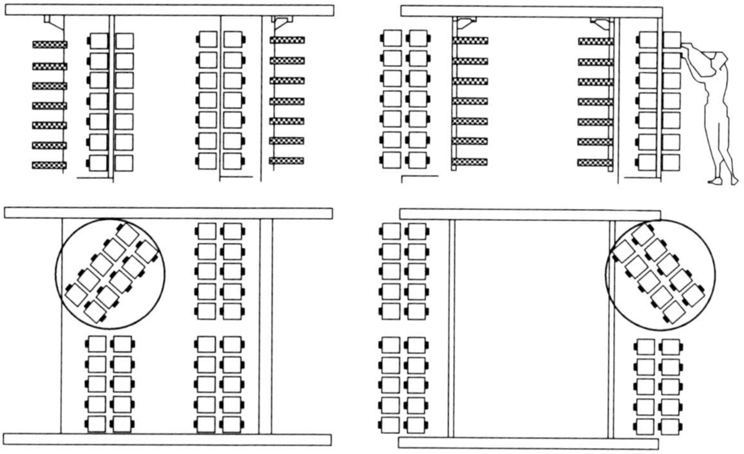

(a)

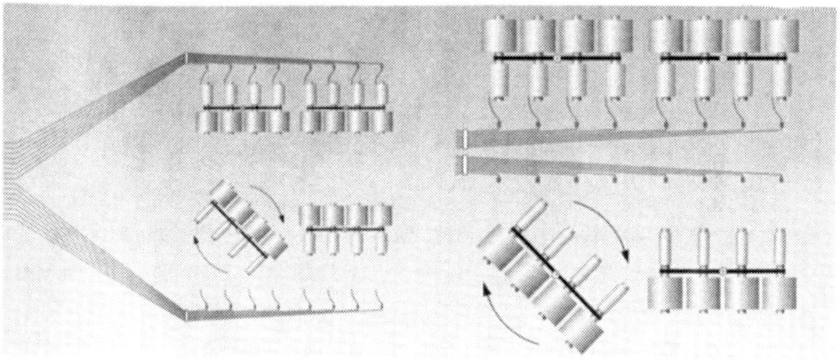

(b)

3.11 (a) Swivelling frame creel and (b) swivelling frame creel GP-D.

3.5 Tensioning units of creels

All modern creels are equipped with a tensioning unit for each package. These can be mechanical or electronic. A number of different tensioning units are available, covering a large range of yarn types and counts. The simplest is the disk tension device[7] as shown in (Fig. 3.12). The thread is drawn from the package, through an eyelet and led around two small pins before going forward from the creel. If an increase in tension is needed, weighted disks are placed on top of the yarn as shown in Fig. 3.12(b). This method is generally suitable for cotton and wool. However, for sensitive yarns such as filament and silk, the steel posts or pins are fitted with porcelain adapters.

There are various types of tensioning device:

- self-compensating tensioner type DTR
- adjustable disc tensioner type DT

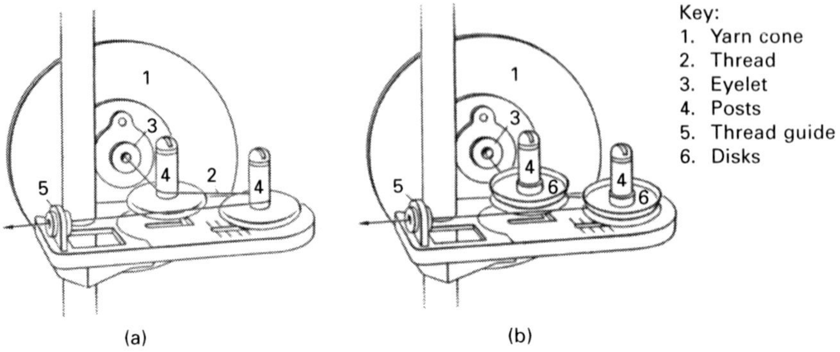

3.12 Disc tension device.

- mechanical self-compensating tensioner
- electronic tension units
- Tsudakoma new tensioner device.

3.5.1 Self-compensating tensioner type data terminal ready (DTR)

Figure 3.13(a) shows a self-compensating tensioner type DTR fitted on a warping machine.[4] It is designed to maintain a constant yarn tension on a given setting with cops, conical or cross-wound packages and to maintain a constant yarn tension from full to empty packages, regardless of speed or package size. The decrease in diameter of the package during warping is compensated by the tension unit so differences in tension are kept to a minimum. Tension peaks are absorbed by the hydraulic damping action of the compensating wing, enabling smooth running of the yarn sheet. Different types are available to suit a large range of yarn counts.

3.5.2 Adjustable disc tensioner type DT

This tensioner DT (Fig. 3.13) has been developed for technical applications with high tensions.[4] The yarn is deflected over a ceramic roller with a large deflecting radius of 12 mm. It is ideal for fibreglass, industrial, or heavy filaments and spun yarns.

3.5.3 Mechanical self-compensating tensioner

Two different mechanical self-compensating tension units, V001 and V002, are offered by VTA, Belgium (Fig. 3.14(a),(b)).[5] These units keep the tension constant, independently of the various colours on the creel, the different

(a) (b) (c)

3.13 Self compensating tensioner, (a) DTR-1000 (b) DTR-Z and (c) DT-tensioner.

types and counts of yarn or the difference in diameter between the various packages. When at standstill, the tension units return to their starting position, thus keeping the yarn stretched.

3.5.4 Electronic tension units

Most modern warping machines are equipped with different types of electronic tension units for packages on the creel. Figure 3.15 shows an electronic tension unit[5] Model EL001 from VTA equipping a modern warping machine creel. It has the following features:

- fully programmable tension control
- stores 999 styles
- custom application for quick installation
- automatic profile of tension from front to back
- constant tension across the yarn sheet
- constant tension regardless of speed
- computer calibrated
- motion sensors for detecting broken ends
- individual end identification
- programmable response time to reduce lost ends
- touch screen.

3.5.5 Tsudakoma new tensioner device

Tsudakoma have introduced a new type of tensioner unit (Fig. 3.16(a)) on their creel TVC-e Creel.[8] The figure shows the two-rod type, but three-rod and

(a)

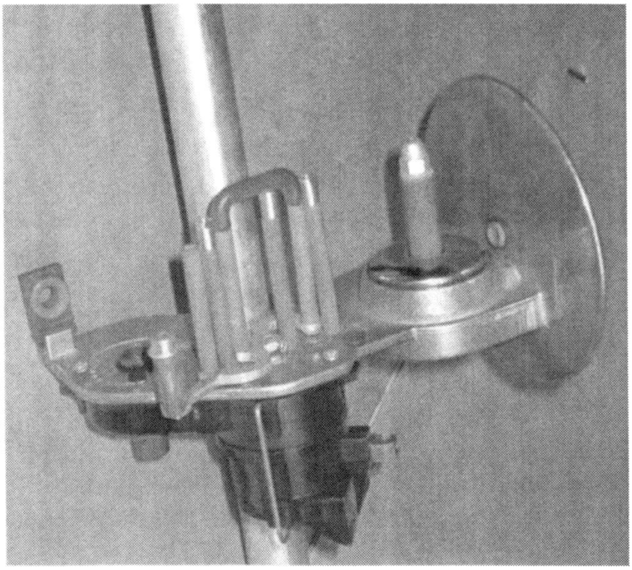

(b)

3.14 (a) Mechanical tensioner V001 and (b) mechanical tensioner V002.

four-rod types are also available. A rod tensioned system with a movable rod is used. By setting three positions for the movable rod at stopping, running and opening, the tension is equalized and yarn breakage may be easily repaired. This device breaks the yarn according to the machine stop/start setting to prevent the yarn from loosening in acceleration and deceleration. An optional tension control device is also available. The creel tension is detected on the warper's main body section. From the beginning to the end of the package, the combined use of the rod position control and the speed control draws out the yarns at a stable tension for winding on the beam.

3.15 Electronic tension unit.

The Tsudakoma tensioner unit also includes a device for automatic adjustment for ballooning distance. This unique device calculates the optimum distance between the yarn package and the tension rod based on the yarn package diameter, which is detected by the yarn package diameter sensor (Fig. 3.16(b)). The tension rod is then brought close to the yarn package.[8] As the distance between the package and the tension rod is maintained so as to prevent multiple balloons, yarn breakage is decreased.

3.6 Thread stop motion

Every warping machine is equipped with a thread stop motion. In the event of a thread breaking, the stop motion immediately stops the machine before the broken thread becomes lost under subsequent layers of yarn on the beam. A light at the stop motion unit illuminates to identify the exact location of the break and assists the operator in repairing the broken end. Whenever an end breaks, the operator locates the broken end from the cone, passes it through the tension unit and thread stop motion and takes it forward to the beam, tying it with the broken end from the beam. The same process is carried out when any cone or cheese is finished and the machine stops.

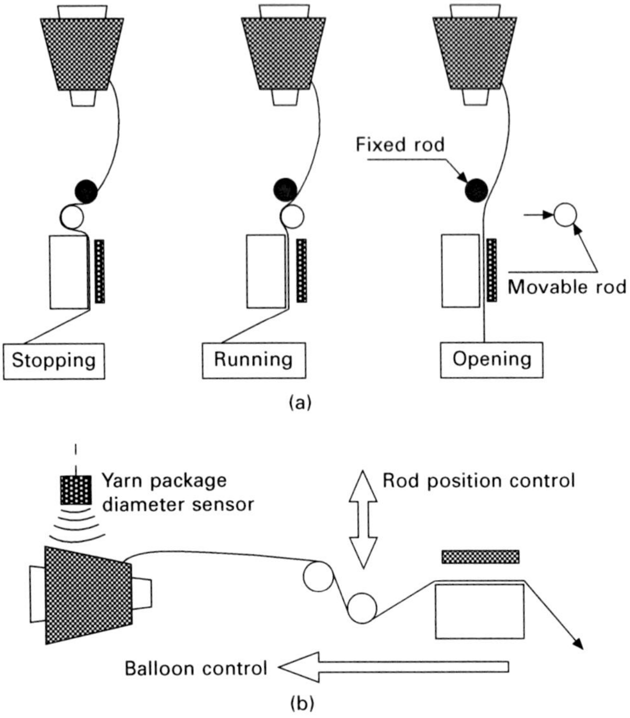

3.16 (a) Two-rod tensioner and (b) automatic adjustment for balloon distance.

Huys (Belgium) offer two types of modern warp stop motion/detection systems:

- FRD-O
- ECD -16.

The first type uses a mechanical dropper to generate the stop motion signal. A LED-lamp on each stop motion rack indicates the vertical position of the yarn break. The ECD-16 type (Fig. 3.17) is based on magnetic reed-contact. Because this contact is hermetically closed, the sensor is not sensitive to dust or other environmental conditions.[4]

Another modern yarn break electric stop motion, with drop needles and dust-proof contactors or electric stop motion with ceramic eyelets and a vertical drop, is offered by VTA Belgium (Fig. 3.18).[5] Benninger offers a unique broken yarn detection device called FILGARD (Fig. 3.19) on their Ben-Creel.[6] When an end breaks, two prominent LEDs flash on the outside of the stop motion assembly, clearly indicating the tier with the broken end. A digital display on the electrical panel shows the tier and the number of the broken end. Repeated end breaks at the same position are specially signalled

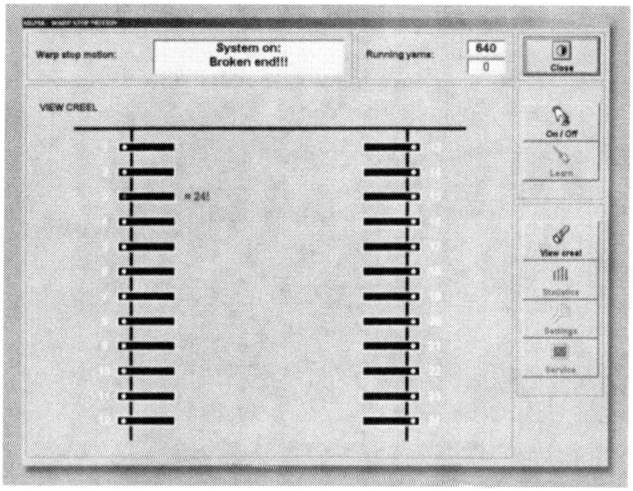

3.17 Warp stop motion.

3.18 Yarn break detector.

3.19 FILGARD yarn detector.

which enables troublesome packages to be identified with certainty and replaced if necessary. The minimum tension detection is only 3 cN, which allows even the finest yarns to be monitored. The detector needle can be adapted to cope with divergent yarn counts.

3.7 Single-end warping machines

Traditional warping methods all have their strengths and weaknesses. Even after many years of technological improvement in the design of warping machines, the basic principles and ways of producing warps have not changed. It remains necessary to prepare and use a considerable number of packages, whether in the form of bobbins, cones or cheeses, which constrains the process.

Customer requirements are changing rapidly for a variety of reasons. For example, the demand for small quantity orders of specialized products at short notice is becoming very common. New production techniques with high flexibility are continuously being introduced into the industry. Firms engaged in the production of fancy and patterned fabrics cannot afford to follow the traditional method of warping to produce 10 to 20 metres of fabric as a sample. The answer to this problem is single-end warping technology which has come into the market in the last few decades.

This requires a minimum single package to prepare a sample warp of the desired density. It is equally possible to produce a patterned warp in any sequence according to the design of the fabric. The process is highly automated and dramatically increases the efficiency of sampling and small volume production. In addition to the use of single-end warping by the fashion and design industry, the equipment is ideal for academic institutions actively engaged in running courses in the field of woven textile design.

A number of manufacturers are marketing single-end warping machines. Figure 3.20 shows a LUTAN[9] model which is available in two versions: v 3.6 and v 5.0 from CCI TECH INC in Taiwan. This can produce warp lengths from 11 to 200 m with a working width of 2,200 mm on Model v 3.6, and 15–500 m on Model v 5.0. In the simplest version, a stationary creel accommodates one package of each colour. Changing of the colour is automatic and allows any possible warp arrangement to a maximum of eight colours. Figure 3.20 shows a patterned warp wound on a drum which is then transferred to the weaver's beam positioned in front of the drum.

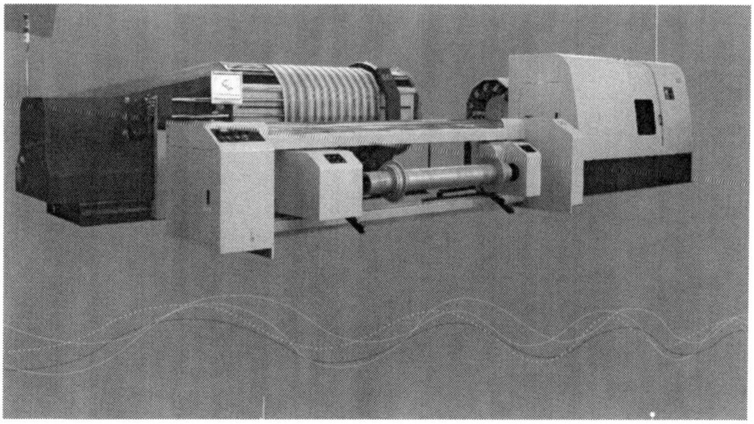

3.20 LUTAN single-end warping machine.

3.8 References

1. Schlafhorst warping machine, MZD brochure.
2. G.A. Bennett, *Introduction to Automatic Weaving*, Harlequin Press Co. Ltd, Manchester, 1948.
3. Schlafhorst warping and beaming machine, DSB brochure.
4. Huys & Vanhevel n.v., Tiegem, Belgium, Brochure – A new generation of section warping machines.
5. V T A, Ingooigem, Belgium – Creels, brochure – Creels and tension units.
6. Benninger AG, Switzerland, brochure BEN-*CREEL*.
7. A.L. Landau, *Warping and Slashing, Sizing Part 1*, International Text Book Co., Scranton, PA, 1949.
8. Tsudakoma Corporation, Kanazawa, Japan, brochure A03QVH05HE.
9. LUTAN single-end warping machine brochure, CCI TECH INC, Taiwan.

4
Yarn preparation for weaving: sizing

K. L. GANDHI, Chartered Consultant, UK

Abstract: The chapter begins with the basic concept of sizing spun cotton and filament yarns. Yarn's hairiness considered to be the most serious hazard in weaving single spun yarns in their unsized state is illustrated with full width weaving trials carried out on single unsized cotton and single unsized singed cotton yarns. It then covers the characteristics of different size materials as well as of a good size mixture. The chapter then deals with the importance of sizing weaving curve and the effect of size on adhesion between fibres in yarn structure.

Sizing machines using cylinders drying and hot-air drying are illustrated in detail along with the factors that affect the percentage of size applied to a warp. The chapter finally covers briefly the automation controls of modern sizing machines.

Key words: size ingredients, modified starches, abrasion resistance, squeezing pressure, sizing weaving curve, cylinder drying.

4.1 Introduction

It is almost impossible to weave single spun warp yarn from cotton, viscose, polyester blends and acrylic/wool blends without the application of size (starch). Most single spun yarns are therefore sized to enable successful weaving of warp yarns. The application of starch, an adhesive material, is carried out in a process called sizing (slashing). Although sizing is a complex process which is key to successful weaving, it is also considered to be a non-value-added process. First, the single spun warp yarns are sized and, after the yarn has been woven into fabric, the protective coating of starch is later removed by a de-sizing process. This enables the selected finishing cycle to be completed without problems, but obviously adds to the cost. Weft yarn is never sized. Correct sizing can help in making cloth from low grade cotton, even where older machinery is used. However, even with new machinery, inadequate sizing will cause serious trouble during the weaving process with resultant losses in material, quality and production time.

In this chapter the author covers only the basic information related to the technology of warp sizing. No attempt is made to cover the chemistry and technology of sizing in any depth. For detailed information on this technical subject, the reader is referred to the books: *Textile Sizing* by Bhuvenesh C. Goswami, R. Anandjiwala and David. Hall, *Sizing of Spun Yarns* by Henry Hacking, *Technology of Warp Sizing*, edited by J.B. Smith, *Warp Sizing Mechanisms* by T. Ramsbottom, *Textile Warp Sizing* by Paul V. Seydel and

Weaving: Technology and Operations by Allan Ormerod and Walter Sondhelm, along with other references given at the end of this chapter.

It is generally thought that sizing is carried out in order to increase the strength of the yarn, but this is not entirely true. There is no correlation between tensile strength alone and the end breaks occurring during the weaving process. The efficiency of sizing treatment cannot be judged by an increase in the breaking strength of the yarn. Flax yarns have a higher strength than cotton and wool but are difficult to weave. The strength of the yarn is many times greater than the normal tension imposed on it by the loom parts during the weaving process and yarn does not usually break because of a lack of tensile strength. On the loom, the warp yarns are subjected to a wide range of mechanical stresses, the most significant being bending, fatigue, oscillating tension and frictional resistance.

4.1.1 Reasons why single spun yarns in their unsized state cannot be woven

The efficiency of the sizing machine at the weaving stage is largely judged by a single criterion, which is the warp breakage rate. A warp break on the loom will occur only when the resistance of the weakest spot in the yarn is less than the tension generated in it. High tension is caused mainly by imperfections in the yarn which create obstructions to the crossing of the warp ends during the shedding process and in their passage through the drop wires, healds and reed. Among the imperfections which are induced during the process of yarn manufacture, the presence of loose fibres called hairs, which protrude from the yarn surface, is considered to pose the most serious difficulties in weaving spun yarns in their unsized state. Yarn hairiness is affected by factors such as fibre properties, spinning parameters and yarn properties.

Hairiness arises from three classes of fibres:

1. Fibres whose ends project from the yarn.
2. Fibres which reach the outer layer of the yarn and then arch into loops emerging from the yarn.
3. Fibres which are only attached to the yarn core, and in the majority of cases, are wound round outside it. These fibres are classified as wild fibres.[1,2]

The length of these protruding hairs can vary from 1 to 5 mm or above. The greater the number of hairs per metre, the more hairy the yarn will be. Yarn hairiness, measured by the number of hairs per metre, can be tested on the Shirley Hairiness Meter. This equipment does not give a figure representing total hairiness but counts the number of hairs exceeding 3 mm in length when the standard setting is used. It appears that hairs emerging more than

3 mm in length from the yarn surface are more susceptible to entanglement with adjacent yarns during weaving. Figure 4.1(a),(b) shows the difference in appearance between the same yarns with and without hairiness.

Properties such as abrasion resistance, breaking strength and elongation and hairiness in two-fold spun yarns are considerably better than in single spun yarns. Sizing is not therefore essential when weaving two-fold spun yarns, although in many cases even these yarns may be sized for better performance.

4.1.2 Weaving trial on unsized warp yarns

A research project conducted a full-scale weaving trial on unsized warp yarns. A beam containing 1,600 ends of 38 tex yarn was prepared. The ends were drawn with one end per heald eye and two ends per dent, weaving plain cloth with 16 ends and picks/cm on a standard Picanol loom running at 220 picks/min (rpm). Fewer than 1,500 picks had been woven before the effect of yarn-on-yarn and reed-on-yarn abrasion was clearly noticed as follows:[3]

- Behind the heald shafts, protruding hairs from adjacent ends were interlocked and prevented the formation of a clear shed (Fig. 4.2(a),(b)). Figure 4.2(b) shows the fibres being pulled out from the yarn structure.
- Accumulations of detached fibres from the yarn structure as a result of yarn-on-yarn abrasion and the reed-on-yarn abrasion were forming beads in between the front heald and the reed (Fig. 4.2(c)).
- Behind the healds, a cloud of detached pieces of fibres were thrown into the air by the upward movement of the warp yarns, forming a reservoir

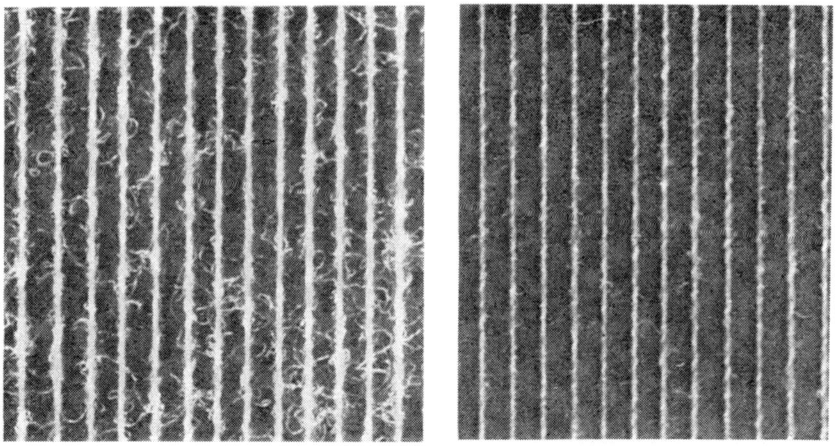

(a) (b)

4.1 (a) Yarn hairiness before and (b) after singeing.

for further obstructions. These detached (wild fibres) and broken fibres were seen to be collecting on the loom parts and on the healds. Some fibres fell on the warp and were quickly wrapped around the yarn (Fig. 4.2(d),(e)).

As the weaving progressed, a considerable degree of obstruction was caused by the beads formed on many of the warp ends across the entire width of the warp (Fig. 4.3). Beads also formed at the back of the healds where ends had

(a)

(b)

4.2 (a) Protruding hairs interlocking with warp during shedding and preventing clear formation of shed, (b) fibres being pulled out during shedding, (c) formation of beads on warp between front heald and reed, (d) detached fibres settling on heald parts and intermittently falling on the warp yarns and (e) clouds of detached pieces of fibres thrown into the air during shedding.

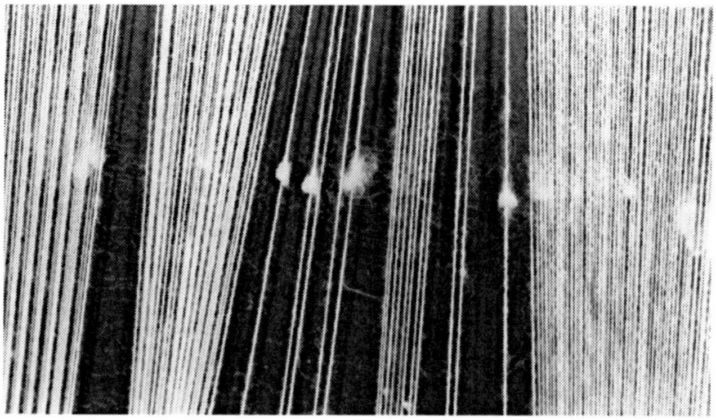

(c)

(d)

(e)

4.2 Continued

4.3 Beads formed in front of reed across the whole width of warp as a result of reed-on-yarn abrasion.

interlocked. Obstructions were cleared by hand but these became so numerous that after 4,000 picks, weaving had to be abandoned. It is noteworthy that for 4,000 picks inserted, only one warp break occurred and this was directly caused by one of the beads (Fig. 4.4(b)(i)). The main deduction seems to be that the yarn was able to withstand the normal weaving tensions and did not suffer significantly from fatigue in its passage from the beam to the fell of the cloth, despite being unsized.

During the formation of the beads it was observed that in the initial stage of their formation (Fig. 4.4), they were not firmly attached to the yarn. A bead, while still small. could slide from its place of formation under the influence of the reed, but as the bead increased in its length and width, it became fixed on the yarn and travelled forward along with the parent yarn. The size of the particular bead and the movement of the reed put the end under high tension which resulted in it breaking. Beads as big as 15 × 3 mm in size were formed on the yarn.

Some of the beads were seen to behave in a typical manner, i.e. due to yarn-on-yarn abrasion, the ends adjacent to the one carrying the beads detached some of the loose fibres from the outer surface of the bead. This resulted in the loose fibres being discarded, either falling under the loom, or, as was observed in some cases, wrapping themselves around the yarn itself or interlocking the two ends. Figure 4.4(b)(iv) shows the appearance of a typical bead from which the fibres had been detached. This bead initially had more or less the same appearance as the one shown in Fig. 4.4(b)(ii) or (iii). The structure of the beads formed in the course of weaving unsized warp was analysed to prepare a Baer-sorter diagram in order to find out the effective length of the fibres. This was found to be 20 mm (Fig. 4.4).

This analysis clearly showed that these long hairs must have been pulled out from the structure of the yarn and not merely broken off. It became clear that the protruding hairs from adjacent ends interlocked and, during

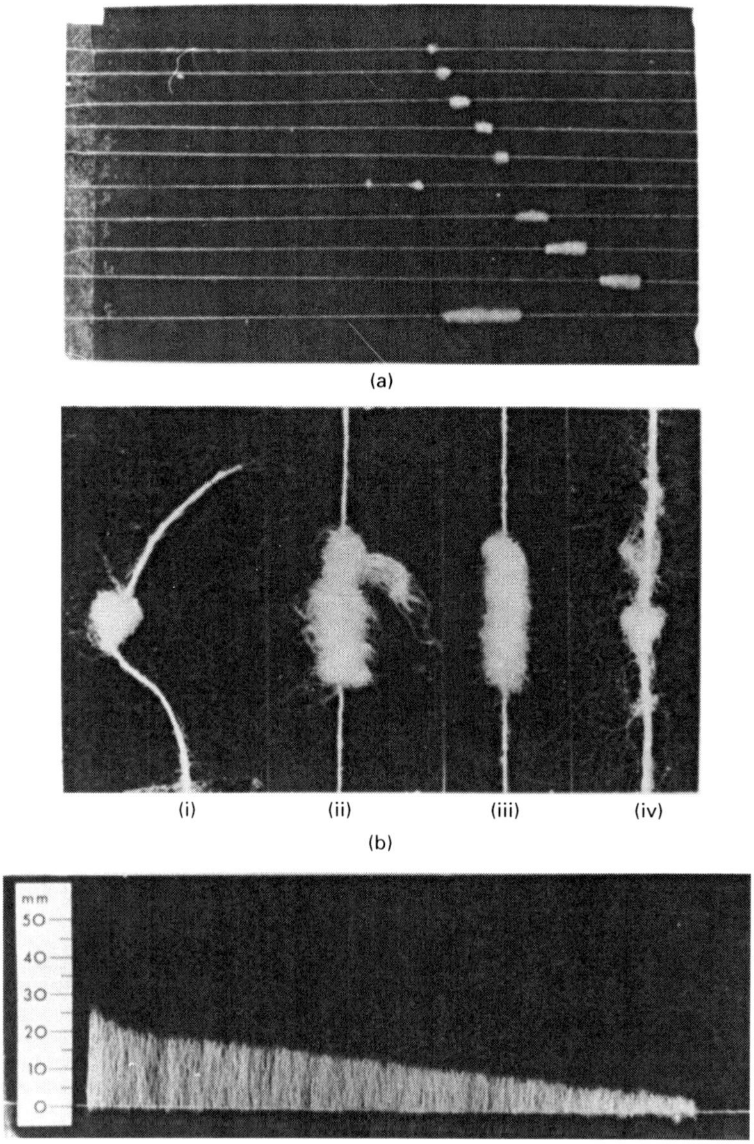

4.4 (a) Different stages of bead formation (b) various beads and (c) Baer-sorter diagram of fibres removed from beads.

the crossing of the warp (shedding), threads were put under tension which either pulled the hair completely from the yarn structure, or after a certain length had been pulled out, caused it to rupture. The severity of entanglement depends on the amount of yarn-on-yarn abrasion, reed-on-yarn abrasion and heald-on-yarn abrasion during the weaving process and the number of ends and picks/cm.

Thus it was concluded that because of the hairiness of the yarn, unsized single spun warp yarns, cannot be woven unless they are treated to improve their hairiness, yarn-on-yarn and reed-on-yarn abrasion resistance. After the treatment, the protruding hairs will not become entangled with each other nor will they be peeled off from the yarn surface during the reed-on-yarn abrasion, which causes the formation of beads.

4.1.3 Effect of singeing on yarn hairiness and weavability

Singeing sewing threads and fabrics burns the protruding hairs to achieve a better appearance in the end products. It is considered desirable to carry out this process to remove the protruding hairs from the yarn and then study the weavability behaviour. As an example, a 37-tex break-spun yarn was singed at a speed of 130 metres per minute. Figure 4.1(a),(b) shows the hairiness of the yarn before and after singeing. The decrease in yarn strength and breaking extension was found to be insignificant.

Weaving trial on singed unsized yarn

A large-scale trial was conducted in which a beam with 1,600 ends from 38-tex singed yarn was prepared to produce a plain cloth with 15.7 ends and picks/cm. A standard Picanol loom running at 220 rpm was used. It was observed that no difficulty was experienced in weaving the warp. Even in the unsized condition, the warp sheet divided clearly and in the absence of hairiness, yarn-on-yarn abrasion did not cause any entanglement of the adjacent warp ends (Fig. 4.5(c),(d),(e)). There was no formation of beads on any of the warp ends across the entire width of the warp. A total of 50,000 picks were inserted and only five warp breaks recorded. Compared to this performance, the trial on unsinged and unsized warp as described earlier had to be abandoned after only 4,000 picks.[3]

In another weaving trial, an unsinged sized warp had only 0.8% of size applied to it. During weaving, a significant number of fibres were detached from the warp as a result of yarn-on-yarn abrasion (Fig. 4.5(a),(b)). This significant undersizing caused many problems during weaving. The trials confirmed that it is the initial hairiness of the yarn which impedes the weaving of spun yarns in their unsized state. An ideal percentage of size must be applied to single spun yarns to improve their abrasion resistance and fatigue properties if weaving results of a high standard are to be obtained.

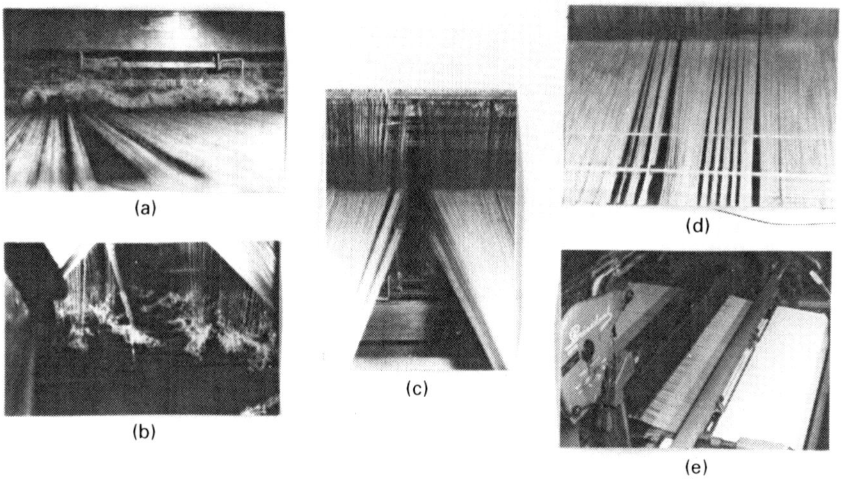

4.5 (a) and (b) Fibre removed from unsinged sized yarn with only 0.8% of size collected on top and bottom of the heald shafts respectively (c) and (e) heald shafts free from fibre deposit when weaving singed unsized yarn (d) singed unsized warp at the back of the loom free from entangled fibres.

4.1.4 The sizing process

Preventing the protruding hairs from entangling with each other and improving the yarn's abrasion resistance is achieved by applying an adhesive (starch) paste around the yarn surface. This is prepared by boiling suitable starch material in water along with other ingredients which are used to enhance the film properties. The process, known as 'cooking' prepares the size paste to a viscosity which will have a significant effect when applied to the yarn.

The application of the adhesive involves immersing the warp yarns in the size paste. The surplus is squeezed out, the warp sheet is dried, then the yarns are separated and wound onto the weaver's beam. After drying, a film of starch is left around the yarn surface and it is this film which improves the abrasion resistance of the yarn. It also enhances its tenacity and fatigue properties as well as the elasticity to withstand the stresses of the weaving process.

4.1.5 Size ingredients for yarns

The main ingredients used in sizing cotton warps are adhesives (starches) and lubricants/softeners.

Starch adhesives

Most adhesives used in sizing are starches. Starch is a hydrophilic polymer which is readily influenced by moisture or water. The adhesion capacity of the sizing product is important. Starch, a natural product obtained from several sources, is still the major sizing ingredient in the formulation of size paste along with other materials, the use of which depends on many factors. Starch is used for fibre-to-fibre cementing action and for laying the protruding hairs firmly on the yarn surface. A number of natural starches are available, i.e. sago, potato (farina), maize, tapioca, wheat and rice. Starches vary widely in their physical properties, exhibiting differing adhesion, viscosity and other characteristics. The most commonly used are sago and maize. In general, sizes of low viscosity are chosen as they are easier to handle, the take-up of paste is smaller and thus the drying of the warp sheet is easier. It is also more economical and achieves better penetration into the sheet and into individual yarns. However, high viscosity is an advantage in certain applications where only a low, or very low, concentration is necessary and the yarn is of a type in which surface coating rather than a high degree of penetration is of value.

Modified starches

Modified or water soluble starches have a number of advantages over unmodified starches. These include uniformity, size consistency, ease of removal and a shorter preparation time. While the modified starches as a whole have adhesive properties similar to those of the natural starches from which they are derived, they fall short in their film-forming properties. To compensate for this, other film-forming substances such as water soluble ethers and polyvinyl alcohol (PVA) are used as film strengtheners and binders with modified starches. Starch ethers are excellent water soluble sizing materials. Solvitoses are important members of this group which require mild de-sizing treatment.

Thin boiling starches

Thin boiling starches are also known as soluble, high-fluidity, or oxidized starches. Fluidity is the reciprocal of viscosity, so a starch with low viscosity has high fluidity. Once the thin boiling starch mixture has reached its required viscosity, it will remain more stable than a raw starch mixture, reducing only slightly in size viscosity. Most of the soluble starches are made from corn starch but other starches may also be used. Starch is added to water and is usually treated with chemical agents which tend to oxidize the starch while it is in a water solution. These starches usually produce a more flexible film and are suitable for the sizing of fine counts of yarn.

Thin boiling starches are converted to more soluble types by acid hydrolyses. These are necessary when sizing heavy yarns for penetration and binding power. In principle, they are used to obtain more weight or a thicker film on the yarn. They need less cooking time than regular starches to stabilize the viscosity.[4–6]

Size materials for synthetic/filament yarns

Filament yarns are composed of continuous filaments aligned in parallel rows which run straight except when they are coiled by the insertion of a twist, giving the yarns an appearance of evenness, smoothness and absence of hairiness. Most of the filament yarns may not require sizing. However, sizing improves their efficiency by a further 15–20%. Sizing improves the frictional resistance of the yarn and bonds the filaments which prevents the peeling back and intermingling of any broken filaments which are present on the yarn before weaving or may break during weaving. In order to achieve this, the selected size must have sufficient adhesion to the fibre to make the filaments cohere.

Requirements for sizes for the treatment of synthetic yarns are different for continuous filament and for spun yarns. Continuous filament yarns (a) have a strong cohesion due to continuity of their filament, (b) possess good elastic behaviour to cyclic extension, and (c) have a higher degree of elongation. Spun yarn has a greater capacity than continuous filament yarn for absorbing size. The traditional starch-based size material suitable for spun yarn has inadequate adhesion to smooth synthetic fibres and cannot be used successfully on synthetic yarn/filament yarn.

Size materials adapted to the specific properties of the fibres are used for synthetic yarns. The choice of the correct size material depends on the type of synthetic yarn, i.e. polyamide, polyester, etc. A satisfactory size recipe for filament yarn, in addition to providing a good adhesion to the yarn, should possess good film-forming properties. It should also impart strength, elasticity, resilience and lubricating properties to the yarn.[7]

The film-forming materials used for filament yarns are generally high polymeric substances capable of being dissolved in water or some other suitable solvent. They are classified into three main groups:

- natural polymer (protein group)
- regenerated polymer
- synthetic polymer.

Table 4.1 lists the classification of sizing materials available for the sizing of filament yarns.[8]

Table 4.1 Classification of sizing materials

Generic name	Examples	Fibre
Natural polymers	Gelatin	Acetate and viscose
	Modified starches	Viscose, glass
	Gums	Viscose
Regenerated polymers	SCMC[a]	Viscose
Synthetic polymers	PAA[b]	Polyamide, acrylic
	Vinyl copolymers[c]	Polyester, acetate, acrylic
	PVOH (PVA)[d]	Viscose, acetate, polyamide, glass

Notes
[a]Sodium carboxymethyl cellulose
[b]Polyacrylic acid
[c]Acrylic copolymers, vinylacetate copolymers, styrene copolymers
[d]Polyvinyl alcohol.

Sodium carboxymethyl cellulose (SCMC)

This is a cellulosic adhesive, water soluble and capable of being easily removed from the cloth during finishing. It is supplied in a range of viscosities and has a high film strength and considerable adhesive power. It can be mixed with PVA and in conjunction with modified starch and PVA, can be used on spun viscose and acrylic fibre.

SCMC produces a tough, clear and flexible film. It tends to absorb and hold moisture so its water binding ability makes the need for high humidity in the weaving room less critical than with other sizes. In solution, the viscosity of SCMC is decreased by raising the temperature and increased by lowering it.[4,5]

Polyvinyl alcohol (PVA)

PVA is synthesized from petroleum and produces size films that are very strong, have good abrasion resistance and are easily de-sized in hot water. It is basically a synthetic polymer resin and can be used satisfactorily on spun yarns in conjunction with soluble starches.

Polyacrylic acid for sizing

This is an excellent adhesive for spun nylon yarns and can be used in the sizing of hydrophobic fibres and their blends such as nylon, acrylics, acetate, polyester, etc. The size gives good binding to these fibres and can also be made water insoluble, permitting their use in water-jet weaving applications.[9]

Lubricants and softeners

When applied on its own, starch tends to make the yarn somewhat harsh and brittle which is detrimental to the yarn during the weaving process. To overcome this, lubricant/softener materials are added to the size paste. Their use reduces friction between yarn-on-yarn, reed-on-yarn and heald-on-yarn, which facilitates the beat-up of the weft. They also reduce the dusting-off due to easier and better separation in shedding. Softeners give a soft handle to the yarn, and help decrease the brittleness of size film, reduce abrasion resistance, increase elasticity, improve size penetration and reduce the sticking of yarns in the warp sheet and on the drying cylinders. Softeners also tend to lubricate a yarn and allow it to slide through the various parts of a loom without chafing. The most common lubricants are tallow, spermaceti and paraffin wax. Emulsified tallow and soluble mineral oils are sometimes used.[5,10]

Tallow is probably more extensively used as a softening agent than any other material. It acts as a lubricant as well as a softener. Its proportion in the size paste is generally between 8 and 10% of the dry weight of the adhesive used. Other common softeners such as castor oil, glycerine, and paraffin wax are also used. Glycerine is an excellent softener as it possesses the desired softening qualities and is hydroscopic in nature, thus maintaining a high regain in the yarn. Caution is required in the use of softeners. If used in quantities in excess of 8–10% in weight of the starch, they will tend to weaken the size film and its strength and abrasion resistance.

4.2 Characteristics of a good sized yarn

The primary characteristics of a good size are:

- tensile strength and elongation
- flexibility
- adhesion
- proper viscosity
- removability
- consistency
- abrasion and sticking resistance.

The secondary characteristics of a good size are:

- transparency
- resistance to mildew and bacteria
- cost
- lack of odour
- softness

- lubrication
- non-corrosive and non-abrasive to machine parts.

The most important of these characteristics which contribute the most to weavability in sized warp yarns are:

- substrate adhesion
- elongation
- flexibility
- viscosity
- desizeability
- abrasion resistance.

4.3 Size mixtures: composition and quality

Size mixtures, their composition and the quantity to be applied to the yarn depend on many factors and conditions and may vary significantly. The most important factors to be considered are:

- Type of yarn (i.e. counts, material, and turns/metre): A size mixture may be required to size a high-twisted, high count yarn efficiently, while another mixture may be required for a soft, low-twist, low count yarn. Each yarn is used for an entirely different purpose and should fulfil all the requirements. The size mixture for each type of yarn is varied for the appropriate characteristics.
- Fabric parameters (ends and picks per centimetre): The higher the number of picks per centimetre in the fabric, the greater is the amount of heald-on-yarn and reed-on-yarn abrasion for each inch of woven cloth. This is because the warp yarns are not drawn forward as rapidly as they are in cloth having a low number of picks per centimetre, and the reed moves forward and backward a greater number of times during the beat-up operation. During this operation, the reed dents severely abrade the warp yarns. Similarly, the higher the number of ends per centimetre in the warp, the greater is the yarn-on-yarn abrasion and friction during the shedding process when the healds change their positions. Therefore, if yarns are to withstand the strain and abrasion forces under such conditions, they will normally require a greater amount of size.

There is no single sizing 'recipe' which is valid for all processes. A recipe suitable for the sizing of all cotton-warp yarns, or blends or synthetic yarns cannot be formulated because of the multiplicity of the fabric constructions and yarn compositions and the different types of weaving machinery. The size recipe and the amount of size deposited on a yarn will vary according to the above mentioned conditions, along with the type of loom used.

4.3.1 Quantity of size mixture applied to the yarn

It has been established that it is impossible to weave single spun yarns without the application of size. The question arises as to the quantity of size applied to the yarn. If too little is applied, the warp would perform better than in an unsized state but problems would remain which prevent successful weaving. If too much size is applied, it will seriously affect breaking extension and flexibility, so leading to an increased number of warp breaks and a loss of productivity. This suggests that there is an ideal amount of size application for optimum results in weaving.

If the warp breakage rate were studied in similar warps sized to varying percentages, it would be found, as shown in Fig. 4.6(a) that the warp breakage rate is reduced as the percentage of size is increased.[6] This is due to reduction in yarn hairiness and a significant improvement in the yarn abrasion resistance due to the coating of size film as well as to an increase in yarn strength.

The reduction in warp breakage rate with an increase in the addition of size, reaches a point beyond which any further addition of size will not show a significant improvement in yarn performance on the loom. In fact, it will tend to increase the number of warp breaks. An excessive size add-on leads to an increased penetration of the size inside the yarn structure, which significantly decreases the yarn's breaking extension. Between the two extremes, a certain size percentage gives the minimum breakage rate and this percentage is called the optimum or ideal size percentage for application to the yarn. The curve plotting the breakage rate against the percentage of size on the warp is defined as the 'sizing-weaving curve'. Warp breaks are generally expressed as breaks/1,000 ends/10,000 picks.

The weaving efficiency, which is inversely proportional to the warp breakage rate, reaches its peak when the warp breakage rate is at its minimum. The

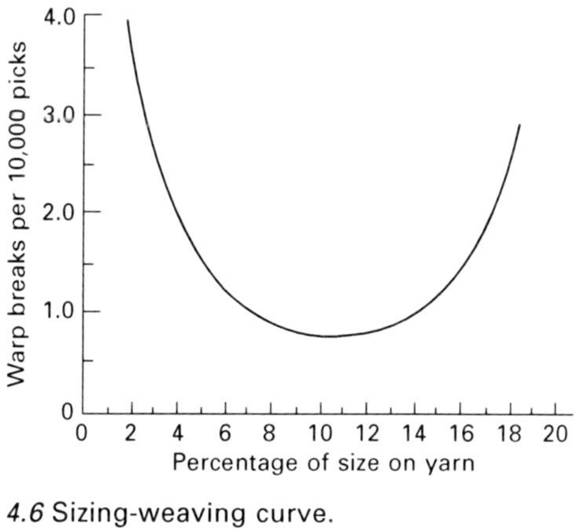

4.6 Sizing-weaving curve.

highest weaving efficiency consistent with the optimal addition of size is usually achieved by trial and error.[11] In practice, it is not simple to identify the ideal size percentage for achieving optimum weaving results for a particular fabric style. The relationship between warp breaks and the amount of size on the yarn will vary greatly, depending on the composition of the size, the type of yarn, the fabric particulars, pick-up of size paste from the size box, type of loom and many other factors.

4.4 Effect of size on adhesion between fibres in the yarn structure

On the basis of information gathered while weaving unsized yarn, it is obvious that sufficient inter-fibre adhesion is required to prevent the hairs being pulled out. This is achieved by adding a suitable percentage of size to the yarn. The size forms bonds between the fibres (Fig. 4.7(a)) and the strength of the bonds depends on the type of size used. A breakdown between fibre and size or in the size itself reduces the adhesion between fibres, making them liable to be pulled out from the yarn structure. This causes yarn breakage. An investigation into the value of this function of sizing was carried out. The amount of force required to extract individual hairs from unsized yarn and from yarn with different degrees of sizing, using sago as the adhesive, was taken as the criterion to evaluate the effect of size on adhesion between fibres. This force was measured on the Instron Tensile Strength Tester, as shown in (Fig. 4.7(b)).

A single hair, over 3 mm in length, was taken at random from a 20 cm length of yarn surface and gripped by the jaws. As can be seen from the figure, when the hair is pulled, it does not remain perpendicular to the yarn axis as the force increases. The angle between the hair and the yarn axis changes constantly and is difficult to observe at the instant when the hair either breaks or is detached. Therefore, the force required to pull the hair is defined as the component at right angles to the yarn of the tension in the hair.

Results from 100 tests carried out are summarized in Table 4.2. Some hairs were pulled out completely, but others broke only when partially extracted. With unsized yarn, some hairs, between 5 and 20 mm in length, were pulled out without registering any force. It is presumed this could happen with loose hairs (wild fibres). The number of hairs completely pulled out from the unsized yarn with a force of 0.1 to 1.0 gf was found to be 33%. With an addition of 7% size, all hairs were anchored to the yarn and none of the hairs were pulled out or broken from the sized yarn in the above range of force. Although a significant increase in the average force required to pull out or break a hair from sized yarn was observed when compared with unsized yarn, it was found that this increase was not proportional to the amount

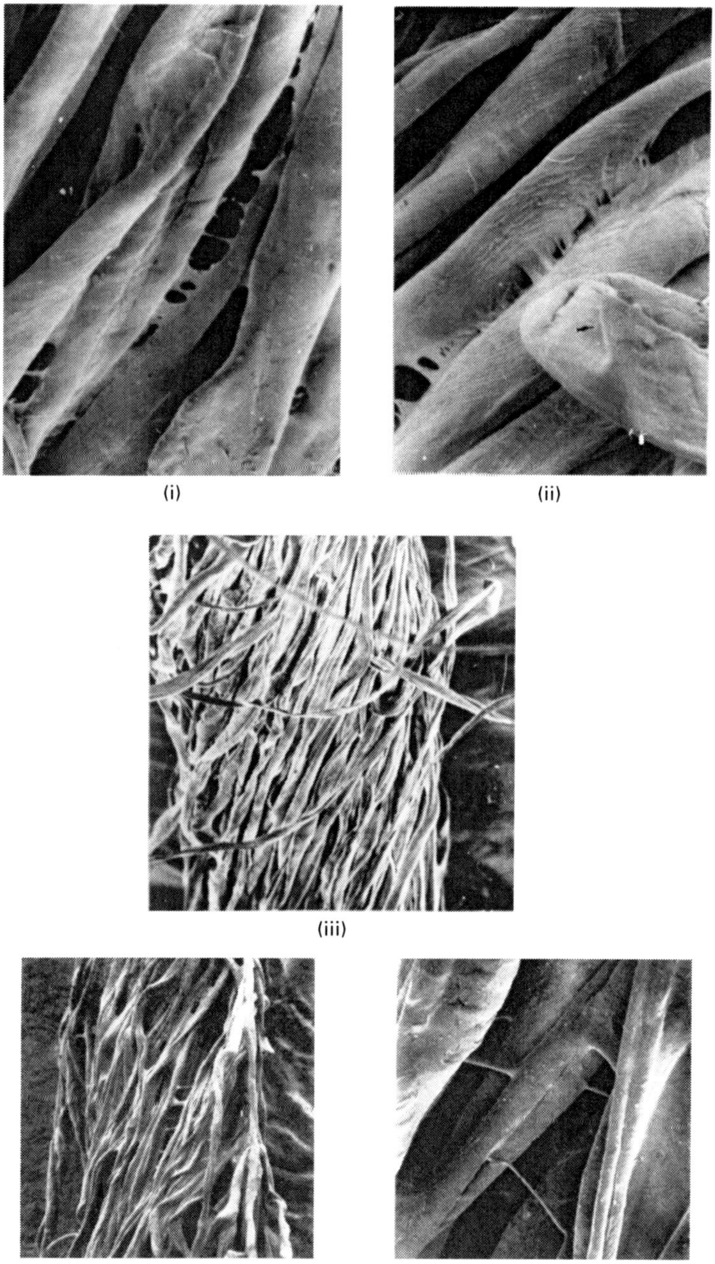

(i) (ii) (iii) (iv) (v)

(a)

4.7 (a) Size film bonds being formed between fibres and (b) hair being pulled out from the yarn.

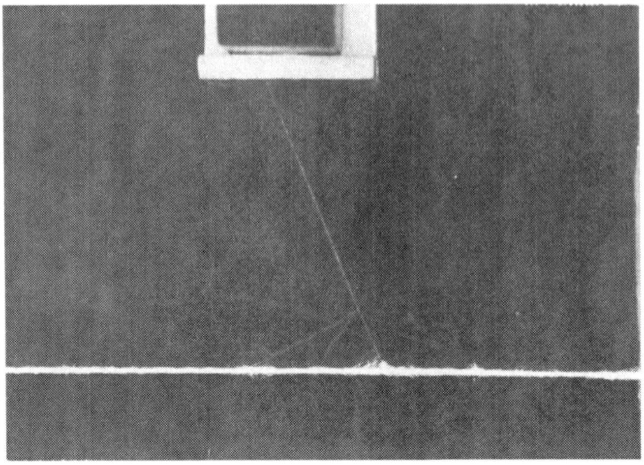

(b)

4.7 Continued

Table 4.2 Force required to pull or break a hair from the yarn structure following single-end sizing.

	Force required to		% hairs		Average length of hairs removed from yarn when	
Size added to yarn (%)	Pull out the hair (gf)	Break the hair (gf)	Pulled out	Broken	Pulled (cm)	Broken (cm)
0	2.0	3.7	54	46	1.30	1.23
7	2.6	4.18	16	84	1.16	1.0
14	3.0	4.0	7	93	1.0	0.78
21	4.4	4.78	5	95	0.9	0.7

Size formulation: sago/tallow (10:1). Yarn: Cotton, open-end spun, 38 tex.

of size added to the yarn. No significant difference was observed between yarn with 14% of added size and that with 21% in respect of the ratio of the number of extracted to broken hairs or to their average length.

The results suggest that beyond a certain point, the further addition of size does not significantly increase adhesion between fibres. Thus any additional amount of size added to the yarn in excess of that ideally required will not serve any useful purpose. It is also likely to decrease the yarn's breaking extension, so leading to a higher warp breakage rate during weaving.

4.5 Size paste preparation: cooking

Size preparation (also known as cooking) is carried out in large wooden or stainless steel kettles fitted with suitable types of agitators to stir the paste.

Depending on the concentration of the size paste, the required quantity of water and size materials are mixed in the kettle and the solution is cooked by injecting steam. The viscosity is controlled and when the size is cooked to the required degree, it is transferred to storage tanks from where it is then pumped into the size box to the sizing machine.

Pressure cookers are also used for cooking size paste. These substantially reduce the cooking time for starch, especially for unmodified starch. Water and size ingredients are added to the vessel and boiled. The steam inside is allowed to build up the pressure to a desired level which is maintained for the required period. When the cooking is complete, the paste is transferred to a storage tank from where it is pumped to the sizing machine.

A number of size preparation systems have been developed to allow quick cooking of starches to a reasonably low and stable level of viscosity. These systems have been automated to a certain extent and are effective not only with starch, but with ingredients which are not easy to mix.

4.5.1 Preparation of size paste when supplied in liquid form

Some sizes are supplied in liquid form for special usage. These are not heated during preparation and the vessels required can be very simple. From a main storage tank of size, specified quantities are passed to a preparation vessel, mixed with water and (if necessary) a suitable lubricant, and then after agitation the size will be ready for use.

4.6 Sizing machines

Figure 4.8(a) shows a typical two-cylinder sizing machine which was in use a few decades ago but is now obsolete. Cylinders of different dimensions were used but the typical arrangement was a large cylinder about seven feet in diameter and a smaller one with diameter of about four feet. The width varied according to the width of the sizing machine and the production requirements of the individual mill. The cylinders were steam heated and the drying temperatures varied from 100 to 105°C, depending on the yarn count, density of the warp being sized, the machine speed and the percentage of size applied, etc.

Figure 4.8(b) is a schematic diagram of a modern seven-cylinder sizing machine for spun yarns.[12] Most modern machines, depending on various factors, are capable of working at 100–120 m/min. On any sizing machine, the warp has to pass through five zones or sections: the creel, the size box, the drying cylinders, the leasing section and the headstock.

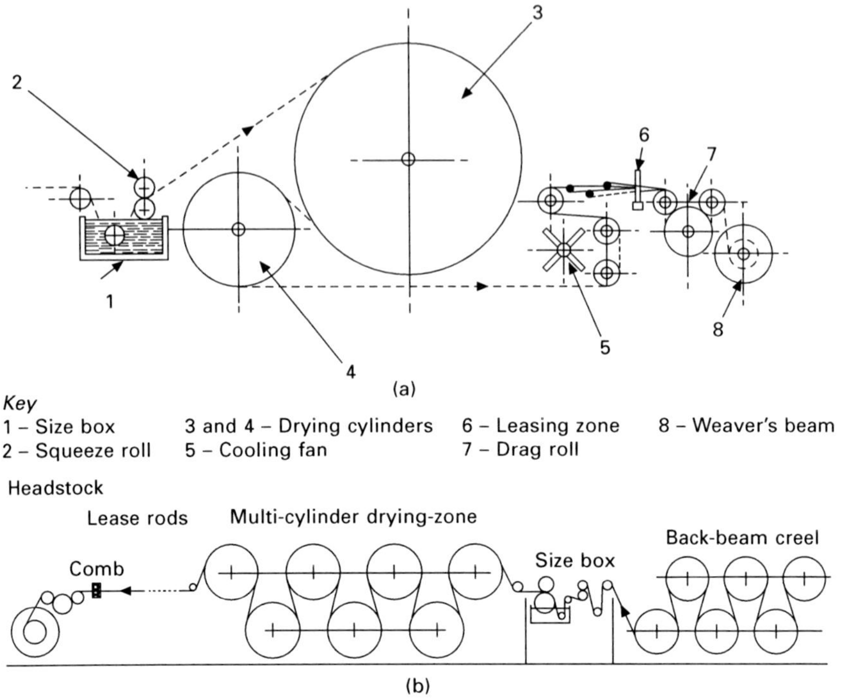

4.8 (a) Schematic layout of two-cylinder sizing machine (b) schematic layout of a modern seven-cylinder sizing machine.

4.6.1 Creel section

This is the first section of a sizing machine and is located at the back of the machine to accommodate the warper's beams. There are various creel designs commonly in use on modern sizing machines. In the over/under creel design, the beams are positioned as shown in Fig. 4.9(a). This type of creel is preferred when the warper's beams commonly used for sizing of spun yarns of cotton and synthetic fibres are exactly the same length.[6] Warp ends in sheet form are drawn from the first and second beams, then drawn forward and combined with the third beam. The process is continued until all the ends have been collected from all the beams in sheet form which is then fed into the next section of the machine, which is the size box.

When the warp is drawn forward from the back beams, it is leased. Lease strings are placed between the yarn sheets from each beam and the yarns are drawn forward until the lease strings reach the headstock of the machine. At this point, the lease strings are removed and split rods (lease rods) are inserted in their place.[13] Figure 4.9(b) shows other designs[6] of creel, methods of collecting warp ends from the back beams and feeding all the yarns in a sheet form to the size box.

4.6.2 Size box

The size box (Figs 4.10(a) and 4.8(b)) consists of an immersion roller, steam heated pipes and a pair of rollers known as the size roller and the rubber covered squeeze roller.[12,14] The configuration of the size box may vary, but its basic function is to apply the size paste to the yarn. As the warp sheet enters the size box, it is passed under the immersion roller which impregnates the warp sheet with the size solution. The immersion roller can be raised or lowered as required. The depth of the immersion in the size solution and the speed of the machine govern the actual time the yarn takes to pass through the size box. The longer the yarn is in contact with the size solution, the more size it will absorb. In general, the greater the speed of the machine, the higher will be the take-up of size paste.

The size paste is kept at a desired temperature by means of steam passing through the perforated steam pipes. Usually, the higher the temperature of the size paste, the greater the degree of size penetration into the yarn structure. The warp loaded with size paste is then passed through the nip of the size and squeeze rollers.

In a size box there may be either one pair of squeezing rollers or a double squeezing roller. In the double squeezing process, the yarn is immersed twice. Some modern sizing machines are equipped with several size boxes. The purpose of extra immersion and squeezing is to facilitate as full a penetration of the warp sheet as possible. The squeeze rollers on modern

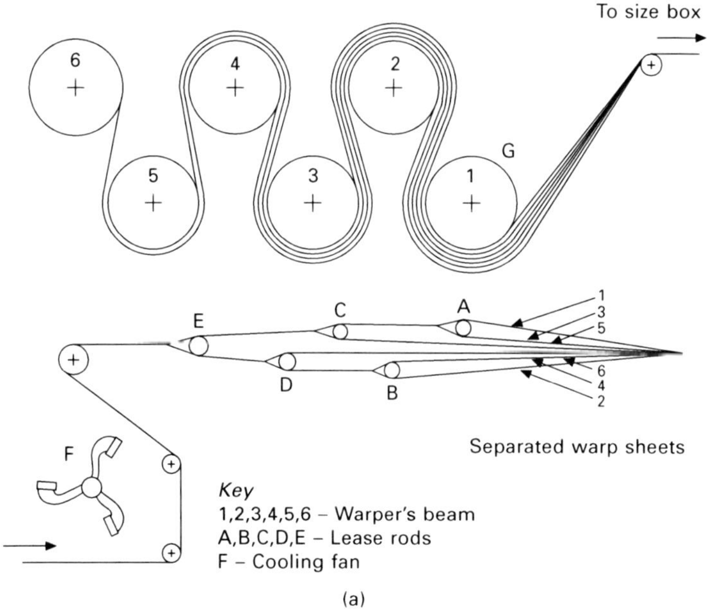

4.9 (a) Over and under (zigzag) creel (b) different creel designs.

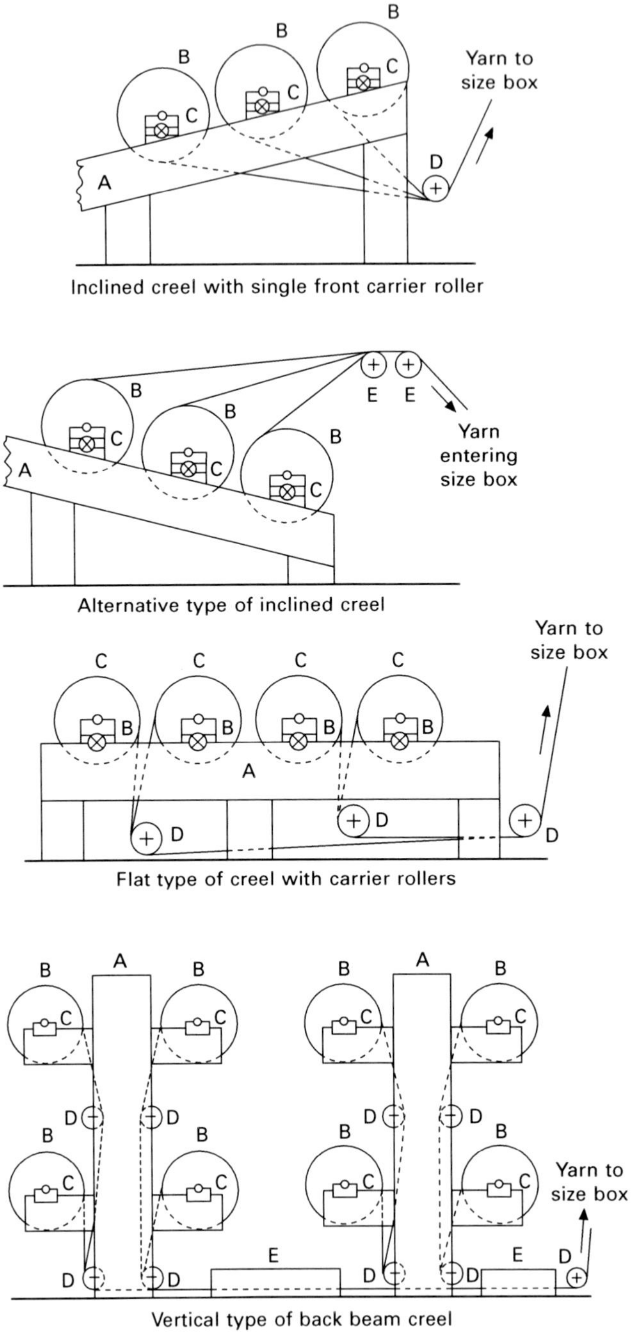

(b)

4.9 Continued

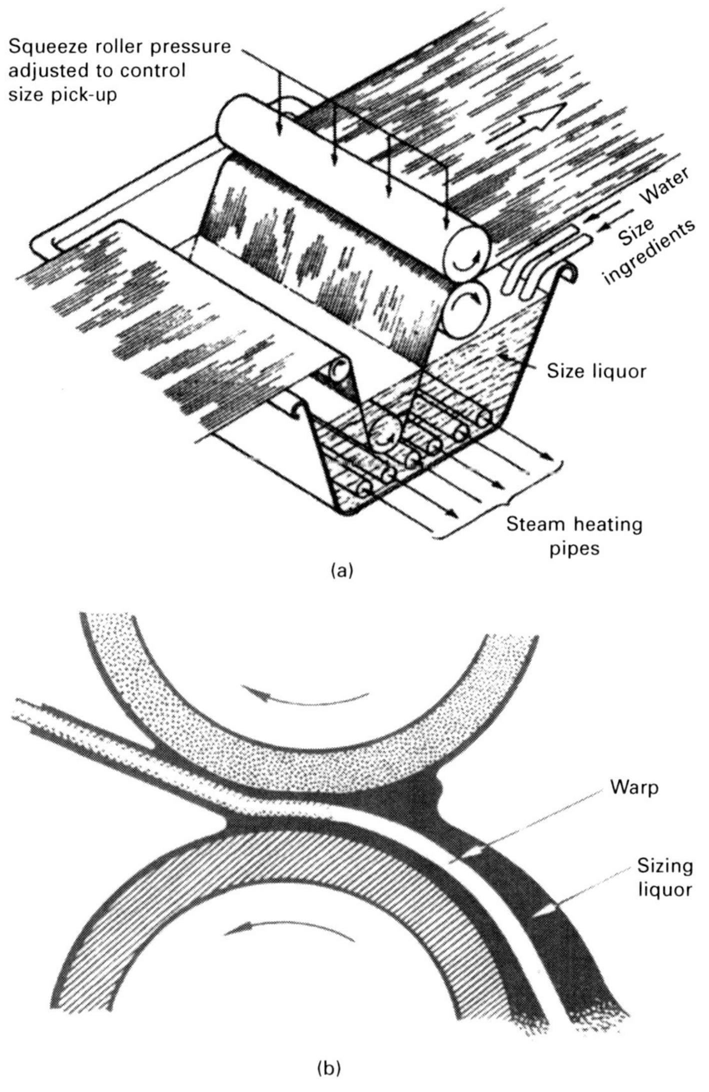

4.10 (a) Size box of a sizing machine (b) Schematic representation of squeezing roller action.

machines are covered with rubber of shore hardness 45°. Rubber rollers on modern machines often have a basic weight of 230–320 kg with extra loading facilities to increase the effective weight well beyond 450 kg.[5]

The squeeze roller exerts a defined amount of pressure on the warp sheet so that most of the excess size is squeezed off the surface of the yarn and a certain amount of penetration into the yarn also takes place (Fig. 4.10(b)).[5] Thus a sized yarn has a coating of size film and penetration of adhesive into the yarn structures. The amount may vary. Penetration of the adhesive into the yarn structure prevents disintegration on occasions when the outer coating fails to withstand the forces of abrasion. It also binds the fibres and prevents them being pulled out. Coating helps to lay the fibres, reducing the hairiness of the yarn and affording initial protection against abrasion.

The pressure exerted by the squeeze roller has a direct effect on the amount of size picked up by the yarn. The heavier the pressure, the lower the pick-up. A slight increase in penetration is more than counterbalanced by the extra size removed from the surface of the sheet of threads by the higher pressure. Increasing the weight or pressure of the squeezing roller tends to decrease the amount of size added to the yarn. By using a higher concentration of size and a heavier roller, the same percentage of size can be applied to the yarn as with a lower concentration of size and a lighter roller. Less water will be picked up by the yarn so it will require less drying and the machine speed can be increased accordingly. An increase in roller weight from 90 kg to 227 kg reduces the percentage of size required in pure sizing by about one-fifth.[6]

4.6.3 Assessing size box performance: size concentration, pick up and percentage of size on yarn

There are a number of variables to consider in assessing size box performance:

- *Size concentration* is defined as the percentage of oven-dry sizing material in a given quantity of size or paste.
- *Pick-up of paste* is the number of kilograms of size paste on 100 kg of oven-dry yarn as it leaves the squeeze rollers. The most accurate way to find the pick-up is through laboratory determinations of concentration and percentage of size. Then:

$$\text{Pick-up} = \frac{\text{Size percentage on yarn}}{\text{Size concentration in box}} \times \frac{100}{1}$$

- *Percentage of size* on yarn is a product of (a) concentration of the sizing paste and (b) take-up of paste as the yarn leaves the squeezing rollers. The percentage of size on yarn can be calculated as follows:

$$\%\text{ of size on yarn} = \frac{\text{Size box concentration of size (\% solids)}}{1} \times \frac{\text{Pick-up of paste}}{100}$$

If the size concentration is 9.0% and after squeezing, 110 kg of paste is left on 100 kg of unsized yarn from the beams, the percentage of size applied to the yarn would be: 9 × 110/100 = 9.9%. If the amount of size paste picked up by the yarn is 150 kg, then the percentage of size on the yarn will now be 9 × 150/100 = 13.5%. This shows how important it is to keep the desired percentage of paste pick-up by the yarn if the right percentage of size on the yarn is to be achieved.

The amount of paste picked up by the warp and the final percentage of size on the yarn depend on many factors. These are given in Fig. 4.11.[7] Figure 4.12(a),(b) shows the cross-sections of sized yarns, the penetration of size into the yarn structure and the varying amount of coating around the yarn.[3]

4.6.4 Drying section

Sizing machines are usually defined by the method by which the yarn is dried. Two techniques for drying the wet warp are normally available:

- *Hot air drying*: in hot-air sizing machines, the wet warp is dried using hot air in an enclosed chamber through which the warp sheet is passed. The yarn is dried on all sides at the same time rather than on one side as in the case of a cylinder drying machine. The air is heated either by electric coil or steam. One drawback of the hot-air sizing machine is its lack of heat efficiency. It is claimed that the fuel costs for hot-air drying are at least 40% higher than for equivalent drying by heated cylinders. Also hot-air drying takes up more space and takes longer to cool down. Therefore, the conduction method using cylinder drying is most commonly used.[15]
- *Cylinder drying*: As the wet warp sheet leaves the size box, it is passed around the drying cylinders (Fig. 4.8(a),(b)) which are heated with steam. The warp is pressed against the surface of the heated cylinder under pressure which is controlled largely by the tension in the warp sheet. The drying capacity of the sizing machine is defined as the weight of water which the machine is capable of evaporating in a given time. A nine-cylinder sizing machine, working at 2.8 kg cm^2 steam pressure would be expected to have a capacity to evaporate 570 kg of water per hour. About 75–80% of the total energy used in the sizing process is consumed in drying the wet warp sheet loaded with size paste. The number of cylinders on a cylinder drying machine can vary from two to 32 cylinders, with four or more size boxes to facilitate the application of multiple layers of sizing material to the warp yarn.[5,11]

The following factors affect the rate of drying of a warp passing over steam-heated cylinders:[15]

- number of cylinders and their dimensions
- pressure of the steam in cylinders

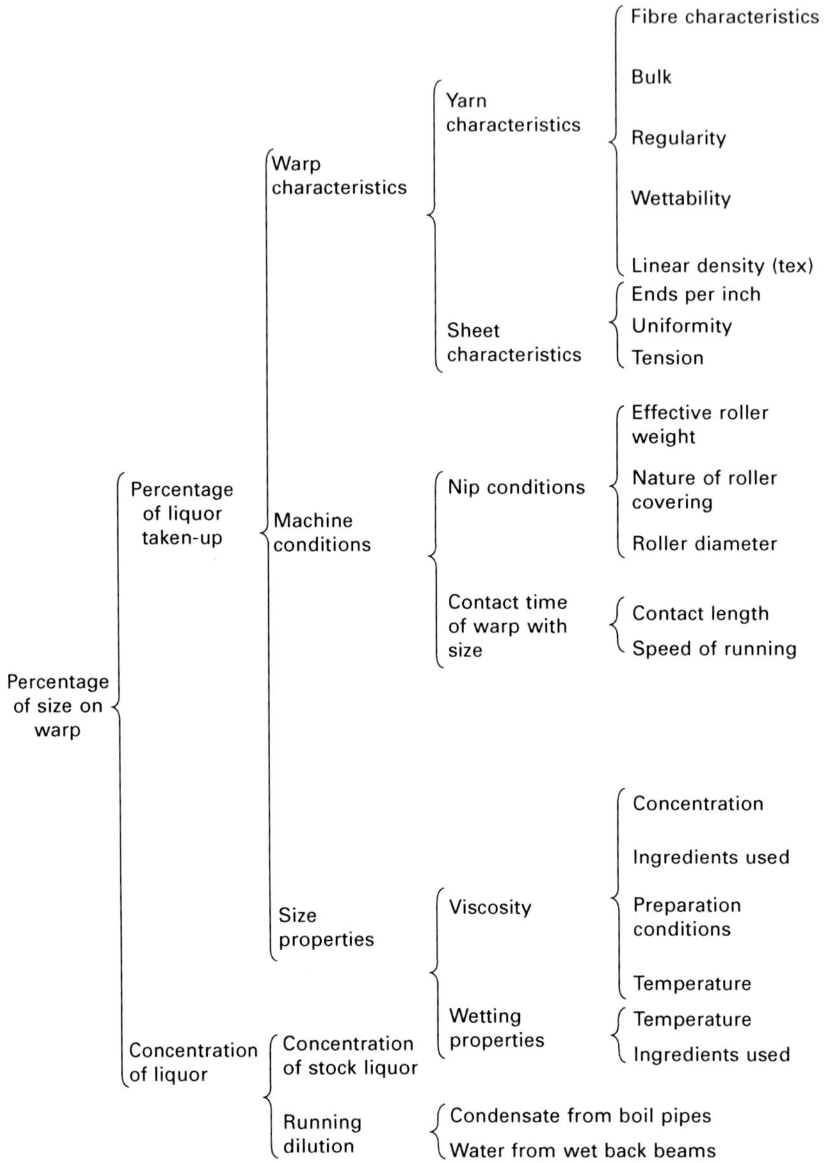

4.11 Factors affecting the percentage of size applied to a warp.

- pressure of air in the steam
- thermal conductivity of the cylinder wall
- warp tension (pressure of warp on cylinders): the higher the tension, the greater the rate of drying

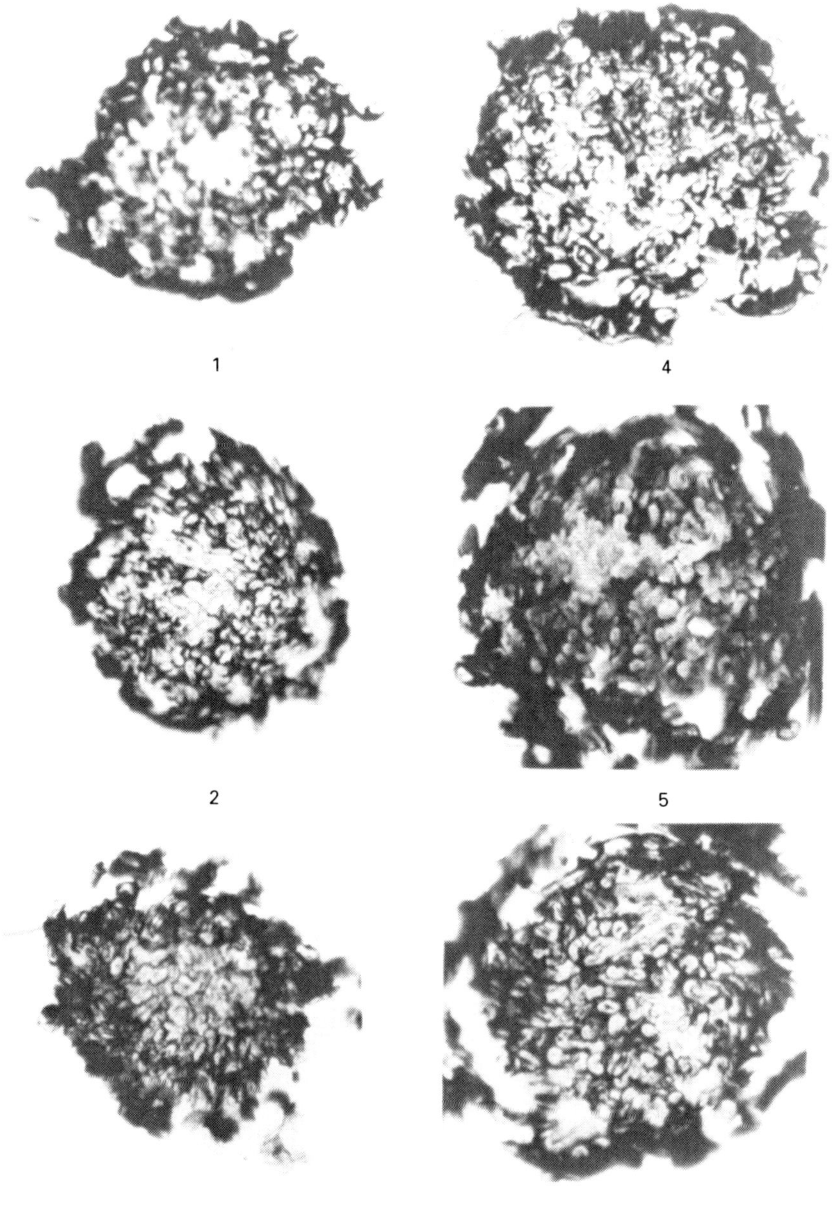

Key:
1,2,3 – Ring-spun yarns
4,5,6 – Open-end yarns

(a)

4.12 (a) Cross-sections of sized yarns. Size paste cases have penetrated the yarn structure (b) cross-sections of sized yarns with varying degree of coating (in black) of size film rather than penetration of size into the yarn structure.

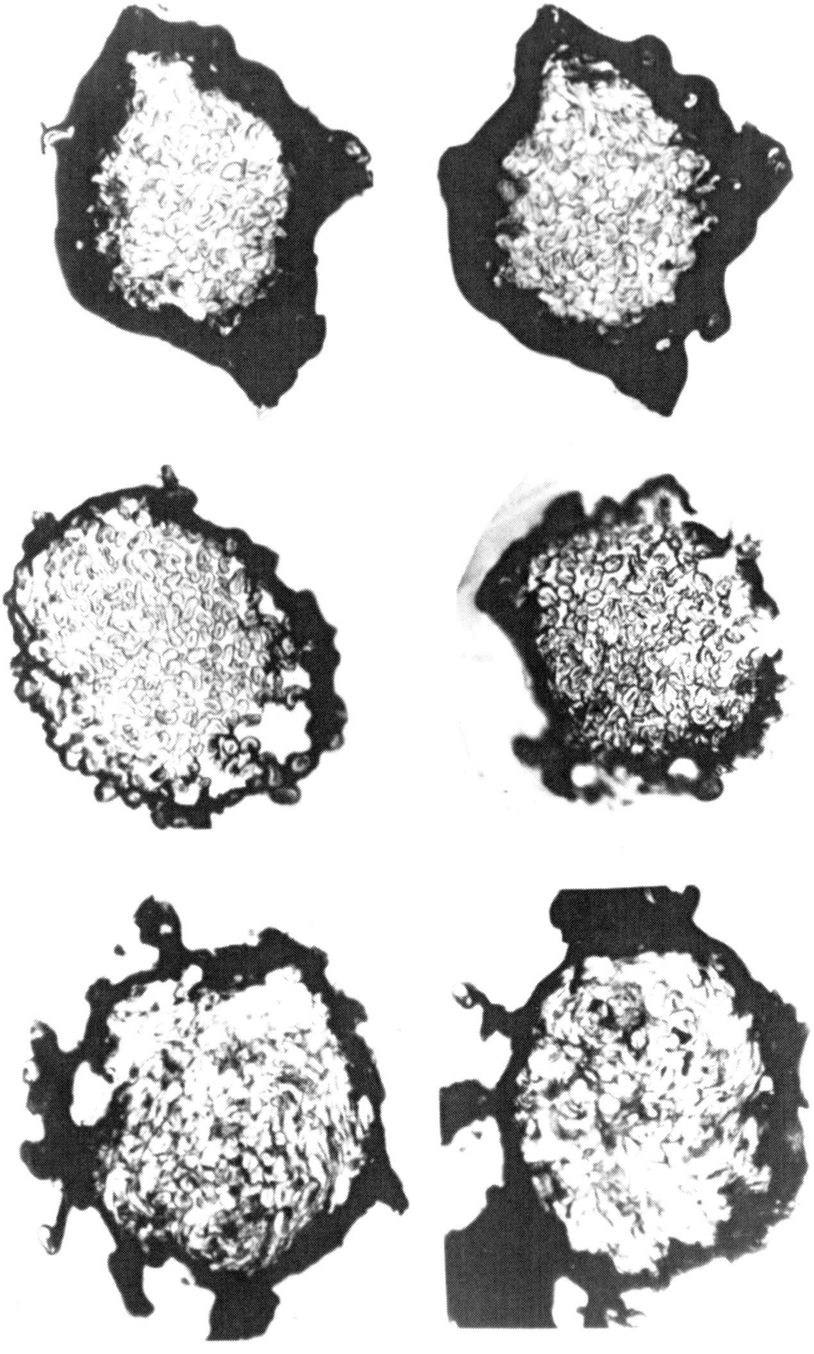

(b)

4.12 Continued

- degree of drying of the warp
- size concentration
- number of ends in the warp and the material.

The drying process governs the properties of the size film – its elasticity, plasticity and flexibility. Drying the warp needs careful attention as an over-dried warp would tend to become brittle and harsh, whereas under-drying would produce a sticky warp and increase the number of warp breaks during weaving. Some multi-cylinder machines are provided with individual temperature controls for each cylinder. In others, temperature control over the whole range of cylinders is obtained by fitting a special type of pressure regulating valve. A moisture regain of nearly 7% is recommended in the sized yarn. All modern warp sizing machines are installed with moisture measuring equipment. Typical cylinder temperature figures for a seven-cylinder filament and spun yarn sizing machine are shown in Table 4.3.[10]

4.6.5 Leasing section (splitting)

As the warp sheet is dried, there may be adhesion between individual yarns. It is of the utmost importance to separate each sized end of yarn before the warp can be used in weaving. This is done in the leasing section, using a series of rods of varying diameter. The number of lease rods is generally one less than the total number of warp beams in the creel. The arrangement is shown in Fig. 4.9(a) where the back beams are numbered 1–6 respectively. The sheets of yarn from each beam, after being divided by the lease or splitting rods A, B, C, D and E, are numbered 1, 3, 5, 6, 4, 2 to indicate their positions at the headstock.[6] The splitting rods extend the full width of the machine and are housed in brackets for easy removal. The separated

Table 4.3 Cylinder temperatures for filaments and spun yarn sizing (°C)

Yarn type	Material	Cylinder number							Mean temp. (°C)	Max. temp. (°C)
		1	2	3	4	5	6	7		
Filament	Polyester	75	95	115	115	115	95	75	98	115
	Viscose	80	90	115	110	110	90	80	96	110
	Acetate	70	80	100	100	100	80	70	86	100
	Nylon	85	95	100	110	110	100	90	99	110
		Group 1		Group 2		Group 3				
Spun	Cotton	140	140	140	140	140	140	140	140	140
	Viscose	120	120	120	120	120	100	100	114	120
	P/Cotton	110	110	120	120	120	100	100	111	120
	P/Viscose	110	110	110	110	110	100	90	106	110

Note: Cylinder No. 1 situated next to size box.

yarns are passed through an expandable comb to maintain the division of the sheet into separate ends and to position them for winding.

4.6.6 Headstock

The final section of the machine is the headstock. This consists of two small guide rollers and the main draw (drag) roller as well as a unit to support weaver's beams of varying widths (Fig. 4.8(b)). The warp sheet is guided around the draw roller as it emerges from the comb. It is then wound onto the weaver's beam, the surface speed of which is kept constant so the yarn can be wound at a consistent tension from start to end of the beam. When the required length of warp has been wound, the warp is cut and a new empty weaver's beam is positioned in its place to continue the process until the desired number of beams has been produced. The warp yarn wound on the weaver's beam is then ready for the drawing-in process. This consists of drawing the yarns from drop-wires, healds and reed, after which they are taken to the back of the loom, ready for weaving.

4.7 Yarn stretch during sizing

During sizing, the yarns are kept under tension, even on a positively driven sizing machine. This results in a slight permanent stretch in the yarn with a resultant decrease in the extensibility or elongation at the breaking point of sized yarns. A useful conception of stretch is that of the amount of elongation which occurs in, or is imposed upon, a given length of warp as it passes through the various zones in a sizing machine.[15]

Stretch is usually expressed as a percentage and the quantity can be either positive or negative. A negative value of stretch implies the opposite of an elongation, i.e. a shrinkage. The greater the yarn stretch, the higher the loss in extensibility of the yarn. This causes a higher rate of end breakage during weaving. The extensibility of the sized yarn is controlled by ensuring that the stretch during sizing is 2% or less. Loom performance deteriorates significantly if this is exceeded.

There are five main zones where stretch is applied:

- creel zone – between the last warper's beam and the first guide roll in the size box
- size box and first drying cylinder – wet stretch
- drying zone
- splitting zone – from the last drying cylinder to the drag roll.
- winding zone – between the draw roller and the weaver's beam.

Stretch in the creel zone should be no more than 0.5%. Wet stretch should be as low as possible and up to 1.5% in the drying and splitting zone. Tension control in the drying area is very important since it governs the elastic properties of the sized warp. There should be no stretch in the winding zone.

4.8 Automation controls of sizing machines

A modern sizing machine may supply sized warps for 300–500 looms and just one hour of sub-standard work in sizing may produce hundreds of metres of poorly sized warp. This may result in weeks of poor work, loss of production, waste of material during weaving and consequent dissatisfaction for both labour and management. Therefore on grounds of economics alone, the preparation of efficiently sized warps is essential.

To meet this objective, most modern sizing machines are equipped with the following automatic controls:[10]

- pneumatic creel loading to control tension
- temperature of size box
- size box level
- squeeze roller pressure and stretch control
- speed of drying cylinders relative to draw roll
- temperature of drying cylinders
- moisture content of sized yarn
- beam tension between beam and draw roll
- beam pressure – roll loading.

4.9 References

1. A. Barella, *JTI*, 1966, pp 477–78.
2. W.E. Morton, *JTI*, 1952, p. 60.
3. K.L. Gandhi, 'A comparison between Break-spun and Ring-spun yarns with particular reference to Sizing and Weaving, PhD thesis, University of Manchester, 1975.
4. Paul V. Seydel, James R. Hunt, *Textile Warp Sizing*, Long & Clopton, Inc., Atlanta, GA, 1972.
5. H. Hacking, *The Sizing of Spun Yarns*, Scientific Era Publications, Stamford, UK, 1980.
6. T. Ramsbottom, *Warp Sizing Mechanisms*, Columbine Press, Manchester, 1964
7. J.B. Smith (ed.), *Technology of Warp Sizing*, Columbine Press, Manchester, 1964.
8. V. Heap, Sizing material for continuous filament yarns, *Textile Institute & Industry*, July 1973, p. 172.
9. D.M. Hall, (ed. S, Adanur), *Wellington Sears Handbook of Industrial Textiles*, Technomic Publishing, Lancaster, PA, 1995.
10. Allan Ormerod, Walter S. Sondhelm, *Weaving – Technology & Operations*, The Textile Institute, Manchester, 1995.
11. B.C. Goswami, R.D. Anandjiwala, D.M. Hall, *Textile Sizing*, Marcel Dekker, New York, 2004.

12. J.E. Booth, *Textile Mathematics, volume 3*, The Textile Institute, Manchester, 1977.
13. A.L. Landau, *Warping and Slashing, Sizing Part 1*, International Text Book Co. Scranton, PA, 1941.
14. P.R. Lord, M.H. Mohamed, *Weaving: Conversion of Yarn to Fabric*, Merrow Technical Library, Watford, UK, 1973.
15. E.H. Jones, in *Technology of Warp Sizing*, ed. J.B. Smith, Columbine Press, Manchester, 1964.

5 The fundamentals of weaving technology

K. L. GANDHI, Chartered Consultant, UK

Abstract: The chapter starts with a brief introduction to woven fabrics categorized into different types and then explains all the primary, secondary and auxiliary loom mechanisms involved in the weaving process. It then covers principles of different shedding mechanisms and the type of sheds that are formed by different mechanisms: different type of shuttle looms are reviewed. The chapter then critically describes the principles of shuttle-less looms (air-jet, water-jet, rapier and projectile) and finally discusses, in brief, the principle of multiphase weaving, its economic advantages and limitations, etc.

Key words: shed, tappet, dobby, jacquard, warp yarn, weft, multiphase, weft insertion rate.

5.1 Introduction

Whether carried out on a simple hand loom or a power loom, weaving consists of interlacing two sets of yarns at right angles to each other. The yarns that run the length of the woven fabric are called warp yarns and the yarns that run across from side to side are called weft yarns (also known as filling). Warp yarns are generally stronger than weft yarns because warp yarns experience a lot of stress and strain during weaving as a result of the different loom motions during the weaving cycle. The manner in which the warp and weft threads interlace with each other is known as the weave. Broadly speaking woven fabrics can be categorized into the following three types:

1. Fabrics in which warp yarns (also referred to as ends) and weft yarns (also referred to as picks) intersect one another at right angles.
2. Fabrics in which some warp yarns interweave to the right and left of the adjacent warp. Such fabrics are called gauze and leno fabrics.
3. Fabrics in which warp or weft yarns project outwards from the foundation of the cloth and form a loop or pile on the surface of the fabrics. These fabrics are called pile fabrics (examples are carpet, terry towel, velvet and corduroy).

Figure 5.1[1,2] shows the passage of warp yarns through a loom which is basically the same whether it is a table loom (Fig. 5.1(a)), a hand loom (Fig. 5.1(b)) or a power loom. All the warp ends are drawn through the eyes of the heald shafts which are raised or lowered as one unit. A minimum number of two heald shafts are required to weave a fabric. The need for more than two

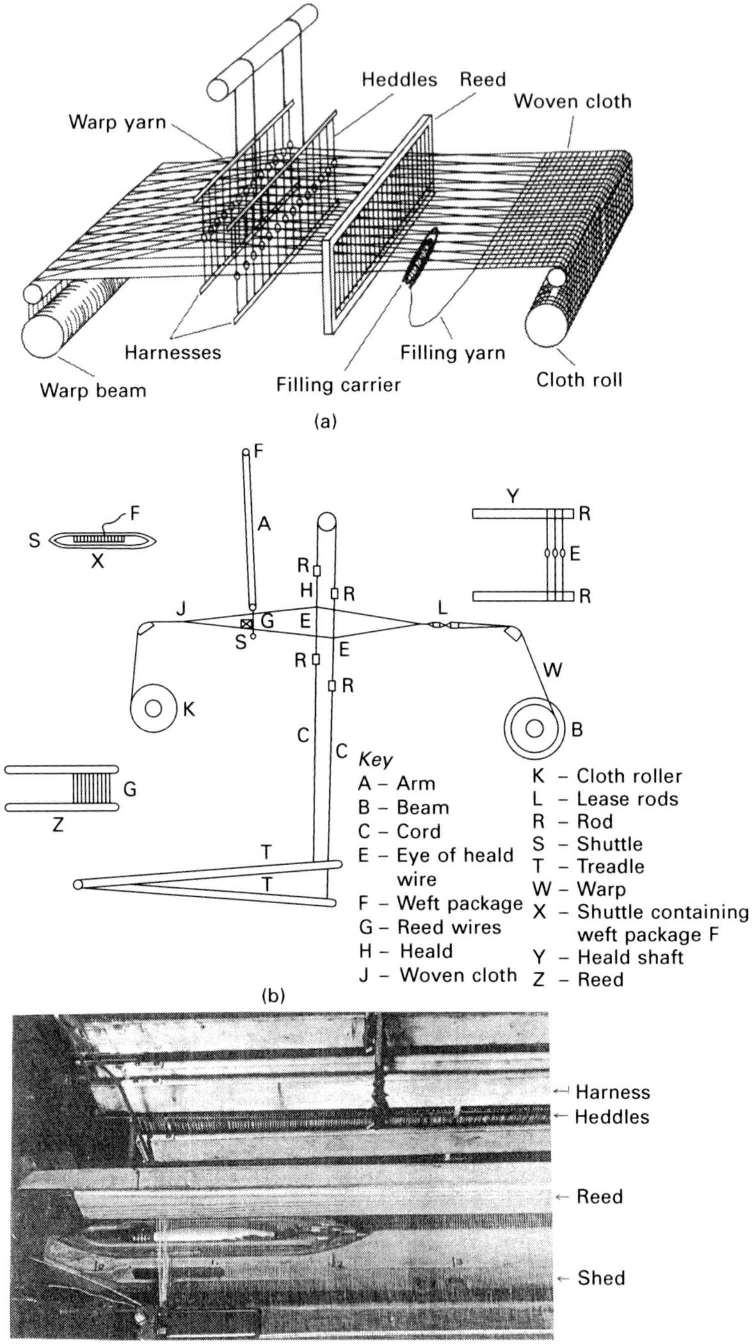

5.1 (a) Passage of warp through basic loom parts, (b) passage of warp yarn from a basic hand loom with treadles and (c) shuttle entering the shed.

shafts depends on the weave structure of the fabric. All the warp ends are then passed through the dents of the reed. The reed is primarily composed of parallel strips of steel wires with gaps (or dents) between them. The reed keeps the warp yarns evenly separated and held parallel as they pass through the dents of the reed. The operation of drawing each warp yarn through its appropriate heald eye and through the dents of the reed is known as drawing-in, before the weaver's beam is positioned at the back of the loom.

5.2 Primary loom mechanisms

In order to interlace the warp yarn with the weft yarn, three primary mechanisms have to take place completing one loom cycle or one complete revolution of the loom. The three primary mechanisms are shedding, picking and beat-up, and these are discussed below. Without any of these mechanisms weaving cannot take place.

5.2.1 Shedding

The shedding mechanism separates the warp threads into two sheets (layers) by lifting some of the heald shafts up whilst lowering others. Since each warp yarn from the weaver's beam passes through an eye of the heald shaft, when some of the heald shafts are lifted and lowered by a shedding mechanism, the corresponding warp ends are raised or lowered thus forming an opening. When this happens, we say a shed is formed and the next primary mechanism, 'picking', can commence. A shed can be formed by different types of shedding mechanisms such as tappet, cam, dobby and jacquard. The minimum number of heald shafts required to form a shed is two. When the number of heald shafts being used is more than two, whether a shaft is lifted or lowered is decided by the weave structure of the fabric being woven on the particular loom. On a table loom, the shed is made by lifting the heald shafts manually by hand, whilst on a simple hand loom it is generally made by depressing the treadle levers (Fig. 5.1(b)).

5.2.2 Picking

On a shuttle loom, when the warp sheet has been divided into two parts and a shed is formed, a shuttle carrying the weft yarn in a package (called a cop or pirn) is thrown through the shed[1] by a picking mechanism (Fig. 5.1(c)). When the shuttle has almost reached the other side of the loom, the shed begins to close as the direction of the heald shafts is changed. On a shuttle-less loom, the weft is thrown through the shed by a projectile (a projectile loom), a rapier (a rapier loom), by air (on an air-jet loom) or by water (on a water-jet loom).

5.2.3 Beat-up

As soon as the weft has been inserted through the shed, beat-up (the last primary motion of the loom) can take place. With the help of the reed, the inserted weft is pushed into the edge of the already woven cloth at a point called the 'fell of the cloth'[3] (Fig. 5.2). The reed gets its forward and backward movement from a crank shaft.

Thus in one revolution (one cycle) of the loom, all three primary mechanisms take place, and are then continuously repeated as weaving continues.

5.3 Secondary loom mechanisms

Although not essential for weaving a fabric, i.e. to interlace warp and weft yarns, there are other mechanisms which assist in the continuous weaving of a cloth on a power loom. These secondary mechanisms are carried out simultaneously.

5.3.1 Let-off

On a power loom, let-off is the delivery of the warp yarn from the weaver's beam at a constant rate, and at a constant suitable tension, by unwinding it from the beam as weaving continues. On a power loom, the position of the fell of the cloth always remains in the same place, whereas on a traditional hand loom, the fell of the cloth position keeps on changing as each loom cycle completes.

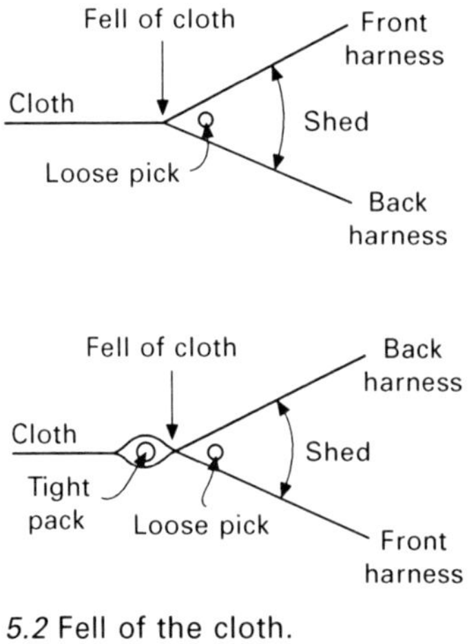

5.2 Fell of the cloth.

5.3.2 Take-up

Take-up is the automatic withdrawing or winding of the fabric as weaving continues, at a constant rate, onto a cloth roller. Take-up motion controls the weft density, i.e. the number of picks/cm in the fabric.

5.4 Auxiliary loom mechanisms

In addition to the primary and secondary mechanisms discussed, there are other auxiliary mechanisms on almost all power looms, whether with or without shuttle. These auxiliary mechanisms are warp stop motion and weft stop motion and, although not absolutely essential, they are nevertheless critical for controlling fabric quality and production on a power loom. Their primary function is to stop the loom when either a weft yarn or warp yarn breaks during the weaving process. The absence or malfunctioning of these mechanisms can lead to a lot of problems, causing serious defects in a fabric.

5.4.1 Warp stop motion

Warp stop motion can be either mechanical or electrical, both of which are widely used. In both types of systems, each warp yarn from the weaver's beam is drawn through the eye of a special drop wire. These drop wires can vary in dimension and weight depending on the quality of yarn (in general, the coarser the yarn, the heavier the drop wire should be). If the warp yarns are to be drawn through the eye of the drop wire at a point away from the loom, the base of the drop wire is closed (Fig. 5.3(b))[4] but if the drop wires are to be inserted onto a warp at the loom, the drop wires (c) are open from the base as shown in Fig. 5.3.

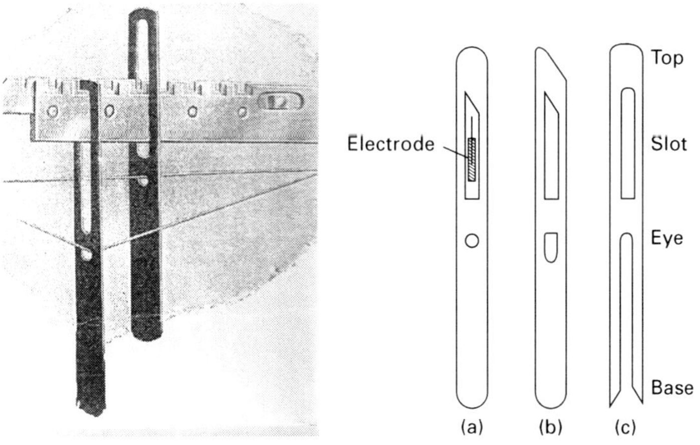

5.3 Warp stop motion.

One of the most common mechanical warp stop motions has the drop wires suspended over three slotted or serrated bars, the outer pair of which are fixed whilst the middle bar moves between them[5] (Fig. 5.3). When a warp yarn breaks, the drop pin falls and interrupts the movement of the oscillating serrated bar. This action (in one way or another) pushes the starting handle from its position and stops the loom. Electrical warp stop motions basically use the same principle, except that there are no serrated oscillating bars and each warp yarn is passed through special drop wires which can make contact across a pair of electrodes. When a yarn breaks, the drop pin falls, thus making contact with the bar and completing a circuit which, through a relay medium, stops the loom.

5.4.2 Weft stop motion

The main function of weft stop motion is to stop the loom when the weft yarn either breaks or when the supply package is completely used and needs replacing. If the loom does not stop immediately when this happens, a fabric defect called a broken pick is created and, if the loom continues running without weft for a number of picks, a far more serious fabric defect is created. In either case, the weaver has to spend some time correcting the fault, resulting in increased workload as well as reducing productivity. On shuttle looms, the weft stop motion is commonly known as the weft fork mechanism and is either located at the side of the loom (between the selvedge and the entrance to the shuttle box) or in a cut-out space.

5.5 Temples

Every power loom is equipped with temples on both the right and left side of the loom in front of the reed and near the fell of the cloth. A few centimetres of fabric near the selvedge on both sides pass through these temples (Fig. 5.4). The function of the temples is to:[6,7]

- control contraction of the fabric due to the interlacing of warp and weft
- keep the fell of the cloth at the same width as the warp in the reed; if this width is not maintained, the reed will abrade against the selvedge ends on both sides and cause breakages during the beat-up cycle
- overcome the drag on the selvedge when the shuttle passes from one box to the other.

There are various kinds of temple:

- *Ring temples* are rings with spikes which are mounted on a shaft. The number of spikes varies, as does the number of rings. Ring temples are used for light, medium and heavy fabrics. If not maintained properly, they can damage the fabric edge.

- *Rubber temples* are temples without spikes and are used for delicate fabrics and on water-jet looms.
- *Steel roller temples* do not have rings but are covered with short spikes. They are suitable for light and medium fabrics and are mostly used in pairs, and are covered with a cap.
- *Full-width temples* are usually used for fabrics that should be free from temple marks, such as nylon, etc. These temples do not have spikes and consist of a steel rod running across the full width of the loom. The fabric passes around it as shown in Fig. 5.4. The temple controls the fabric close to the fell of the fabric.

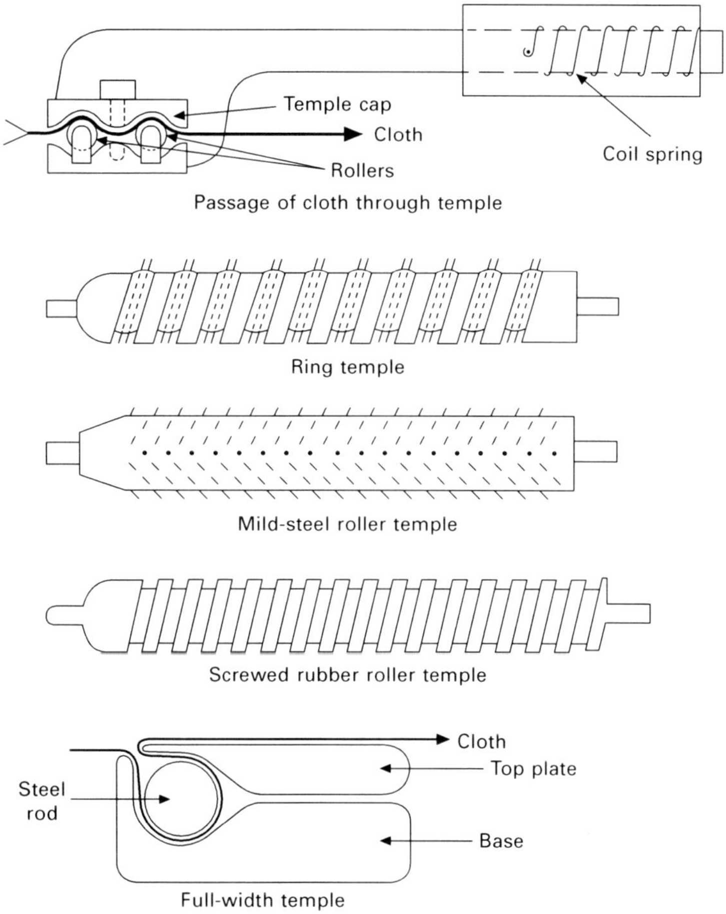

5.4 Types of temples.

For medium and heavy fabrics, the following distinct types of temples are normally used:[7] tapered rings and tapered cap with all the rings inclined; inclined rings, all the same size, with an semi-circular oblong cap; inclined and straight rings on the same shaft, with a semi-circular oblong cap; or roller temples, the rollers being covered with short spikes, and the two rollers being covered with a convenient cap.

5.6 Shedding mechanisms

Shedding is the first primary mechanism that takes place during the weaving operation and is the division of the warp threads into two parts, by the lifting and lowering of selected sets of warp threads on each new loom cycle. Which of the warp threads should be raised and lowered is dependent on the weave structure.

The lifting or lowering of the heald shafts is achieved by using different shedding mechanisms which can be positive or negative. In positive shedding, the heald shafts are both raised and lowered by crank cams or tappets or by a lever system within the shedding mechanism. In negative shedding, the healds are either raised or lowered by the shedding mechanism but then returned by the action of an external device, such as a spring or reversing roller mechanism. Different shedding mechanisms are discussed below, and a summary of the main shedding motion characteristics are presented in Table 5.1.

5.6.1 Negative cam (tappet) shedding

Figure 5.5(b) shows the passage of warp yarn through various parts of a loom fitted with a negative tappet shedding mechanism.[7] The shedding cams (tappets) get their rotary motion through the bottom shaft on which they are fitted. When the tappets rotate, they depress 'followers' (called anti-frictional bowls), which are fitted to the treadle levers, fulcrummed towards the back of the loom. The lower ends of the heald shafts are connected to the treadle levers, whilst the tops are connected through cords to a reversing roller device consisting of two rollers of different diameters. Figure 5.5(a) shows the shape of the tappets and cam.[8] When the cams rotate, they push the bowls down, thus depressing the treadle levers and the corresponding heald shafts. Positive action is achieved because the downward drive given to one heald is used to generate an upward action for the other heald shaft via the roller reversing action fitted at the top.

Figure 5.6(c) shows a negative cam shedding from Toyota[9] which can realise speeds of operation up to 1,000 rpm. Figure 5.6(d) shows one of the Stäubli cam shedding motions[10] designed for precision high-performance

Table 5.1 Summary of main shedding motion characteristics for cam (tappet), crank, dobby and Jacquard shedding

Shedding mechanism		Main characteristics
Cam (tappet) shedding	a	Simple to maintain and operate.
	b	Loom speed is relatively fast.
	c	Limited to designs repeating on 8–10 heald shafts.
	d	Complicated for frequent change of designs and therefore more economical for mass production of the same patterns.
	e	Fewer fabric faults occur during weaving.
	f	Design repeat in the weft direction is limited.
Crank shedding	a	Most characteristics are the same as for cam shedding.
	b	Mainly used on high-speed air-jet and water-jet looms for weaving plain and basic weaves only.
Dobby shedding	a	Can be built to control up to 48 heald shafts but commonly made with 32 levers. Each lever controls one heald shaft.
	b	Offers unlimited picks per repeat in the design (in weft direction).
	c	Easy to change the design. Requires only that the new design be punched on the plastic paper or on to lags with pegs.
	d	Higher initial costs and maintenance costs.
	e	Slightly prone to produce fabric faults.
Jacquard shedding	a	Can produce virtually unlimited designs both in warp and weft directions.
	b	Most jacquard shedding mechanisms, except the most advanced, tend to limit loom speeds compared to cam and dobby shedding weaving machines.
	c	Costly to install and maintain.

machines and ideally suited for all plain weaves, heavy fabrics and technical fabrics.

5.6.2 Positive cam (tappet) shedding

Figure 5.6(a) shows a positive cam shedding mechanism[8]. The cam (B) with a groove (C) is fitted on the cam shaft. A bowl (D) fits in the groove. The heald shafts are connected with appropriate connecting rods to one end of the cam lever (E), which is fulcrummed at (F). When the cam gets its rotary motion from the cam shaft, the bowl moves and follows the path of the groove, transmitting the motion to the link rods (G and J) which then pass on the upward or downward movement to the heald shafts, depending on the position of the bowl D in the grooved cam.

As the upward and downward movement of the heald shafts is controlled by the cam, this type of shedding is classified as positive shedding. There are as many cam units as there are working heald shafts. The tappet or cam

(a)

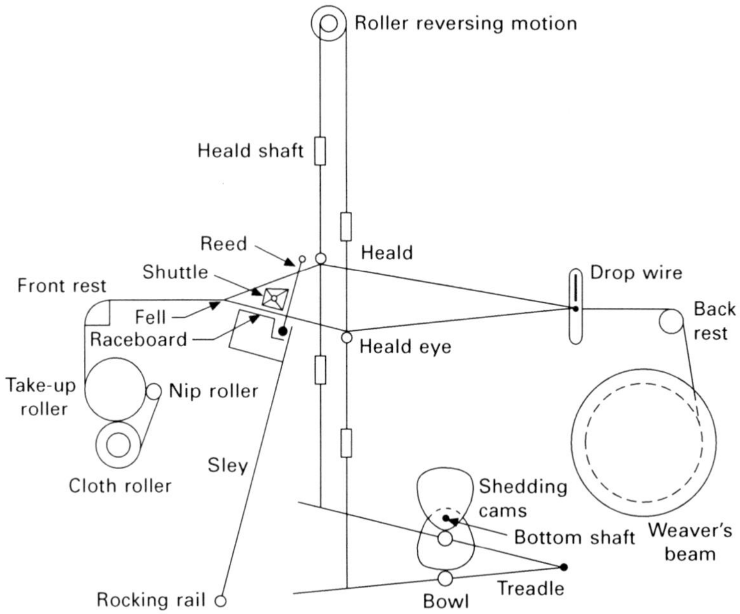

(b)

5.5 (a) Shape of tappet and cam, (b) passage of yarn from the loom parts.

shedding mechanism can be located inside or outside the loom.[5] Figure 5.7(a) shows an inside tappet shedding mechanism for up to eight shafts on a Picanol loom. The action is positive in lowering the heald frames, which are then returned by a clock spring motion. Figure 5.7(b) shows a typical example of an outside positive tappet shedding mechanism on a Saurer loom. The heald shafts are driven positively from below by means of angle levers and connecting rods, ensuring smooth and accurate shedding.

Another type of positive cam shedding, on a Sulzer loom, is shown in Fig. 5.6(b). Such double cam shedding consists of a pair of matched cams mounted on a shaft, with an oscillating link (L) fulcrummed (at F) by two anti-friction rollers, one of which is in contact with each cam face.[7] There are as many double shedding cams and motion transmission mechanisms as there are heald shafts, with each double cam having its own anti-frictional roller. When a cam continuously rotates in a clockwise direction, the oscillating lever swings in clockwise and counter-clockwise directions, and dwells at the end positions when necessary. The oscillating levers are connected through a series of links to the heald shafts and, finally, their clockwise and counter-clockwise movement is transmitted in upward and downward movements to the heald shafts. Adjustment for the depth of the shed is made at A by moving the stirrup (S) up or down, which does not affect the height of the shed. This is adjusted at B.

5.6.3 Crank shedding

Figure 5.8 shows a schematic view of a crank shedding mechanism,[11] which consists of a crank rocker mechanism (A_0ABB_0) and a slider crank mechanism

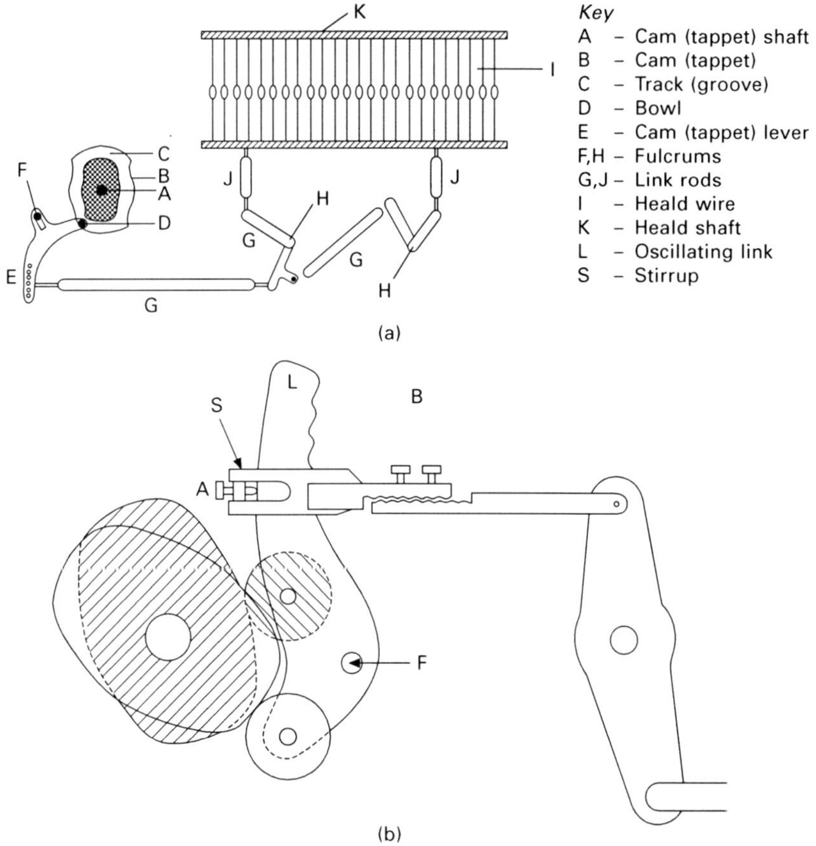

5.6 (a) Positive cam shedding mechanism, (b) positive cam shedding, (c) Toyota negative cam shedding, (d) Stäubli cam motion.

(c)

(d)

5.6 Continued

(B_0CD). The crank link (2) rotates at half of the loom's speed, the crank's continuous rotation being transmitted to link (4) by link (3). During one revolution of the crank, link (4) swings between its foremost and rearmost positions. The slider crank mechanism converts the angular displacement of link (4) to the linear displacement of the heald frame. The foremost position of link (4) corresponds to the bottom position, whilst the rearmost position corresponds to the upper position of the heald frame. Heald frames change position in each loom revolution and, consequently, this mechanism is suitable only for plain weave.

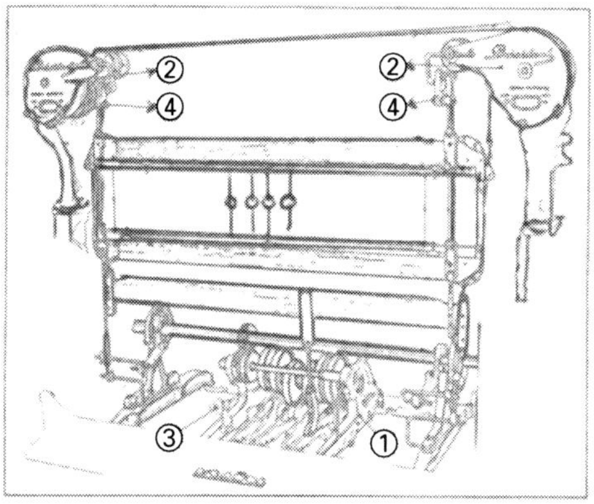

Key
1 – Tappet set
2 – Clock spring motion
3 – Shed adjusting knob
4 – Independent selvedges

(a)

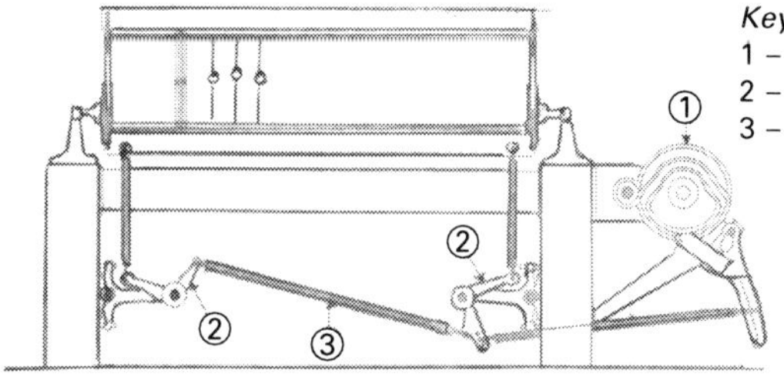

Key
1 – Tappet set
2 – Angle levers
3 – Connecting rods

(b)

5.7 (a) Inside tappet mechanism, (b) outside tappet mechanism.

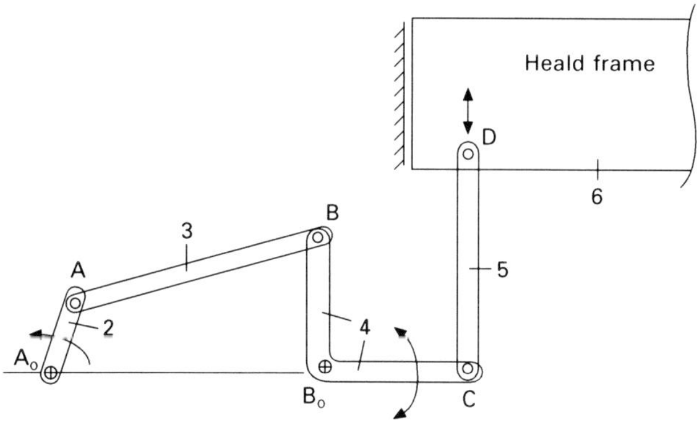

5.8 Crank shedding.

5.6.4 Dobby shedding

Dobby shedding mechanisms have been in use for well over 150 years and, over the years, there have been significant improvements in their design to meet the requirements of modern high speed weaving machines. Dobby shedding offers a considerably greater scope than most other mechanisms and is used in the production of simple figured fabrics. Modern dobby mechanisms are basically divided into single lift and double lift, or positive and negative types. The single-lift dobby provides a bottom closed shed whilst the double-lift dobby creates either an open or semi-open shed. Positive dobbies are commonly used to weave medium to heavy weight fabrics, whilst negative dobbies (which tend to be simpler) are used for light to medium weight fabrics. In negative dobby shedding, the heald shafts to be raised are lifted by a cam by the dobby mechanism but are normally lowered by the action of springs (Fig. 5.9(a)).[12] In positive dobby shedding, the shafts are raised and lowered by the dobby mechanism (Fig. 5.9(b)).[12]

In early versions of dobby shedding mechanisms, despite their different designs, the principle of selecting and lifting the heald shafts for positive or negative dobby is more or less the same[2] (see Fig. 5.10(a)). The dobby is moved by a rod (R) worked from the crank. The hook (D) is given backwards and forwards ('to and fro') motion through the lever (L). The actual selection of the heald shafts to be lifted on each pick is made by the presence or absence of pegs in the lags. The selection of the heald shafts to be lifted on each pick is pegged in the lags[12] (Fig. 5.10(b)). These pegs are joined in the form of a chain, which is put on a barrel (B) which in turn moves into a new position on every revolution of the loom. A series of feelers, or levers (A), are located above the pegs. One end of each lever (A) supports another hook (C). The presence of a peg lifts the filler which, through the leverage mechanism, lowers a hook (C) (as indicated by broken lines in the figure). The hooks (C and D) are engaged and the resulting movement of

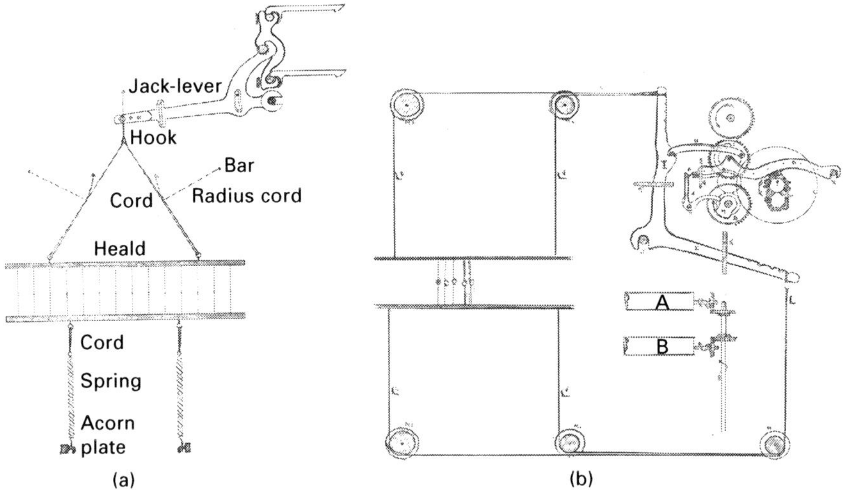

5.9 (a) Negative dobby, (b) positive dobby.

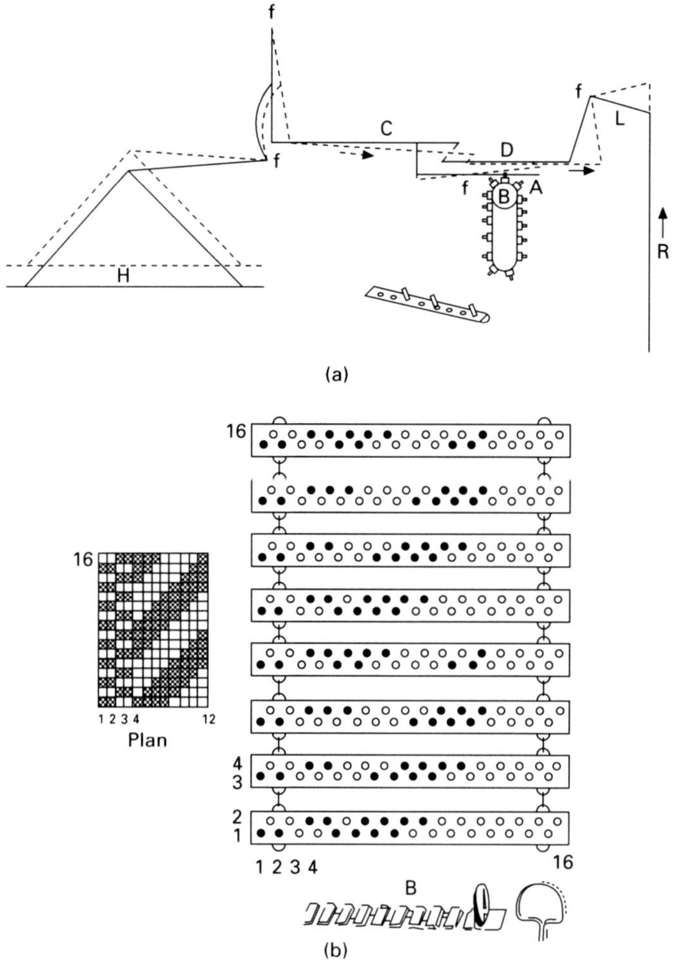

5.10 (a) Working principle of dobby, (b) pegged lags joined in chain form.

the hooks through the levers raises the heald shaft (H). For each heald shaft there is one feeler, one hook and a lever, their total number indicating the capacity of the dobby, i.e. the maximum number of shafts that can be used for a given design. Figure 5.11 represents a modern version[13] of a dobby attachment on a loom, using a plastic punched tape instead of the pegs on lags as described above.

5.6.5 Jacquard shedding

Jacquard fabric is a fabric in which a large number of warp threads, in excess of the capacity of a dobby, weave differently and therefore require a jacquard

5.11 Plastic punched tape that controls the warp shedding.

mechanism.[14] In tappet and dobby shedding, heald wires are not operated singly but are attached to heald frames, or heald shafts, and each wire on a given shaft conforms to the movement of the shaft, rising or falling with it. Both types of shedding system are used to control shedding where, due to the simplicity of the interlacing, only a few heald shafts are required. Both methods have limitations if the design repeat is of more than 32 shafts.

In Jacquard shedding, heald wires are manipulated individually by means of cords (or a harness) and not collectively through the agency of a heald frame. The shedding mechanism is mostly mounted on top of the loom on a gantry. Figure 5.12 shows a modern Stäubli jacquard on a Vaupel weaving machine with the harnesses attached to the top of the jacquard. Jacquard shedding is divided into two categories, single cylinder single-lift and double cylinder double-lift, and is used for high speed weaving machines.

Figure 5.13 shows a schematic diagram of a single cylinder single-lift jacquard.[7] The essential elements of a traditional jacquard are hooks, needles, perforated squared cylinder and a griffe with knives. On a single-lift jacquard, there is one needle for each hook, with one end of the needle resting in a spring box, the coiled spring of which presses the needle towards the right and out of the needle board. All the needles face the holes on the squared cylinder, each needle corresponding to a hole. Each needle has a small bent kink into which the vertical hook is positioned.

A Jacquard design is punched onto strips of card, with one card for each pick. When all the cards have been punched, they are laced together in the form of a chain and placed around the squared perforated cylinder. The holes in the card match the holes in the cylinder, whilst blanks in the card cover holes in the cylinder. For each loom cycle, the cylinder has two movements – away from the needles as well as turning once to bring the next punched

5.12 Stäubli jacquard on top of the Vaupel loom.

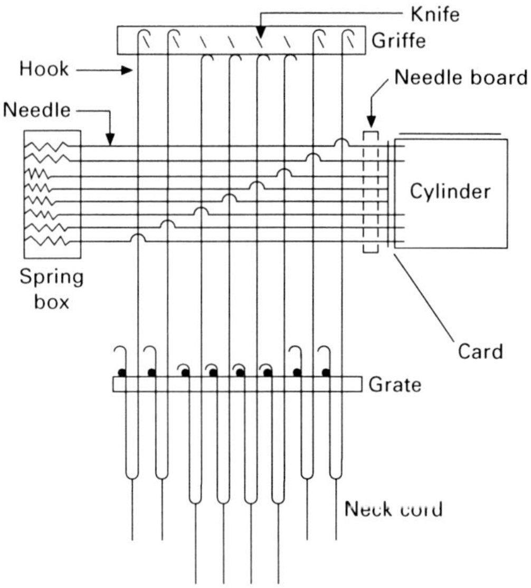

5.13 Single cylinder single-lift jacquard.

card into place and pressed against the needles. If there is a hole in the card opposite the particular needle, the needle can enter the hole in the cylinder beneath and the position of the corresponding hook will not be disturbed.

As this is happening, the griffe on the top with knives fixed into it reciprocates vertically once every pick. If the hook has not been disturbed from its position, the knife gets engaged with the hook and, while moving up, takes the hook with it. At the bottom of the hook the harness cord is attached by a mail through which the warp end has been passed. Thus, when the hook goes up, the warp is raised with it.

If there is a blank in the card, as the cylinder moves inward and the card is presented against the needle, the card pushes the needle to the left. Thus, the hook controlled by this needle is moved out of the path of the knife as it goes up and the hook is not engaged; the hook and the warp ends tied to it are therefore not lifted. Each hook controls a number of harnesses depending on the number of repeats of the design in the given width of the warp. This type of jacquard forms a bottom closed shed and limits the loom speed. The figuring capacity of the jacquard is limited by the number of hooks it has. On a single-lift jacquard, there is one needle for each hook.

Single cylinder double-lift jacquard

Figure 5.14 presents an outline of a single cylinder double-lift jacquard.[6] In this type of jacquard, each needle is made with two kinks to control the movement of two hooks. It has two sets of knives mounted in two griffes that move up and down in opposition over a two-pick cycle. The cylinder reciprocates and turns every two picks. The principle of selecting the ends to be raised is almost identical to the single-lift jacquard. The mechanism

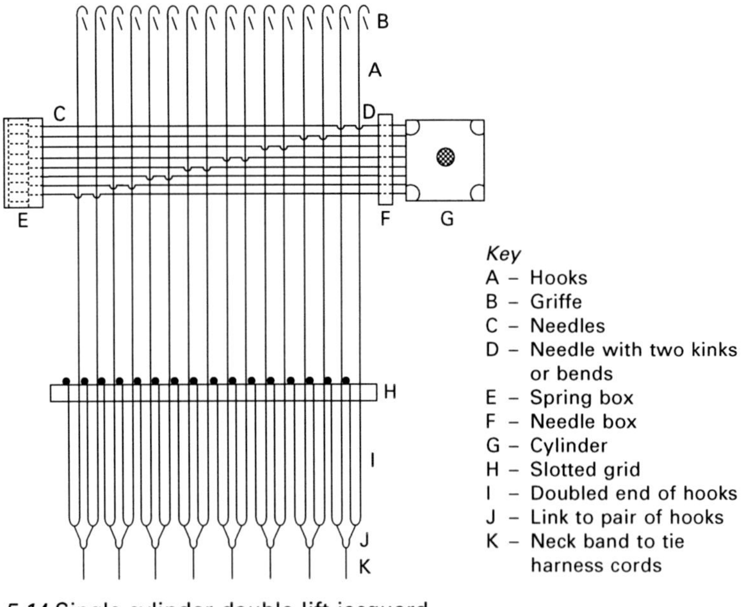

5.14 Single cylinder double-lift jacquard.

offers the advantage that if an end is required to remain up for more than two or more consecutive picks, it is lowered half way between the picks and raised again. This reduces the strain on the particular warp ends and the breakage rate also. It produces a semi-open shed.

Double cylinder double-lift jacquard

In a double cylinder double-lift jacquard (Fig. 5.15) there are two sets of needles and hooks as well as two cylinders (one of which carries the even-numbered cards and the other, the odd-numbered cards).[6] Each needle controls one hook. Since the cylinder speed is half that of the single cylinder jacquard, the double cylinder double-lift jacquard is capable of running at higher speeds. Apart from that, it works on the same principle as the single cylinder double-lift jacquard, and produces a semi-open shed. One advantage of this mechanism is that, if an end needs to remain up for more than two or more consecutive picks, it is lowered just half way between the picks before being raised again. This reduces the strain on the warp ends concerned and consequently reduces the breakage rate.

Over the years, various electronically controlled jacquards have appeared on the market and have become very popular, with Stäubli being the leading manufacturer. Although the fundamental principles of jacquard shedding

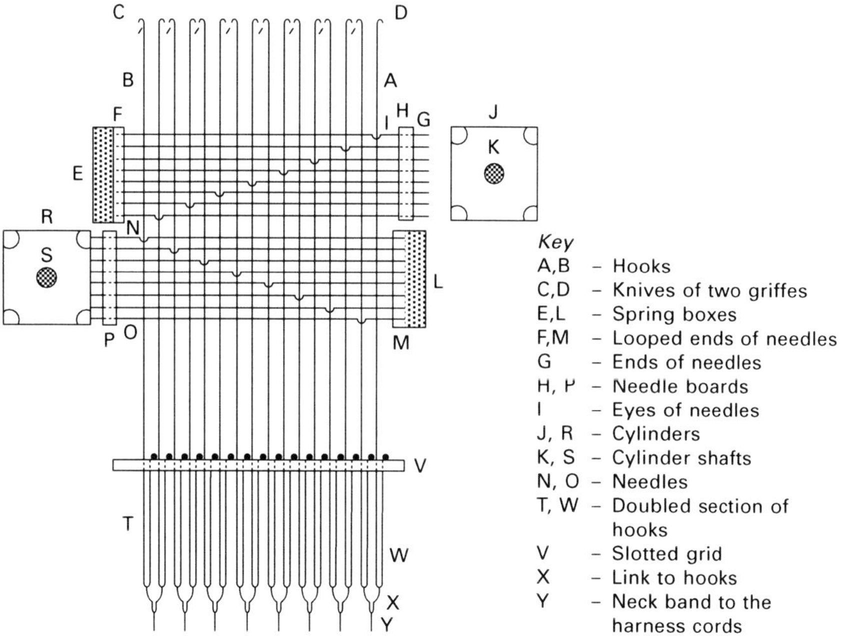

5.15 Double cylinder double-lift jacquard.

have not changed much, the methods of selecting the warp ends to be raised have been revolutionized. Figure 5.16(a) demonstrates some of the operating phases of the functional principle of Stäubli CX 860 type jacquard with electronic control. The double rollers 'a' move the harness cord to the upper and lower shed positions. Figure 5.16(b) shows a jacquard machine with a hook-selection device.[2,15]

5.7 Different types of shed

During the process of forming a shed (see Fig. 5.17), cam, crank, dobby and jacquard shedding mechanisms all give different types of motion to the heald shafts which, in turn, produce different types of shed, which can be divided into two main groups: open and closed.[2,7]

5.7.1 Bottom closed shed

The mechanism used to form a bottom closed shed is very simple and reliable. After every pick, all the yarns return to the bottom line from their top position and the shed closes, even if some of them are required to be raised again on the next pick. As such, some warp yarns move a distance equal to twice the depth of the shed, each time. Hence, there is a great deal of wasted movement and it takes a long time to produce a shed and consequently this type of shed is not suitable for high speed looms. Wear and tear of the loom, and power consumption, are also high.

5.7.2 Centre closed shed

In a centre closed shed, all the yarns return to the middle position (centre line) to form a closed shed after every pick. For the next pick, yarns which are to form the top and bottom line are raised and lowered respectively from the centre line and, after inserting a pick, both the lines meet at the centre. Whatever the weave, every end is either raised or lowered at every pick resulting in wasted movement. Centre closed shed tend to place a little more strain on the warp than open shed, but are employed widely in plain tappet, dobby and jacquard looms.

5.7.3 Open shed

With an open shed, the warp threads form two stationary lines, one at the top and the other at the bottom, and the shed never closes. After the insertion of the pick, changes are made by only some yarns from one fixed line to the other, i.e. some threads are lowered from the top line and some are raised from the bottom line, simultaneously. Since there is no wasted movement,

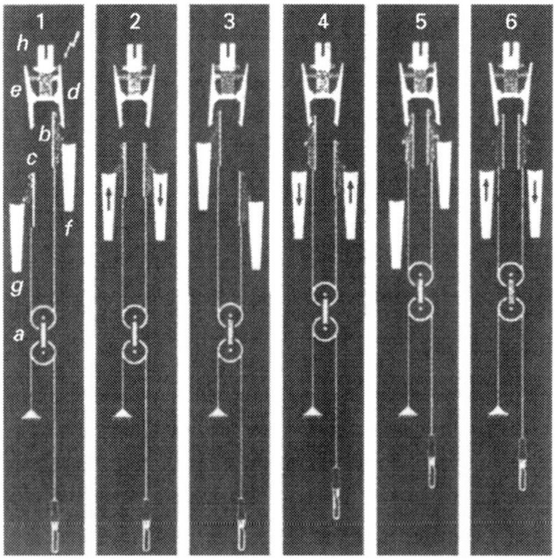

(a)

(b)

5.16 (a) Electronically controlled hook-selection system: 1 – (lower shed position) hook *b* in its upper-most position has placed ratchet *d* against electromagnet *h*. This magnet is activated according to the pattern, briefly retains ratchet *d* and prevents hook *b* from hooking onto ratchet; 2 – (lower shed position) hooks *b* and *c* follow the knives *g* and *f* moving up or down. Double rollers *a* compensate the motions of hooks *b* and *c*; 3 – (lower shed position) by the rising motion of knife *g*, hook *c* has placed ratchet *e* against electromagnet *h*. According to the pattern, the magnet is not activated, which causes hook *c* to hook onto the ratchet; 4 – (shed motion) hook *c* is hooked onto ratchet *e*, hook *b* follows the rising knife *f* and thereby lifts the harness cord; 5 – (upper shed position) hook *c* remains hooked onto ratchet *e*. Hook *b* has placed ratchet *d* against electromagnet *h* by the rising motion of knife *f*. According to the pattern, the magnet is not activated, which causes hook *b* to hook onto ratchet; 6 – (upper shed formation): hooks *b* and *c* remain hooked onto ratchets *d* and *e*. Knives *g* and *f* are in rising and descending motion, respectively. (b) jaquard machine with hook-selection device (indicated by arrow).

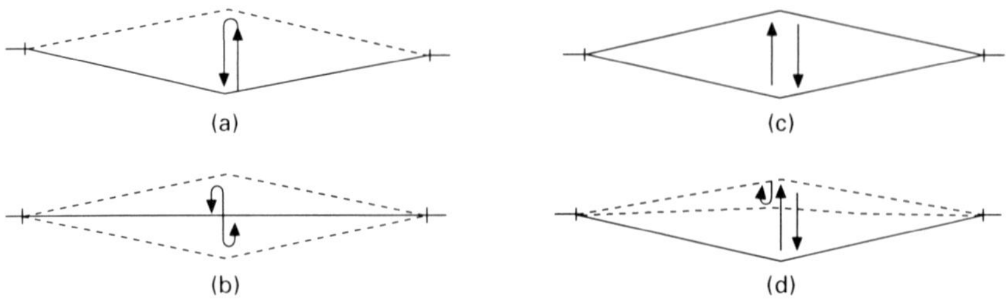

5.17 Types of shed (a) bottom closed (b) centre closed (c) open and (d) semi-open.

the shed is formed relatively quickly. Open sheds offer advantages such as low strain on the warp, low power consumption, less wear and tear of the loom, and can be used on high-speed looms. However, because the warp is not levelled, it is difficult for the weaver to repair any broken ends and a levelling mechanism is provided on most looms to overcome this problem. As the shed is always open, the yarn is subjected to strain and, with weak yarns, increased breakages may result.

5.7.4 Semi-open shed

The semi-open shed combines elements of both the closed and open shed systems. A stationary bottom line is retained. Warp yarns from the top line which are not required to be raised in the next pick are lowered to the bottom line, whilst yarns which are required to stay up for two or more picks in succession are lowered but stopped in the centre position between picks (i.e. not fully lowered to the bottom line), before being raised again to the top position. At the same time, ends from the bottom line which are required to be at the top line are raised. The shed is thus never fully closed or open, hence the name 'semi-open shed'. In this type of shed there is less wasted movement by the warp and higher loom speeds can be offered. However, the strain upon the warp is not so equally distributed as in the open shed type. Double-lift dobbies and jacquard normally use this type of shedding.

5.8 Classifications of plain and automatic shuttle looms

5.8.1 Plain non-automatic shuttle loom

In a plain non-automatic shuttle loom, a weft package called a cop is inserted onto the spindle in the shuttle[6] (Fig. 5.18) and the yarn is drawn from the eye of the shuttle. When the supply package is exhausted, the loom stops and the weaver takes out the empty package, inserts a new one in the shuttle, draw

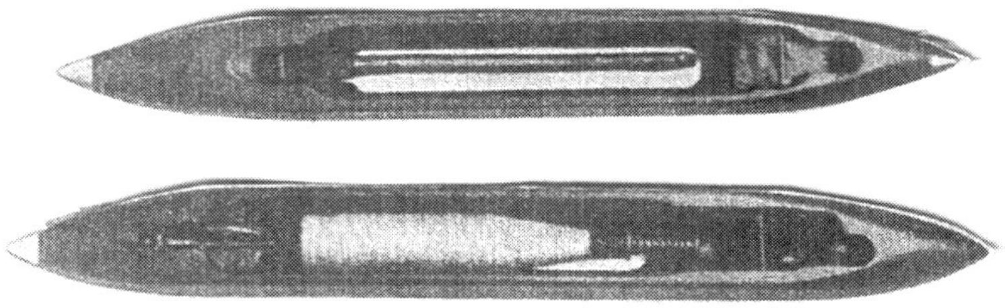

5.18 Shuttle for non-automatic loom.

the yarn from the shuttle eye, and re-starts the loom. This procedure puts a serious limitation on the number of looms a weaver can attend to.

5.8.2 Pirn changing automatic loom

This type of loom requires a specially designed shuttle without a spindle, as well as a weft package (Fig. 5.19). When the supply package is finished, the exhausted pirn is ejected from the shuttle and a new one is fed automatically, whilst the yarn is self-threaded from the thread guide. Automatic pirn changing mechanisms can be fitted to tappet (cam) looms, dobby looms and jacquards. Figure 5.20(a) shows a Northrop pirn changing loom.[16] The weft, in the form of ring pirns with three metal rings at the base (Fig. 5.19), is placed in a circular magazine (sometimes called a battery), which can hold 24 packages (Fig. 5.20(b)). The supply is kept topped up by battery filler. The shuttle used on this type of loom is specially designed with a pair of grooved spring jaws to hold the pirn with the three rings (Fig. 5.19). The circular magazine is mounted on one side of the loom. When the running weft supply package is finished, a weft feeler mechanism senses the absence of weft and triggers a mechanism which in turn raises a change lever. When the sley moves forward to the beat-up position, the change lever is then pressed back by a striker fixed to the sley. As a result of this action, a transfer hammer is depressed, forcing a new pirn from the magazine into the two grooved spring jaws in the shuttle and, at the same time, ejecting the empty one (Fig. 5.20(b)).[6]

5.8.3 Automatic shuttle changing looms

Automatic shuttle changing looms still use conventional non-automatic shuttles, and the weft package (provided in the form of either cops or pirns) is therefore interchangeable with those used for non-automatic looms. This is an advantage when, as is often the case, both automatic and non-automatic looms are used in the same factory, since no change in the spinning equipment is called for (as is the case, for example, for a pirn changing

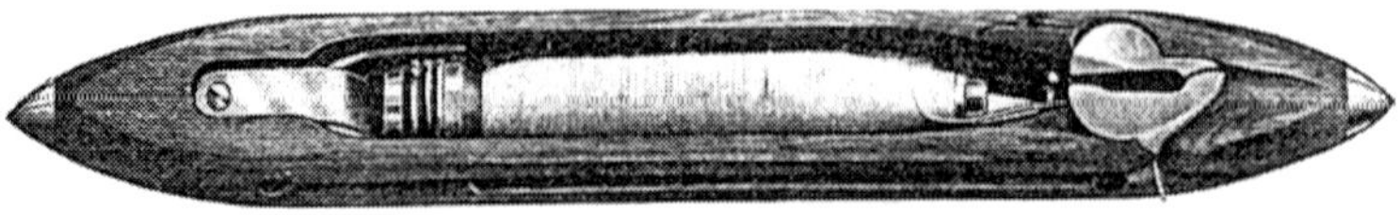

5.19 Shuttle for automatic pirn changing loom.

loom). Shuttle changing looms are well suited to the use of weft of a fine or delicate nature.

Shuttles are changed by a number of different methods, though the most successful looms appear to be those which are automatically stopped for the change and then automatically re-started when the change has been made. Figure 5.21 shows an automatic shuttle-changing loom.[16] A vertical stationary magazine, with the capacity to hold about 9 or 10 shuttles, is located on the opposite side of the loom to the weft fork and weft feeler, either of which can be used to set the shuttle change mechanism in motion. When a change of shuttle is required, the loom is automatically stopped and the changing mechanism controls the various movements. The shuttle box front is lifted and the spent shuttle pushed out of the box; the bottom shuttle in the magazine is lowered onto a conveyer which places it in the box, afterwards returning to its position below the magazine. The box front is lowered and the loom is automatically re-started. The whole operation takes 3–4 seconds.

5.9 Drop box looms

Drop box looms are used for weaving fancy fabrics requiring different colours or kinds of weft yarn in any sequence; the boxes are moved up and down to bring the correct shuttles into position, as required. There are basically two types: (a) pick-and-pick and (b) pick-at-will (Fig. 5.22(a),(b)).[6] In the case of pick-and-pick looms, an even number of picks of any colour or kind of weft yarn can be inserted as desired, whilst on pick-at-will looms, either an odd or even number of picks can be inserted as desired. To achieve this, there must be the desired number of shuttle boxes on either side of the loom. Drop box looms are generally classified as shown in Table 5.2.

5.10 Weft insertion on shuttle looms

Weft insertion involves throwing the weft yarn (the pick) through the shed formed by the shedding mechanism. On a shuttle loom, the weft package is inserted in the shuttle and the shuttle is then thrown through the shed with the help of the picking mechanism. On each side of the shuttle loom there is a box called a shuttle box in which the shuttle rests after being thrown from one box to the other by the picking mechanism. The picking mechanism located at each side of the loom propels the shuttle from the box, through

(a)

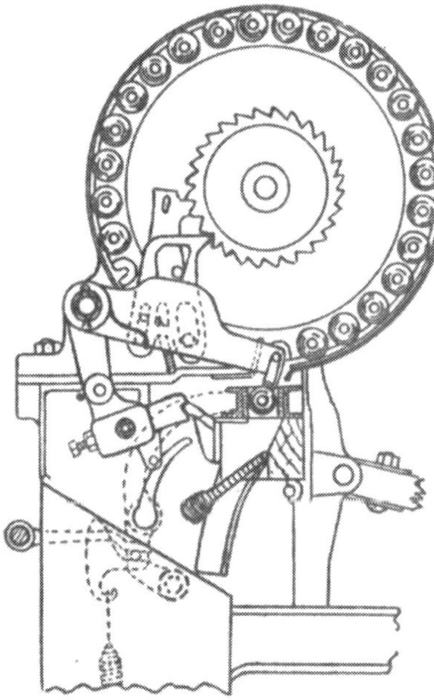

(b)

5.20 (a) Northrop pirn changing loom with dobby on the top left-hand side of the loom and under-pick motion on the right-hand side, (b) pirn changing mechanism showing the battery in the transfer position.

5.21 Northrop shuttle changing loom.

(a) (b)

5.22 (a) Hattersley 4 × 1 drop box loom, (b) Hattersley 4 × 4 drop box loom with dobby on top of the loom.

the shed to the other shuttle box. In the next revolution of the loom, the picking cycle throws the shuttle back to the other side; this process is then repeated, as required.

Usually there are two types of picking mechanisms: (a) over-picking; and (b) under-picking (see Fig. 5.23 for a schematic representation).[7] In both cases, the rotary motion of the bottom shaft is converted into rectilinear motion to project the shuttle through the shed.

Table 5.2 Classification of drop box looms

No. boxes			
Left side	Right side	Maximum shuttles that can be used	Category of the drop box loom
2	1	2	Pick and pick
2	2	3	Pick at will
4	1	4	Pick and pick
4	3	6	Pick at will
4	4	7	Pick at will
6	6	11	Pick at will

5.10.1 Over-picking mechanism

With an over-picking mechanism, a so-called picking-tappet is fitted on the bottom shaft (Fig. 5.23(a)) which, during its rotation, hits a cone fitted on a vertical rod (called a picking shaft). On top of the shaft is a picking stick, on the free end of which a leather band is wrapped. The other (free) end of the leather band is fastened to another part, called a picker, which slides freely on a spindle. As it revolves, the picking tappet hits the cone, which turns the vertical picking shaft sharply. This energy through the picking stick and leather band is transferred to the picker which hits the shuttle tip and throws it out of the shuttle box. The shuttle passes through the shed and is received in the shuttle box on the other side of the loom. In the next loom cycle, picking then takes place from this box and the process alternates from side to side.

5.10.2 Under-picking mechanism

The principle of operation for an under-picking mechanism is almost the same as for an over-picking mechanism, except that the picking mechanisms are located on the sides of the loom; the parts of the mechanism are almost the same, i.e. a picking cam fitted onto the bottom shaft (Fig. 5.23(b)), a cone, picking stick and the picker which slides on a spindle. The picking stick hits the picker directly which throws the shuttle from the box on one side of the loom to the box on the other side.

5.11 Weft insertion on shuttle-less looms

The principle of weaving for both shuttle and shuttle-less looms is exactly the same, i.e. the primary motions – shedding, picking and beat-up – take place in one cycle for both types of loom. The main difference between shuttle and shuttle-less looms is the method of weft insertion: on a shuttle loom, the picking (i.e. weft insertion) unit is located on both sides, whilst

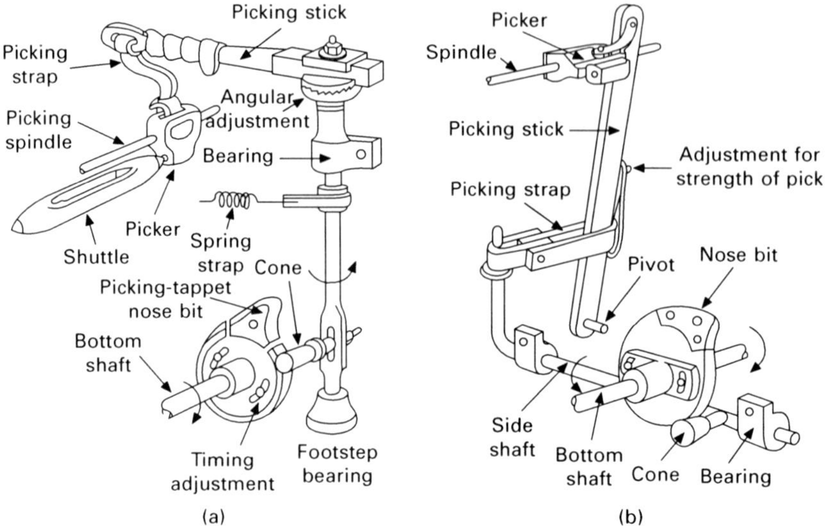

5.23 Over- and under-picking mechanisms: (a) cone-overpick, (b) cone-underpick.

on shuttle-less looms, the picking always takes place from the left side of the loom.

The revolutionary difference between a shuttle and a shuttle-less loom is the method of weft insertion across the shed. In the case of a shuttle loom, the shuttle (itself weighing around 500 g or even more) is thrown across to insert one pick length of yarn, weighing just 4–5 g (depending on the yarn quality), which clearly consumes a lot of energy. In the case of shuttle-less looms, the yarn is thrown not by a shuttle but by any of four different methods: by rapier, projectile, or by a jet of air or water.

5.11.1 Rapier looms

Rapier looms are considered to be the most flexible weaving machines and produce a huge variety of fabric styles and designs. There are basically two types of rapier looms: (a) single rapier and (b) double rapier looms. With single rapier looms, a single rigid bar (a 'rapier'), of circular cross section equal to the width of the reed, enters the shed from one side, picks up the tip of the weft yarn on the other side and then passes it across the width of the warp as it retracts. Consequently, the rapier only carries the yarn in one direction and half of the rapier movement is wasted, limiting the loom speed significantly and, as such, this type is not very popular (though they are suitable for yarns that are difficult to control).

In a double rapier loom, two rapiers are used: a 'giver' and a 'taker'. The giver takes the yarn from the supply package on the left of the machine and takes it to the middle of the shed where it is met by the taker, which retracts and transfers the weft yarn to the other side of the warp. Double rapier looms can be either rigid or flexible (and in both cases, the second rapier, i.e. the taker, completes the weft insertion). These two types are considered below:

- *Double rigid rapier looms*. Here, the weft yarn is transferred in one of two ways (Fig. 5.24): either by tip-to-tip transfer or loop transfer.[17] In tip-to-tip transfer (the Dewas system), the giver grips the tip of the yarn and takes it to the middle of the warp where it transfers to the other rapier head (the taker), which retracts and carries the weft to the other side of the warp. This method inserts each pick separately and both edges of the fabric are identical. In loop transfer (the Gabler system) the yarn is not gripped from the tip. Instead, the weft yarn is hooked by the giver (i.e. the rapier which extends the yarn in the form of a 'U' shaped loop to the middle of the warp). The yarn is then transferred to the taker which takes it to the other side of the warp by straightening it. The yarn for each full pick is withdrawn from the supply package in the first half of the pick cycle, which puts tension on the yarn and sometimes restricts the range of weft yarns (i.e. fine or delicate yarns) that can be woven using this method. This method of weft insertion is not currently very common. However, the method allows for convenient selvedge formed at one side of the fabric. Figure 5.25(a) shows a rigid rapier sequence.[18]

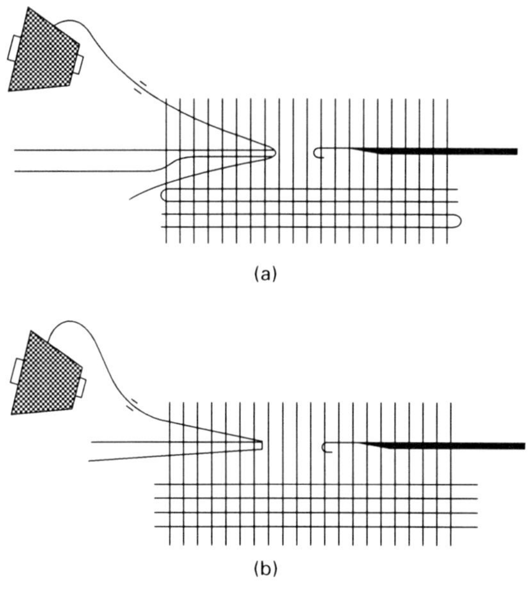

5.24 (a) Gabler and (b) Dewas system of weft insertion.

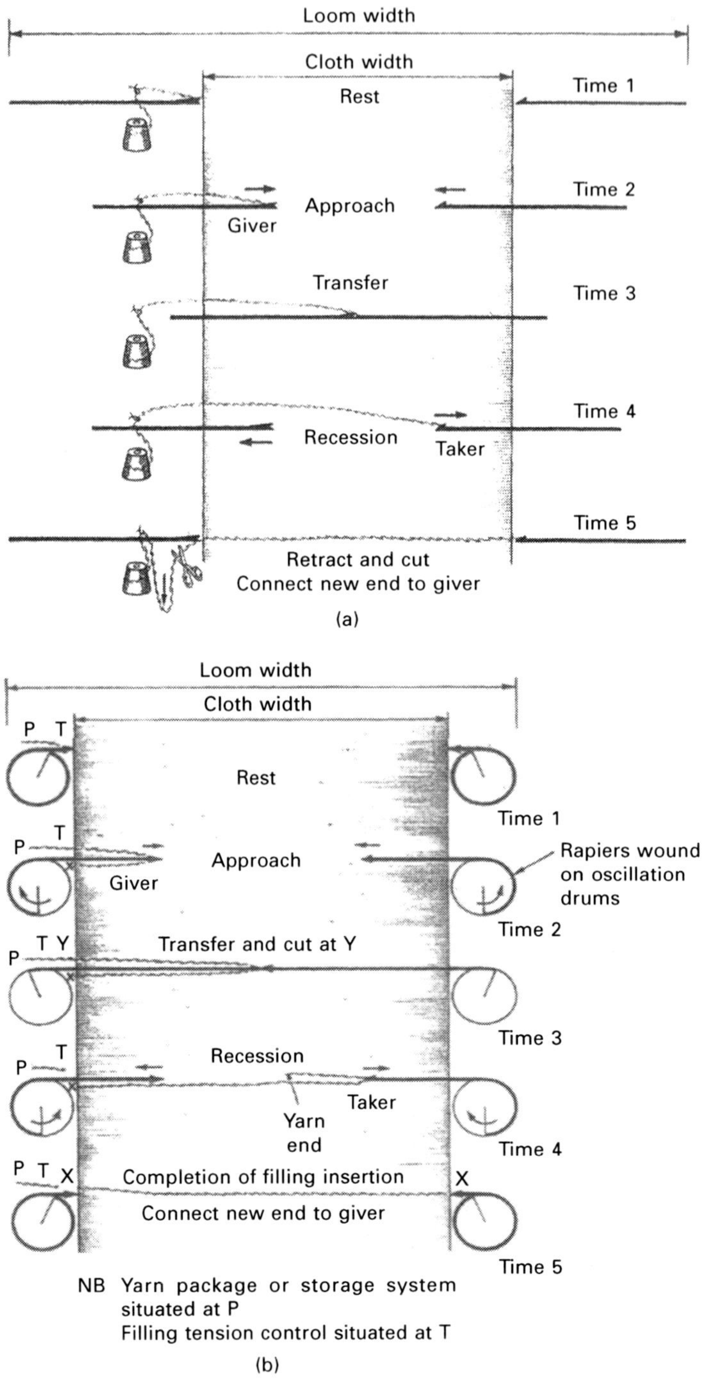

5.25 (a) Schematic diagram of a rigid rapier sequence, (b) schematic diagram of flexible rapier sequence.

- *Flexible rapier looms.*[18] Here the principle of weft insertion is almost the same as on a rigid rapier Dewas system, except that the rapiers have a tape-like structure that can be wound on a drum as shown in Fig. 5.25(b). This design saves space and consequently flexible rapier looms are more common than rigid rapiers. The yarn is gripped by both the giver and the taker.

5.11.2 Two-phase weaving machines

Many years ago, Saurer introduced their Saurer 500 two-phase rapier machine[17] in which the rigid rapier is centrally driven (as shown in Fig. 5.26) and the weft is inserted from the centre from weft packages adjacent to the inside selvedges of the two fabrics (the Saurer 500 produces two fabrics simultaneously). The rapier inserts picks alternately in the left-hand and right-hand fabrics: with the left-hand reed on back centre, the rapier first inserts a pick from the left-hand rapier gripper; as the rapier is withdrawn, the right-hand reed recedes to the picking position and the right-hand rapier gripper picks up the weft end from the other weft package and inserts it into the right-hand fabric. At the same time the left-hand reed forces the previous pick to the cloth fell where it is severed very close to the two selvedges.

5.11.3 Projectile looms

The manufacturer Sulzer first introduced the projectile method of weft insertion to the market in 1952, and then exhibited it at International Textile Machinery show (ITMA), in Brussels, in 1955. Over the decades, significant developments have gone into its design and the company continues to totally

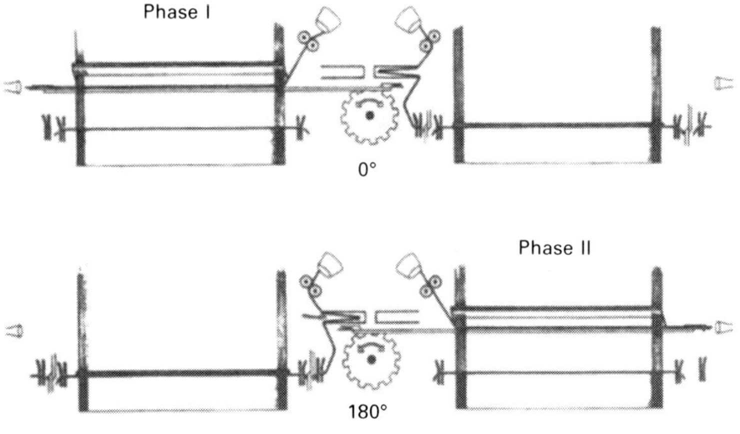

5.26 Saurer 500 two-phase rapier loom.

monopolize the market because of the following advantage offered by this method of weft insertion:[19]

- flexibility to weave more than one width of fabric at a time, i.e. 33–540 cm
- low power consumption
- reduced waste of weft yarn due to its unique, clean tucked-in selvedge
- its ability to weave multicolour weft for any sequence of up to six different weft yarns
- low spare parts consumption and easy maintenance.

For projectile looms, the principle of weft insertion[2,17,20,21] is such that the weft yarn is inserted in the shed by means of a projectile[21] weighing only 40–60 g. (Fig. 5.27). The gripper (projectile) carrying the weft is thrown from the picking side to the receiving side across the machine. The energy used to throw the gripper holding the weft yarn is derived from the shear strain energy stored in a torsion bar which is twisted a predetermined amount and triggered off to provide the means of propulsion. During this journey, the gripper projectile glides through the shed in guides and draws the weft yarn into the shed.

The picking action is illustrated[20] in Fig. 5.28. The projectile (missile) (1) is about to be picked by a metal picker (2) joined by a short connecting rod to the picking arm (3). This arm is attached to the rotatable end of a steel torsion rod (11), which has its other end fixed at (12); the rod is twisted by the anti-clockwise rotation of the arm (13) attached to the sleeve (10). This arm is turned when the toggle formed by the arms (8 and 9) is straightened out by the action of the cam (4) which, rotating in a clockwise direction, moves the lever (8) to the right. The diagram illustrates the situation immediately following the straining of the torsion rod: the rod is being held at its position of maximum torsion and cam (4) has ceased to operate on the cam follower

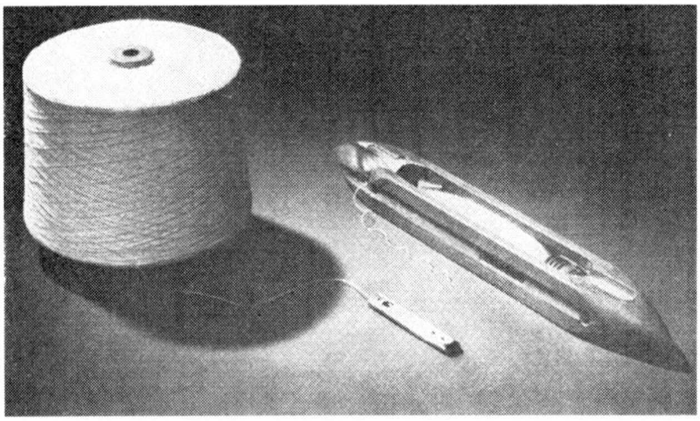

5.27 Sulzer package and gripper shown for comparison with a conventional shuttle and pirn.

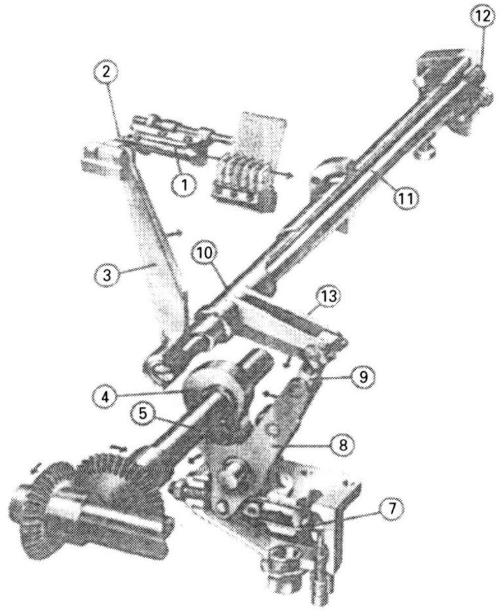

5.28 Picking mechanism of Sulzer weaving machine. The strain energy in the torsion rod (11) is about to be released by further movement of the cam follower (5), causing parts (8), (13), (3) and the missile (1) to move in the directions shown by the arrows (courtesy of Sulzer Bros Inc.).

mounted on the arm (8). The cam (4) is itself provided with a cam follower (5) which, in the next phase of the operation, initiates picking by pressing on an extension of arm (8), which moves the arm to the left and abruptly nullifies the holding action of the toggle. The torsion rod is now free to untwist and the pent-up energy in it is released, causing the picking arm to rotate and launch the projectile at high speed, typically at about 30 m/s. The residual energy in the picking system is absorbed in the hydraulic buffer (7).

The energy available for picking depends on the angular displacement of the free end for a particular design of torsion bar; the strength of pick is completely independent of the speed of the machine. At the receiving side, the weft-end gripper is positioned to grip the weft after reception. It is braked in the receiving unit and is then transported by a conveyor belt (situated under the shed) to its original position on the picking side. It is thought that the gripper projection energy is about one-third to one-half of that normally needed in conventional shuttle picking.

The weft package is situated outside the machine on the picking side, all the picks being delivered from the same side. The weft is withdrawn from the package through a tension device, a weft tensioner, a feeder, scissors, and the weft-end gripper. Figure 5.29(a) shows the weft insertion system in detail on a Sulzer projectile weaving machine, whilst Fig. 5.29(b) shows a projectile with weft yarn entering a shed.[1]

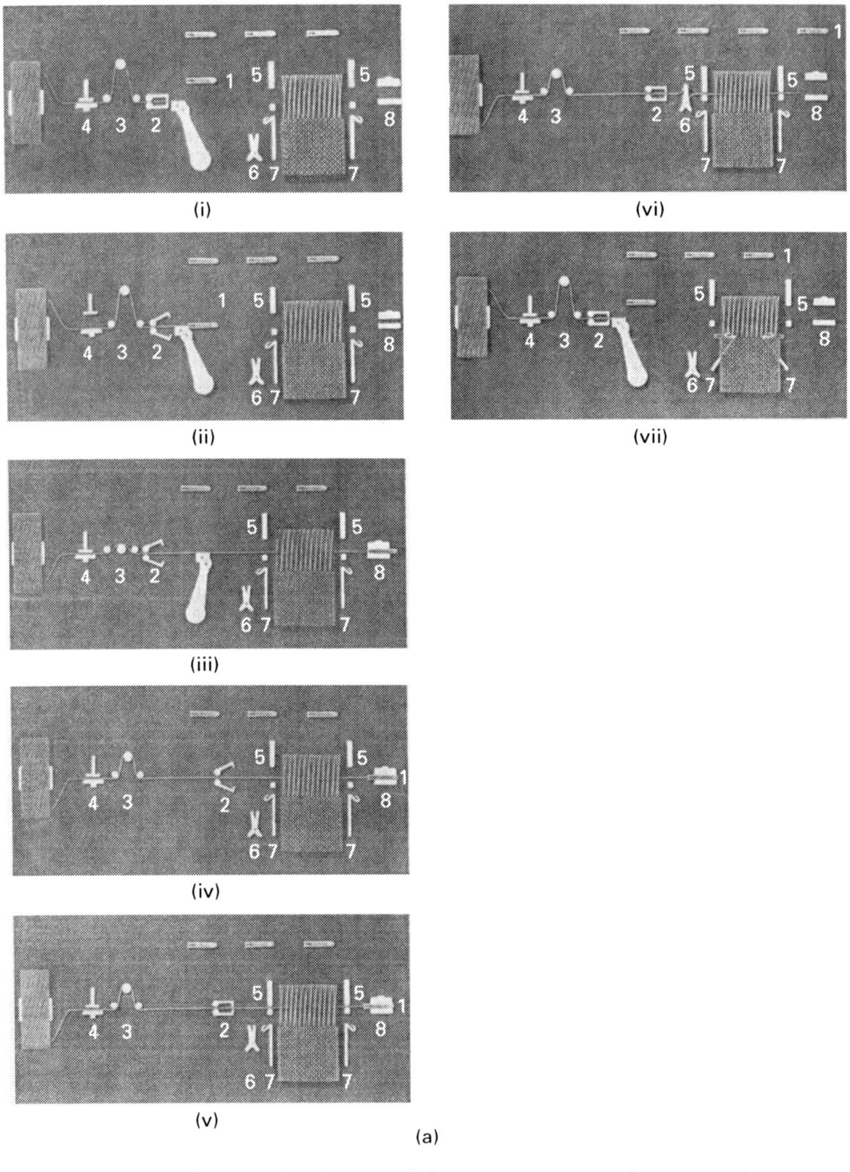

5.29 (a) Schematic of the weft insertion system of a projectile loom. (b) Sulzer projectile entering the shed. Every pick is cut off after complete insertion. The ends are then tucked in the shed and woven in with the next pick (courtesy of Sulzer Bros Inc.).

(b)

5.29 Continued

A description of the weft insertion system corresponding to the seven stages in Fig. 5.29(a) is as follows:

(i) The projectile (1) moves into the picking position.
(ii) The projectile feeder (2) opens after the projectile has gripped the end of the weft thread presented to it.
(iii) The projectile has drawn the thread through the shed during which time the weft tensioner (3) and the adjustable weft brake (4) act to minimize the stress on the thread at the moment of picking.
(iv) The projectile (1) is stopped by the brake (8) and pushed back inside the receiving unit housing. Feeder (2) moves close to the edge of the cloth.
(v) The feeder (2) grips the thread, while the selvedge grippers (5) hold the weft at both sides of the cloth.
(vi) The thread is severed by the scissors (6) on the picking side and released by projectile (1) on the receiving side. The ejected projectile (1) is then placed on the conveyor belt which carries it outside the shed back to the picking position.
(vii) The weft has now been beaten up by the reed. The needles (7) tuck the weft ends into the next shed (a tuck-in selvedge).

5.11.4 Weft insertion on jet-looms

On a jet loom, the weft is inserted by means of a nozzle and carried across the shed by a jet of a working substance, i.e. a fluid. The relative velocity between the jet and the weft thread produces a force on the weft which results in its insertion in the shed. As the tractive force applied to the weft is not very high, the weft thread cannot be unwound directly from a cross-wound package but must be prepared for picking by a metering device.

In the majority of jet weaving machines, the picking system is fitted only on one side of the machine. In all types of jet weaving machines, the picking system is fixed firmly to the machine frame so that the beat-up mechanism carries only the reed or the air duct. Hence the moment of inertia of the beat-up mechanism is small and, therefore, the operation of the jet weaving machine is smooth and uniform.

Air-jet weaving machines are supplied with compressed air from a central source or equipped with individual built-in compression units. Water-jet weaving machines are equipped with individual injection pumps and water is supplied to them from the water main; waste water is discharged into a drain.

Figure 5.30 shows the working principle of jet looms, in the following stages:[22]

(a) The weft (2) is unwound from the cross-wound cone (1) and passed through the guide eye and tensioner (3). The metering device (4) measures the necessary pick length of the weft yarn and the holders (5) serve to retain the weft thread after picking. The water or air is supplied to nozzle (6) through tube (7).
(b) The reed (A) moves backwards. The metering device (4) completes the preparation of the necessary pick length of the weft thread.

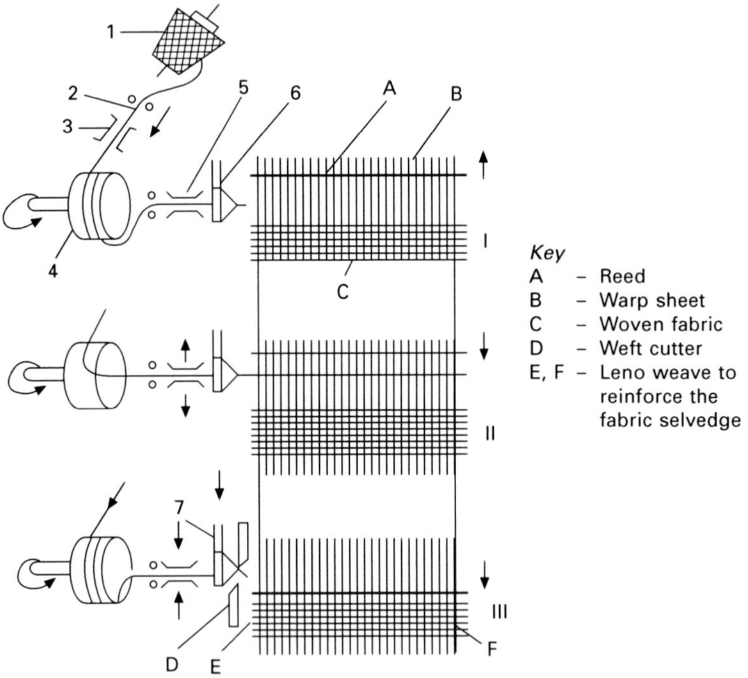

5.30 Principle of sequence of operation of jet weaving machines.

(c) Holders (5) open and, simultaneously, water or air is released into the nozzle (6) to take the weft yarn across the shed.
(d) When the weft insertion is completed, holders (5) clamp the weft thread. The reed (A) beats up the inserted weft to the fabric fell and the cutter (D) cuts it off near the nozzle. Simultaneously the weft thread is secured by the leno weave (E, F). Before each weft is beaten up, the metering device (7) starts preparing the next pick length.

5.11.5 Air-jet looms

Air-jet weaving is considered the most efficient and productive way of inserting the weft across the shed, for the production of light to medium weight fabrics, terry towels, furnishing fabrics and denims, etc. Compared to the rapier and projectile methods of weft insertion, the mass of the insertion medium in air-jet weaving is very small. This novel feature allows high weft insertion rates to be achieved. Although the first commercial single-nozzle air-jet loom was exhibited at ITMA in 1958, the real breakthrough came in 1967–1968 when a 165 cm wide loom with relay nozzles running at a weft insertion rate of 530 m/min was exhibited. Over the last few decades, there have been further technological advances in air-jet weaving, as a result of which weft insertion rates from 2000 to 2900 m/min have been achieved, weaving fabric widths from 190 to 340 cm.

The principle of weft insertion[17] is shown in Fig. 5.31(a). The weft is withdrawn from the weft package(s) by the constant-speed weft feeder units prior to delivering to the two main jet nozzles (3), whilst air is delivered from the pressure regulating valve (7) to the two control valves (6). The velocity of the weft is maintained by the relay nozzles (4) which are integral to the sley and located at intervals across the reed space. The main nozzles provide the initial acceleration, whilst the relay nozzles provide high air velocity to carry the weft across the width of the warp. The relay nozzles are controlled along with the main nozzles from the central processor (1), which is controlled from the operator's terminal (2). The warp let-off (8), the warp tension sensor (9), the shed formation control (10), the weft detector (11), and the warp detector (12) are all controlled through the central processor (1). The stretching nozzle (5) is shown outside the receiving side selvedge. The weft is projected into a tunnel[17] profiled in the reed (shown in Fig. 5.31(b)) by relay jets in the picking position. The profiled reed provides guidance for the air and separates the weft yarn from the warp. A cutter is used to cut the weft yarn when the insertion is complete.

5.11.6 Water-jet looms

For nearly two decades in the 1970s and 1980s, water-jet looms were regarded as superior due to their impressive weft insertion rates of up to 2,700 m/

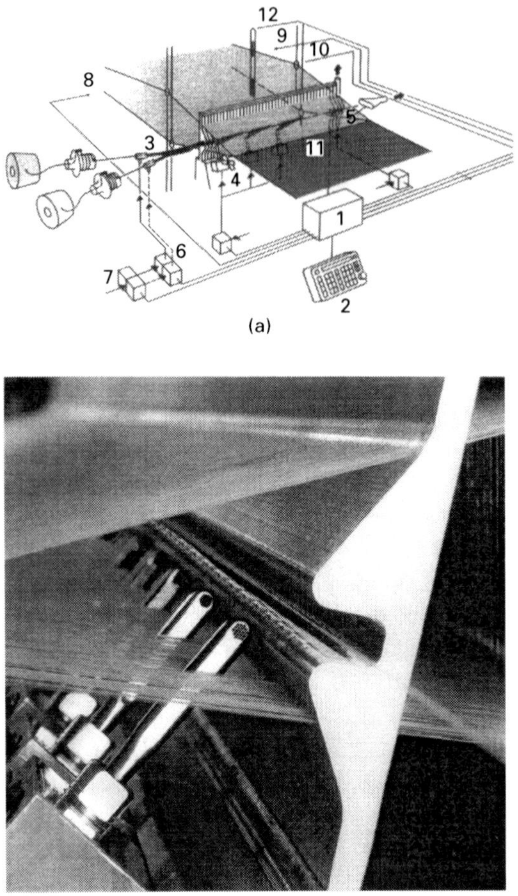

(a)

(b)

5.31 (a) Sulzer Rüti L 5100 (air-jet) weft insertion system and (b) tunnel reed and relay jets in picking position on Sulzer Rüti L 5100.

min.[23] However, they are confined to handling only 100% hydrophobic yarns and, from the 1980s onwards, dramatic improvements in the productivity of air-jet and rapier weaving machines started to overshadow their popularity such that, at ITMA 2003, not a single water-jet loom was exhibited. However, at ITMA 2007, one manufacturer (VUTS from Liberec) exhibited their Camel W water-jet machine, weaving a reinforced fabric 178.6 cm wide at a speed of 600 ppm and with a weft insertion rate of only 1,071 m/min. The trend has continued though, and air-jet and rapier machines have taken a major share of the fabric production previously undertaken by water-jet looms which, although still in commercial use in some parts of the world (and running with weft insertion rates higher than 1,200 m/min), are now decreasing in popularity.

Although similar in principle to an air-jet loom (i.e. it uses water as a fluid, instead of air), water-jet looms differ in construction and operating conditions in a number of ways. Each loom has its own miniature pump to feed water under pressure to the nozzle and, to get the best results, the water must be filtered, of a specified hardness, and supplied to the loom at a constant temperature.

Figure 5.32(a) shows the working principles of a water-jet loom.[24] Weft yarn, from the cone package (7) is drawn off by a measuring device (2) and

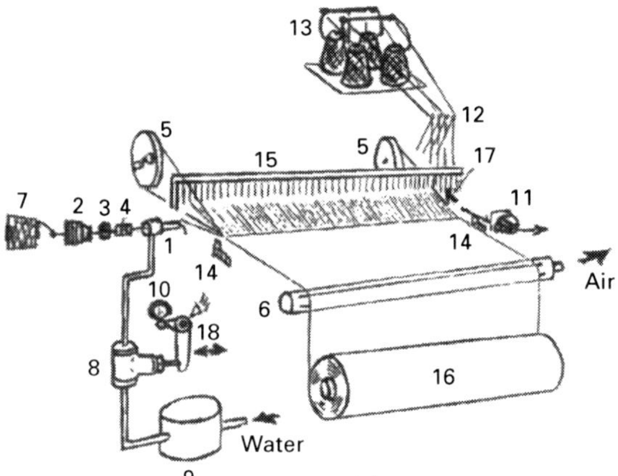

(a)

Key

1 – Nozzle
2 – Measuring (feeding) device
3 – Tension regulator
4 – Weft clamp
5 – Leno mechanism
6 – Front beam
7 – Cone package
8 – Pump
9 – Water container
10 – Cam
11 – Guide to take waste
12 – Guide to yarns from cones (13)
13 – Cones and yarn clamping device
14 – Thermal knives
15 – Reeds
16 – Cloth roller
17 – Leno ends
18 – Pressure device

(b)

5.32 (a) Principle of weft insertion on water-jet loom, (b) water-jet outlet with weft from the nozzle.

passed through a tension regulator (3) and the weft clamp (4). When the pick has to be inserted, the weft clamp (4) loosens its hold and the weft, inserted inside the nozzle (1), is struck by a jet of pressurized water and launched through the shed. The weft is held flat by the leno threads near the entry and exit points of the shed on both sides near the selvedges. The thermal knives (14) enter into action on the launch side to cut the weft, and on the opposite side to trim the fabric. A yarn clamping device (13) holds the weft waste which is cut off by the right- handed thermal knife (14). The water from the container (9) is conveyed by a pump (8) controlled by a cam (10) and a device that controls the pressure (18). A drying device removes the humidity absorbed by the fabric, sucking it through grooves produced in the front beam (6). The loom completes its one cycle and this continues all the time. Figure 5.32(b) shows the water jet outlet along with the weft yarn from the nozzle.[22]

Water-jet looms are normally compact, mechanically simple and require much lower energy than air-jet picking systems. They are best suited for weaving fabrics made from water-insensitive continuous filament yarns, flat polyolefin and glass. One of their drawbacks is their limitation on fabric width. The jet of water which is used to propel the weft yarn disintegrates increasingly into droplets the further it is from the nozzle, and this characteristic (together with the effect of gravity) limits the loom width appreciably compared to air-jet looms. Another limitation is that water-jet looms are not readily adaptable to deal with multi-coloured or multiple types of weft. Currently, Toyota offer one of their models which can be equipped with three-colour pick-at-will and w-FIS which allows the insertion of fine or coarse and heavy yarns without changing the pump and nozzle, thereby offering high versatility.

Weft insertion rates on shuttle-less looms

Weaving machine productivity, based on the weft insertion rate (WIR), measured in meters per minute (m/min) is directly proportional to loom speed (expressed by picks per minute) if no loom stoppage occurs. WIR indicates the amount of weft that is converted into fabric, or inserted, in one minute during weaving. This is a product of loom speed and width of warp in the reed. Table 5.3 presents the maximum WIRs that were exhibited at ITMA

Table 5.3 Maximum weft insertion rates (WIRs) exhibited at ITMA 2007 for different methods of weft insertion

Loom	Maximum weft insertion rate (WIR) (m/min)
Rapier loom	1,628
Air jet loom	2,904
Projectile loom	1,497
Multiphase looms	2,160 at ITMA 1987 (on Nuovo Pignone & Textima) 4,777 (on Sulzer M8300 machine at ITMA 2003)

2007 for different methods of weft insertion.[23] Maximum WIRs on a modern shuttle loom can vary from 550 to 650 m/min.

5.12 Multiphase weaving

Weaving machines are classified into two groups: single-phase and multiphase. In a single-phase loom, a single insertion of weft yarn is made in one revolution of the loom, the shed is fully opened across the whole width of warp, and the pick is inserted. In a multiphase loom, not just one but several sheds are formed, and at the same time several weft insertion elements enter the entire width of the warp.

Single-phase shuttle-less looms such as rapier, projectile, air-jet and water-jet are considered to be the second generation of weaving machines in which the primary mechanisms of weaving (i.e. shedding, picking and beat-up) take place in one loom cycle, thereby inserting one pick in a single phase. Only one shed is formed across the width of the warp.

In multiphase weaving, several sheds are formed across the width of warp at the same time (see Fig. 5.33(b)).[2] Multiphase looms are considered to be the third generation of weaving machines. Multiphase weaving can be achieved by employing two different principles, either: (a) in which shedding, picking and beat-up occur across the width of the warp, i.e. in the weft direction, or (b) in which these actions take place along the length of the warp (in the warp direction).

5.12.1 Principle of weft insertion in multiphase weaving

Several weft carriers, one behind the other, traverse the reed space and insert one pick length of weft yarn (stored inside them) at a time.[17] Weft yarn is continuously drawn from weft packages and wound onto the weft carriers one pick length at a time. Their phase number and velocity determine the

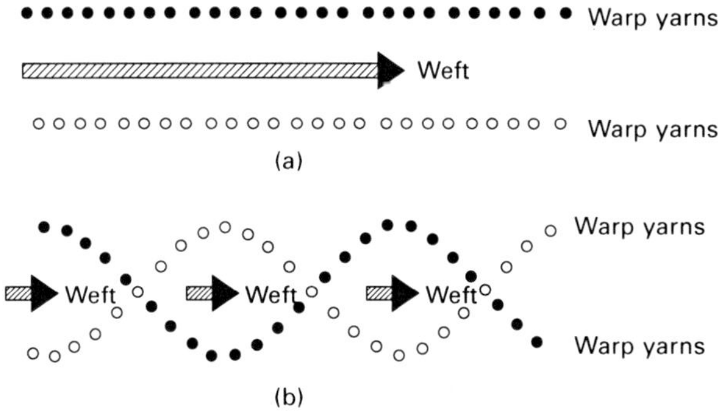

5.33 Principle of (a) single-phase and (b) multiphase weaving.

weft insertion rates. The shed and reed clearance open and close around each carrier as it traverses the reed space, the movement of the upper extremities of the shedding elements giving the impression of a wave motion travelling from picking to receiving side. The special reed motion impels the weft carrier through the shed, each of which inserts weft length as it goes. At their reversal point, the mobile reed dents beat the weft yarn emerging from the weft carriers against the fell of the cloth. Upon reaching the receiving end, the carriers then return under the fabric to the picking side for recharging, the weft from the package is transferred to the carriers and the process continues.

5.12.2 Sulzer *Textil* multiphase loom M8300

Sulzer *Textil* introduced their revolutionary multiphase loom M8300[25] at ITMA 1995, with a working width of 190 cm running at 2,050 ppm with weft insertion rates of 3,894 m/min. The machine has no healds and a non-reciprocating beat-up. The weft is inserted with the aid of compressed air at low pressure and at a uniform speed, with low weft loading. Shedding is based on the multilinker shed principle: sheds are formed positively by shed-forming elements which deflect the warp threads into the upper shed position. The curvature and rotation of the rotor shown in Fig. 5.34(a) and (b) cause the shed-forming elements to open the sheds consecutively. A single movement of the warp positioners positions the warp threads so that they are either picked up and lifted by the shed-forming elements, thus forming the upper shed, or remain in the lower position. As many as 2,800 sheds are formed consecutively per minute. Each warp thread is inserted individually into the eye of a positioned warp.

A number of sheds are arranged in parallel, one behind the other in the direction of the warp, and are opened simultaneously. Four weft yarns are inserted at the same time, though staggered as in a relay race, into the open sheds on the rotor. To open a number of sheds one after another in the warp direction, the warp is led over a 12-channel continuously rotating drum (the weaving rotor), which is provided with combs consisting of numerous individual elements. The combs form the shed as well as a guide channel for the insertion of the weft as soon as the shed is opened.

The rotating movement of the weaving rotor lifts the threads off the shoulders of the shed forming elements and lays them over the weft channel, while the remaining threads come to rest below the weft channel. Warp positioners determine the threads that are to be raised. Once the shed is formed completely, low pressure air carries the weft yarn through a weft insertion channel. During this insertion, further weft yarns start to enter the combs that follow. As soon as the weft has been inserted completely, it is clamped and cut on the feed side. The weft yarn is beaten up by the special beat-up reed that follows each shed forming comb.

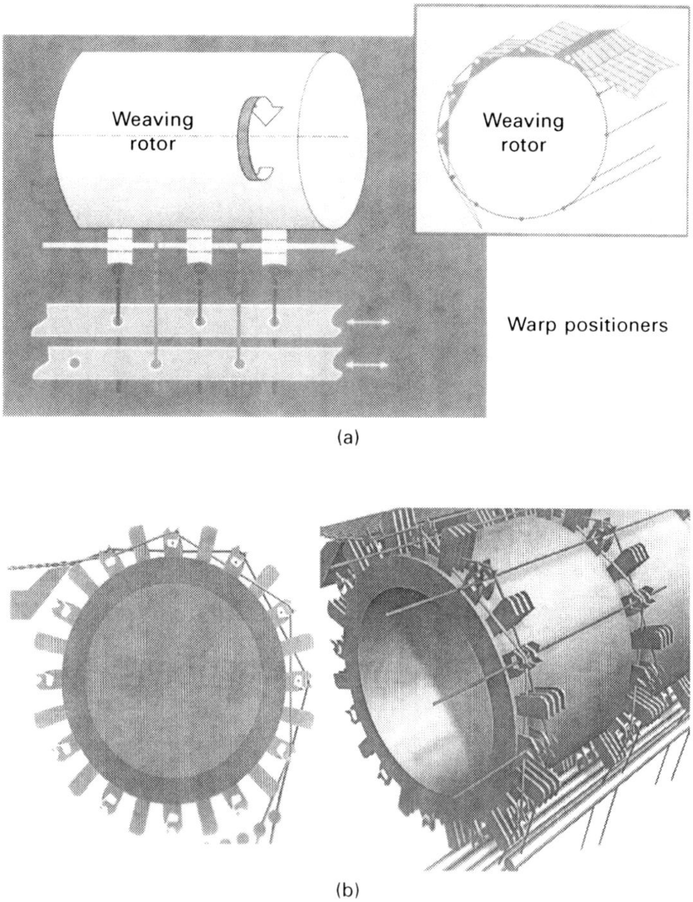

5.34 (a) Sulzer Multiphase loom M8300, (b) weaving rotor.

This innovative technology offers the following immense advantages:

- a reduction in weaving cost by 30–40%
- 3–4 times the productivity
- up to 30–40% lower energy consumption
- 60% lower space requirement than a single-phase air-jet weaving machine
- noise levels around 10 db (A), lower than that of a single-phase weaving machine
- reduced yarn residues and dust.

However, multiphase weaving machines have serious limitations from the point of view of fabric design and their narrow field of application. This has, of course, resulted in the demise of multiphase weaving.

5.13 References

1. Bernard P. Corbman, *Textiles – Fibre to Fabric*, McGraw-Hill, New York, 1983.
2. Teruo Ishida, *Innovations in Weaving Machinery*, Osaka Senken Ltd, Japan, 1994.
3. Ben H. Crawford, *Draper Loom Fixing*, A.J. Showalter Co., Dalton, GA, 1947.
4. R. Marks, P.J. Lawton, D.A. Holmes, *An Introduction to Textiles Vol III*, Comett Eurotex School of Textiles, Bolton University, UK.
5. William A. Hanton, *Automatic Weaving*, Ernest Benn Ltd, Manchester, 1929.
6. J.W. Hutchinson, *The Practical Management of Looms and Yarns*, Bradford, 1949.
7. R. Marks and A.T.C. Robinson, *Principles of Weaving*, The Textile Institute, Manchester, 1976.
8. *Woven Fabric Production – Shedding*, NCUTE. http://www.pdexcil.org/news/51N1203/weaving1.htm.
9. Toyota Brochure for loom LWT 710.
10. Stäubli brochure – Shedding solutions for modern frame weaving.
11. Recep Eren, G. Ozkan, M. Karahan, Comparison of heald frame motion generated by rotary dobby and crank and cam shedding motions, *Fibre & Textiles in Eastern Europe*, October/December 2005, Vol. 13, No. 4, p. 52.
12. W. Wilkinson, *Practical Weaving*, Coulton & Co., Nelson, 1915.
13. Norma Hollen, Jane Saddler, Anna Langford, *Textiles*, Macmillan, New York, 1979.
14. *Textile Terms and Definitions*, The Textile Institute, Manchester, 1995.
15. The 4 of Horgen, Bulletin No. 86, 1988.
16. G.A. Bennett, *Introduction to Automatic Weaving*, Harlequin Press Co. Ltd, Manchester, 1948.
17. Alan Ormerod, Walter S. Sondhelm, *Weaving – Technology and Operations*, The Textile Institute, Manchester, 1995.
18. P.R. Lord, M.H. Mohamed, *Weaving: Conversion of Yarn to Fabric*, Merrow Publishing Co. Ltd, Watford, UK, 1973.
19. Sabit Adanur, Chapter 9 – Projectile weaving, *Handbook of Weaving*, CRC Press, Boca Raton, FL, 2000.
20. J.J. Vincent, *Shuttle-less Looms*, The Textile Institute, Manchester, 1980.
21. Victor Duxbury, Gordon Wray, *Modern Developments in Weaving Machinery*, Columbine Press, Buxton, UK, 1962.
22. Oldrich Talavasek, Vladimir Svaty, *Shuttle-less Weaving Machines*, Elsevier Science, Oxford, 1981.
23. Kim Gandhi, ITMA Fruits of the Loom, *Textile Month*, 2007, Issue 6, p. 8.
24. www.textile school.com/school/weaving/waterjetweavingmachines.aspx.
25. Sulzer *Textil* M8300, Brochure – the reinvention of weaving.

Part II

Woven structures

6
Woven structures and their characteristics

J. WILSON, Consultant Designer, UK

Abstract: This chapter looks at the basics of the weaving process and woven fabric construction. It covers basic weaves including plain weave, twills, hopsacks and satins and explains how these are notated and specified. The chapter also looks at drafting and lifting, repeating pattern, and combining weaves, and briefly covers jacquard, dobby, pile and leno weaving techniques.

Key words: weaving, plain weave, twill, hopsack, satins, fabric notation, dobby, woven jacquard, pile, leno.

6.1 Introduction

Woven fabrics are produced by interlacing two sets of threads, known as the warp and weft, at right angles to each other. The warp threads run parallel to the selvedge (the reinforced fabric edge) down the length of the cloth, and each warp thread is known as an 'end'. The weft threads run across the cloth from selvedge to selvedge and are called 'picks'. (For clarity the terms warp ends and weft picks are used throughout this chapter although warp is synonymous with ends and weft is synonymous with picks.) The threads can interlace in a number of ways limited only by the type of loom and the system in place to raise warp threads. Woven fabrics are classed as constructed textiles as are knits and lace fabrics.

The interlacing pattern of the warp and weft is known as the weave. The weave influences the appearance of the fabric, its handle, drape and its behaviour in use. The simplest weave is plain weave (Fig. 6.1).

A woven fabric is made up of many elements including the weave and the component yarns and fibres. All these elements, as well as any treatments or finishes applied, will affect the final fabric in terms of its appearance, handle, drape and behaviour in use.

6.2 Representing woven fabrics

Weaves can be difficult to describe. They can be drawn out as in Fig. 6.1, but this is very time consuming so they are usually illustrated on special squared paper called weave design paper or point paper. To draw out even a simple weave such as plain weave (Fig. 6.2) can be time consuming. The way weaves are generally notated uses filled squares or crosses to indicate where the warp thread is uppermost. This graphical representation of the

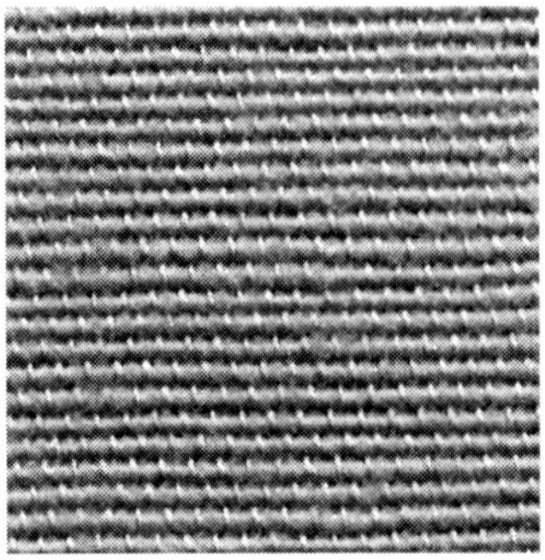

6.1 Plain weave (photograph by the author).

6.2 Graphical representation of a plain weave fabric (photograph by the author).

interlacings is marked on weave design paper which is also known as point paper (Fig. 6.3). The standard weave design paper used is ruled in groups of 8 × 8, separated by thicker bar lines.

6.2.1 Notating weaves

Each vertical row of squares on weave paper indicates a warp end while each horizontal row of squares represents a weft pick. Each square therefore indicates a point where 1 end and 1 pick intersect. There are only two types of intersection possible and these are warp over weft or weft over warp. Warp over weft is indicated by a mark in a square of the point paper: this may be a cross or the square may be completely filled in. Weft over warp is indicated when a square of the point paper is left blank (see Fig. 6.4). In complex structures such as multi-layered fabrics, it is helpful to use several different marks; however, all marks indicate warp up.

Figure 6.5 shows plain weave notated in two different ways: by blocks and by crosses. Either method is acceptable as any mark indicates warp over weft.

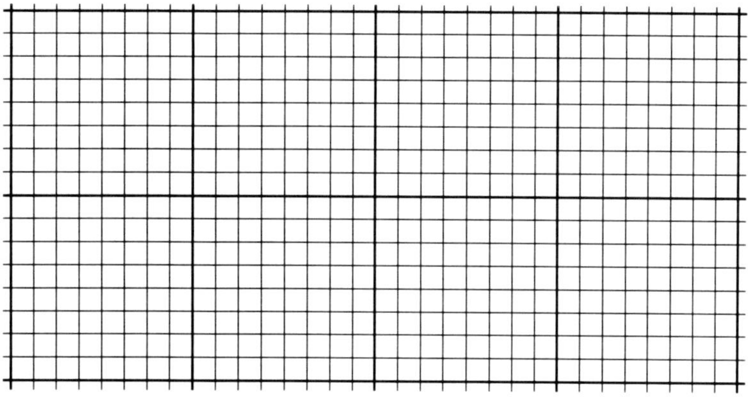

6.3 Weave design paper.

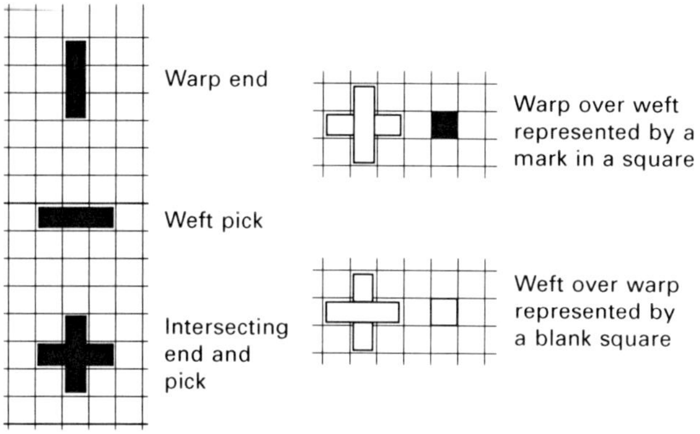

6.4 Notating woven fabrics.

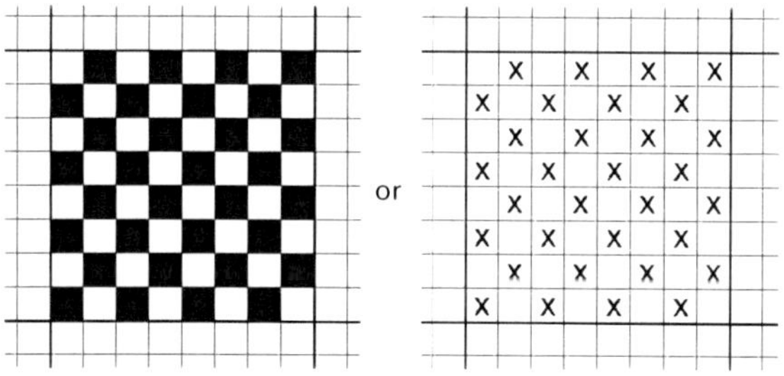

6.5 Notation of plain weave.

6.2.2 Plain weave

Plain weave is the simplest and most frequently used weave of all. Because it has the maximum number of interlacing or binding points possible, it is firmer and stronger than a fabric which is otherwise identical but made in another weave.

Lightweight fabrics are usually made in plain weave to ensure that the ends and picks do not slip over each other and cause the fabric to distort. Chiffon and voile are two lightweight fabrics made using plain weave. Voiles are normally produced from cotton or linen, while chiffons are traditionally made from silk.

A variety of decorative effects can be achieved using plain weave. Stripes or checks (as in a gingham) can be made by introducing coloured yarn. Different yarn types can be used to decorative effect in plain weave structures. Textural stripes can be made by cramming together threads in certain sections to increase the density of groups of ends in these areas.

6.3 Weaving

To better understand woven fabric structures it is necessary to understand the weaving process. A woven fabric is constructed from the interlacing of two sets of threads: the warp and the weft. The warp is a set of threads in a vertical direction wound onto a frame or beam. The weft is the thread interlaced in a horizontal direction to create the weave.

6.3.1 Weaving on a simple hand loom

Warp ends are wound onto a warp (or warper's) beam in the required length, density and width. The first end of the warp is on the left-hand side facing the loom. Lease rods are used to separate the warp ends. These maintain the position of the warp ends within the warp and make it easier to draw the individual ends through the healds on each shaft.

A shaft is a frame with wire (or string) healds. These healds have eyelets in the centre and each warp end is pulled or drawn through one. Each differently interlacing end in the weave requires a different shaft – ends that interlace identically are drawn through the same shaft. The healds are lifted (and in some looms also lowered) via the shafts during the weaving process to form a shed. The shed is the triangular shaped space through which the weft yarn passes.

To achieve the required number of ends per centimetre across the width of the fabric, the ends are spaced out across the width using a reed. A reed is essentially a 'closed comb', made from a series of metal wires and is used to separate the warp ends and 'beat-up' the pick against fabric already woven. The warp yarns are placed within the spaces or 'dents' of the reed. The number of dents per cm and the number of ends in the dents will determine the density of the warp.

As already stated, a woven fabric is constructed by the interlacing of warp ends and weft picks. By raising certain shafts the ends drawn through the healds on these shafts are raised. The shed is formed between the ends

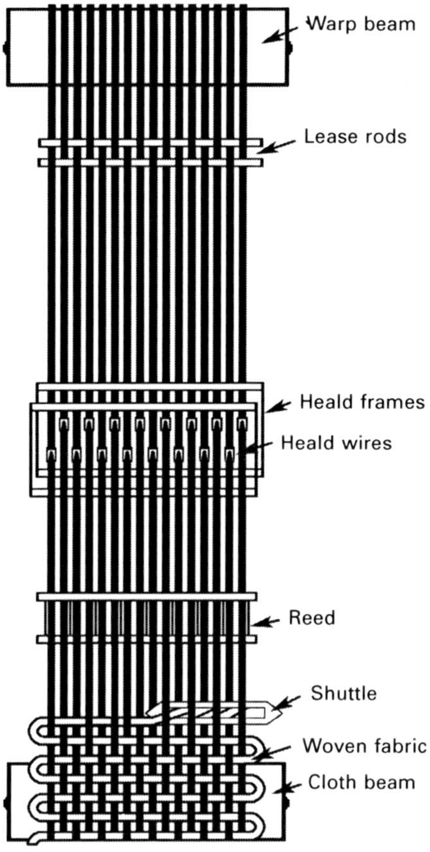

6.6 A simple loom.

raised and those left down. The weft is inserted through this shed using a shuttle. The shuttle is guided across the loom by the reed which is also used to beat up the last inserted pick. The picks are inserted according to the required pick spacing (weft density). The resulting fabric is wound on to a cloth roller.

A simple loom is shown in Fig. 6.6.

6.3.2 The characteristics of a woven fabric

The characteristics shown by a woven fabric are governed not only by the way the threads interlace to construct the fabric (the weave) but also by the yarn used (the colour, count or linear density and the fibre content), the number of ends and picks per cm, the ratio of ends per cm to picks per cm, how close or open the fabric is, and the finish.

6.3.3 Keeping records of woven samples

It is important to keep records of how fabrics have been made so they can be replicated in the future if required. Usually fabric making particulars will be recorded on a standard specification sheet. This will have several sections to record all the details required to make the fabric. Fabric making particulars for a woven fabric would normally include the following:

- a fabric name and/or number; while names are easier to remember, numbers can be less confusing; frequently, both are used as identifiers of a particular fabric;
- a record of all yarns used; the colour, count, quality and supplier of all yarns needs to be recorded, as does the fibre content of each;
- the number of warp ends, the width of the fabric on the loom, and the finished width;
- warp and weft plans with any repeats clearly indicated;
- the weave structure with suitable instructions as to the order the warp threads lift (this is normally the draft and lifting plan and this is covered later);
- the number of ends and picks per centimetre;
- any finishes that have to be applied such as flame-retardant finishes, any brushing or raising, etc.;
- the weight of an appropriate measurement of fabric in grams per square metre or grams per finished running metre.

For initial sampling it is normally enough to use one sheet to record warp details:

- a record of all the warp yarns used in each section; the colour, count and fibre content of yarns needs to be recorded, as do the warp plans with repeats clearly indicated;
- the draft for each section (drafts are covered later in this chapter).

And one sheet to record:

- all the weft yarns used; the colour, count and fibre content of yarns needs to be recorded, as does each weft (picking) plan with any repeats clearly indicated.

For each fabric sample there should be reference to which warp section was used and which weft. For a final fabric sample a proper specification sheet giving full fabric making particulars should be used.

Warping and picking orders

The warp pattern repeat is the smallest number of ends in either colour or count, when repeated and combined with the weave. A warp pattern repeat of 2 black, 2 white, 4 black and 4 white will usually be written as in Fig. 6.7 and is known as the warp pattern, warping plan or warping order. Figure 6.7 shows one way in which a warp pattern can be notated.

The weft pattern repeat (weft plan or picking order) is the smallest number of picks in either colour or count which, when repeated and combined with the weave, up and down the fabric. Figure 6.7 shows a warp pattern repeat of 2 black, 2 white, 4 black and 4 white. Figure 6.8 shows the weft pattern notated in a similar way to the warp in Fig. 6.7.

The warping order is normally given from left to right while the picking order is given from bottom to top although this can vary in different parts of the country and in different parts of the world.

6.4 Colour and weave effects

Combining weaves with different colour patterns in the warp and weft can be used to create different patterns or colour and weave effects. A weave specification sheet, as shown in Fig. 6.9, can be completed to describe the required pattern.

The plain weave fabric in Fig. 6.10 has warp and weft colour patterns of 1 thread black then 1 thread white. A cross indicates the warp end is crossing over a weft pick and a blank square indicates where a weft yarn is on top of a warp yarn. This means that the colour of the warp yarn will be what is seen at any intersecting points where there is a cross and the colour of the weft yarn will be what is seen at any intersecting points where there is a blank square. Figure 6.10 shows the weave notation for plain weave with the colour and weave effect. Figure 6.11 is a photograph of a plain weave fabric woven with warp and weft colour patterns of 1 thread black then 1 thread white. This colour and weave effect is a weft hairline.

A number of different colour and weave effects can be achieved on plain weave and these are shown in Figs 6.12 and 6.13 (warp hairline), and 6.14 and 6.15 (four point star). Note that in Fig. 6.14 the weave repeats on 2 ends × 2 picks while the warp and weft patterns repeat on 4 ends (2 black/2 white) and 4 picks (2 black/2 white) respectively.

Black 2/20s cc	2		4		6
White 2/20s cc		2		4	6
					12 ends

6.7 Warp pattern.

Black 2/20s cc	2		4		6
White 2/20s cc		2		4	6
					12 picks

6.8 Weft pattern.

Loom	Number of shafts	Reed – dents per cm	Width on loom
Ends per 10 cm (loom)	Picks per 10 cm (loom)	Weight per square metre (loom)	Weight per square metre (finished)
Ends per 10 cm (finished)	Picks per 10 cm (finished)	Description	

Count and quality	Colour	Warp pattern

Count and quality	Colour	Weft pattern

Draft

Weave

Lifting plan

6.9 Example of a weave specification sheet.

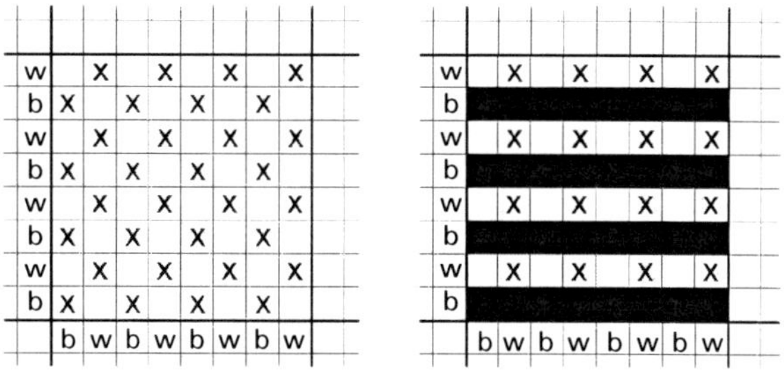

6.10 Plain weave colour and weave weft hairline.

6.11 Weft hairline on plain weave (photograph by the author).

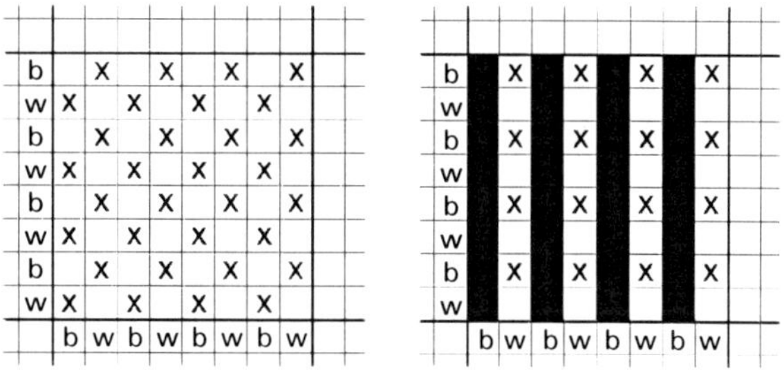

6.12 Plain weave colour and weave warp hairline.

6.5 Sett

A fabric with the same number of ends and picks per centimetre is said to be square sett (see Fig. 6.16). Altering the sett, that is the ratio of ends per cm to picks per cm, can give a very different appearance and handle.

By altering the sett of plain weave and using yarns of different counts (linear densities) for the warp and weft the effect of ribs can be achieved (see Figs 6.17 and 6.18). A plain weave fabric with more ends than picks per cm, where the warp yarn is finer in count than the weft yarn, will give ribs

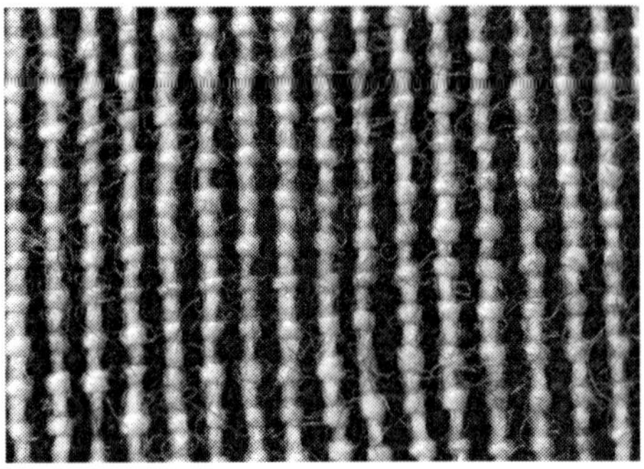

6.13 Warp hairline on plain weave (photograph by the author).

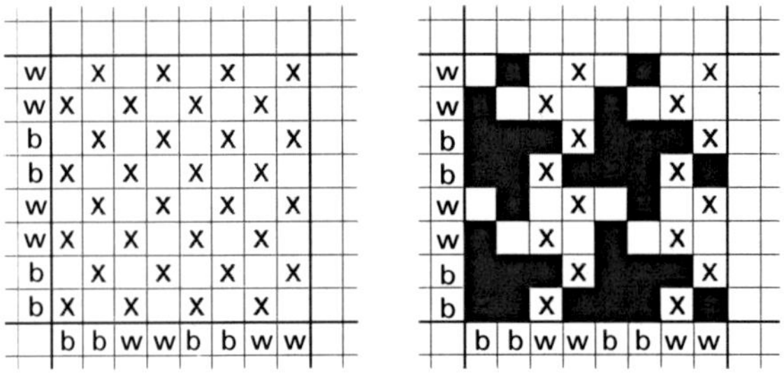

6.14 Plain weave colour and weave four point star.

6.15 Four point star on plain weave (photograph by the author).

running in the weft direction and so the fabric is described as warp-faced with a weft-ways rib. Comparing the threads taken from both directions will show the ends to be more heavily 'crimped' than the picks, i.e. the ends do more bending around the picks while the picks lie across the cloth. Poplin, taffeta, grosgrain are all plain weave fabrics that are warp-faced with a weft-ways rib. Poplins are normally made from cotton or linen, while taffeta and grosgrains are traditionally made from silk.

6.16 Square sett plain weave.

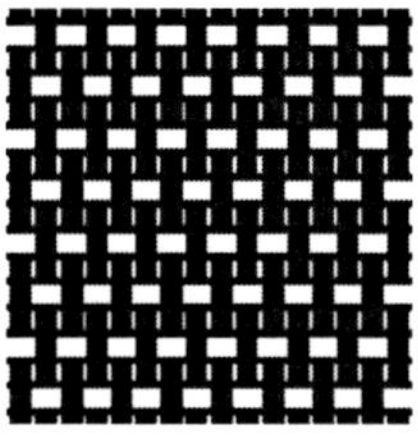

6.17 Warp-faced plain weave.

6.18 Weft-faced plain weave.

6.6 Weaves

An almost infinite number of interlacings can be created. The only real limitation is the number of ends that can be lifted independently. This will depend on the mechanism a loom uses for lifting the threads in the warp. For non-jacquard looms the limitation on the different number of interlacing ends will be the number of shafts that the loom being used has. For jacquard looms this will depend on the number of hooks. The patterning capabilities of jacquard looms are greater than for looms without any jacquard mechanism and as such more complex woven structures with larger repeats can be woven.

6.6.1 Weave repeat

One repeat contains the smallest number of different intersections between warp and weft which, when repeated in either direction, gives the fabric structure (Fig. 6.19).

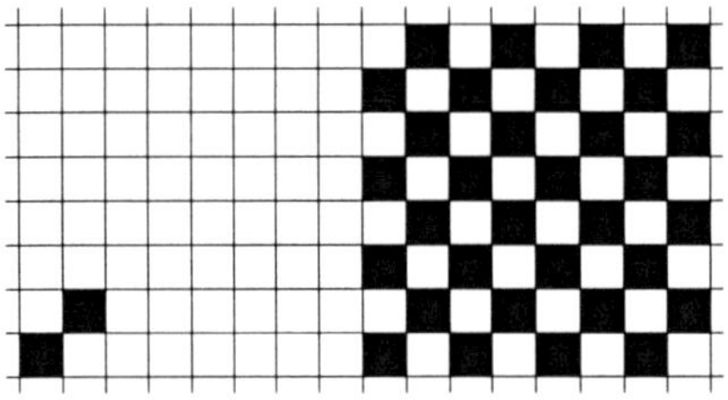

6.19 Weave repeat (a) one repeat and (b) several repeats.

While just one repeat needs to be indicated on point paper it is often helpful to show more than one repeat, particularly with weaves that have small repeats. While a weave can be equal in the number of ends and picks, the weave repeat can also be unequal with more ends than picks or more picks than ends, as shown in Figs 6.20 and 6.21.

6.6.2 Plain weave derivatives

Cords (sometimes called rib weaves)

Plain weave can be extended to form cord effects. In a warp cord the warp float is extended to cover two or more weft picks at a time. These floats on alternate ends have a tendency to come together forming ribs running across the cloth in a weft-ways direction. The floats virtually obscure the weft (Fig. 6.22).

Similarly weft cords are formed by extending the weft float which produces ribs or cords of weft running up and down the fabric in a warp-ways direction (Fig. 6.23).

6.6.3 Hopsack or matt weaves

Hopsack or matt weaves are based on plain weave but with two or more threads working in the same order in both warp and weft (Fig. 6.24). The fabrics have a smooth surface with what is essentially an enlarged plain weave structure. If two fabrics of otherwise similar construction (same yarns and sett) are woven as plain weave and 2/2 hopsack, the hopsack fabric will be less stiff and less stable because of its fewer intersections, and will drape differently.

When weaving hopsacks there is a tendency for the threads to roll over each other. This makes weaving difficult and the appearance of the cloth suffers. This rolling can be prevented by the way the threads are separated in the reed. Where possible, the ends that work together should be separated by the wires in the reed.

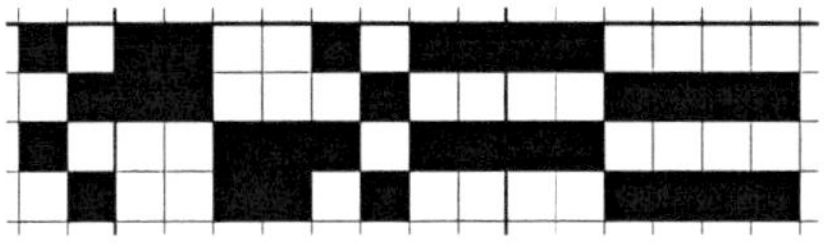

6.20 Weave repeat 4 ends × 16 picks.

6.21 Weave repeat 16 ends × 8 picks.

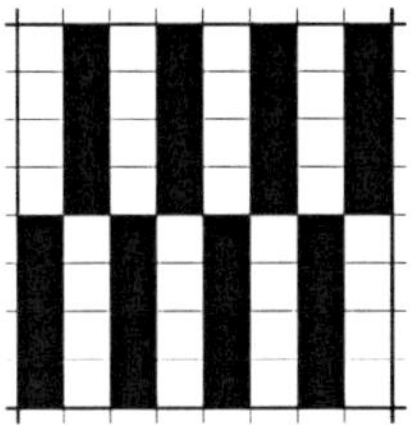

6.22 Warp cord.

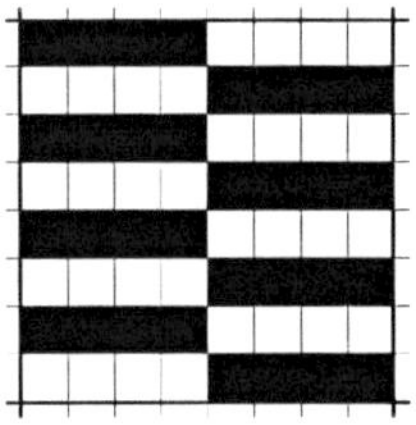

6.23 Weft cord.

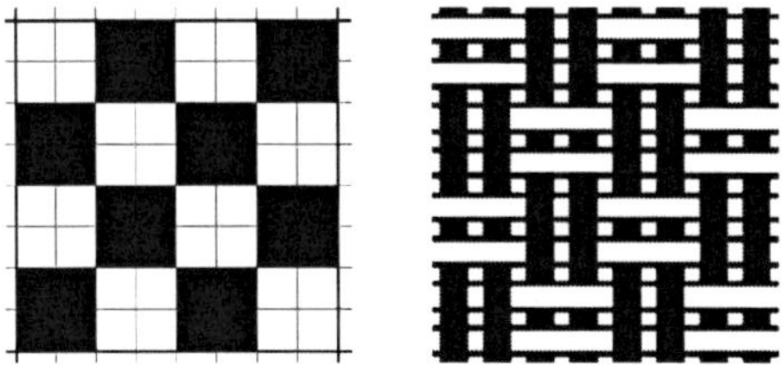

6.24 2/2 hopsack notation and fabric representation.

As hopsacks have two or more consecutive picks with identical interlacing, there is a tendency for picks to pull out from the side. To prevent this, hopsacks are woven with selvedges, small sections of plain weave at either side of the fabric being woven.

A variety of colour and weave effects can also be created on hopsacks. Figure 6.25 shows a 2/2 hopsack coloured so as to produce a four point star effect.

Irregular hopsacks

Irregular hopsacks can be created by using different sized hopsacks together and an example is shown in Fig. 6.26.

Stitched hopsacks

Hopsack structures are liable to slippage especially in coarser weaves or when woven with worsted yarns. To produce a firmer, more stable cloth, hopsacks can be stitched. Stitching with a diagonal line as shown in Fig. 6.27, creates a new weave that is often referred to as a 'barleycorn' weave. Although stitching gives more stability, it interferes with the clarity of the hopsack effect.

6.25 Four point star on 2/2 hopsack (photograph by the author).

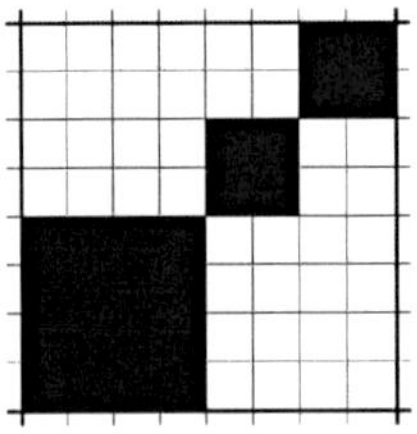

6.26 Irregular hopsack.

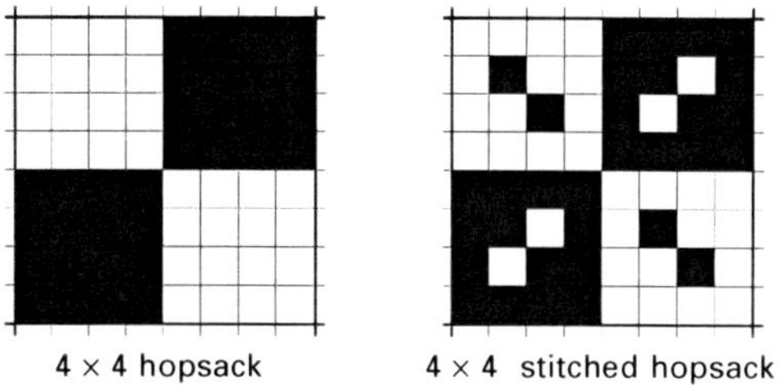

6.27 Stitched hopsack.

6.6.4 Twill weaves

Twill weaves are characterised by continuous diagonal lines created by the way the threads interlace. Twills are used where more weight or better drape is required as they are less firm than plain weave fabrics with the same yarn and ends and picks per centimetre because of the lesser number of intersections. They can also be more closely sett than their equivalents in other weaves. The direction the diagonal runs in (up to the right or up to the left) is given by defining it as Z or S (see Figs 6.28 and 6.29).

The smallest number of threads on which it is possible to construct a twill is three. The repeat of a twill can be expressed in numbers. Those above the line define warp up while those below the line define warp down. The repeat size of any twill is the total of the numbers added together. These numbers can be read in both the warp and weft directions.

$$\frac{1}{2} = 3$$ also written as 1/2 twill right (see Fig. 6.30)

$$\frac{2}{1} = 3$$ also written as 2/1 twill right (see Fig. 6.31)

Square or balanced twills, where the diagonal line lies at a 45° angle, are constructed with the same number of ends and picks per cm, with the same

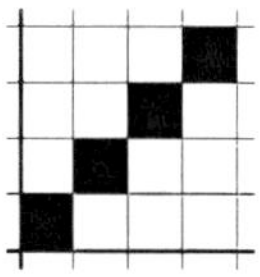

6.28 Z-twill.

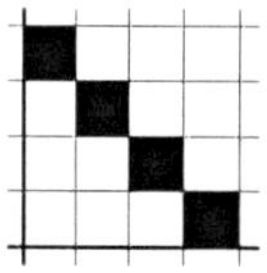

6.29 S-twill.

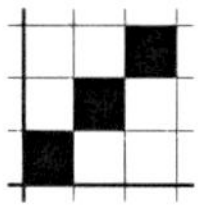

6.30 1/2 twill right.

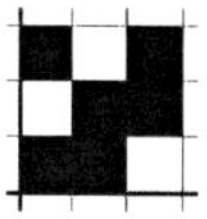

6.31 2/1 twill right.

yarn count and the same interlacing for warp and weft (see Figs 6.32 and 6.33). Changing the ratio of ends per cm to picks per cm will affect the angle of the diagonal line formed by the twill.

Warp-faced twills

Warp-faced twills have a weave structure that has predominantly warp floats showing on the face of the fabric. With an identical number of ends and picks per cm the diagonal line lies at a 45° angle (Fig. 6.34). With a higher number of ends than picks per cm the angle will be greater than 45° (Fig. 6.35). Denim is traditionally a 3/1 warp-faced twill with an indigo-dyed warp and an ecru weft.

Weft-faced twills

Weft-faced twills have a weave structure that has predominantly weft floats. With an identical number of ends and picks per cm, the diagonal line lies at a 45° angle (Fig. 6.36). With a higher number of picks than ends per cm, the angle will be less than 45° (Fig. 6.37).

Although twills can be created with floats over several ends and picks, care should be taken to construct weaves with sufficient intersections to avoid long floats. Any fabrics with long floats tend to be unstable and likely to snag. As a general rule, a float of 7 in either the warp or weft direction is considered to be the absolute maximum.

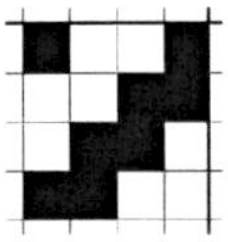

6.32 2/2 twill.

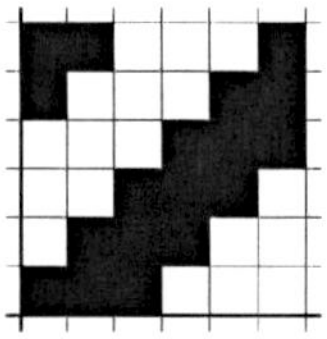

6.33 3/3 twill.

6.34 Warp-faced twill square sett (with an angle of 45°).

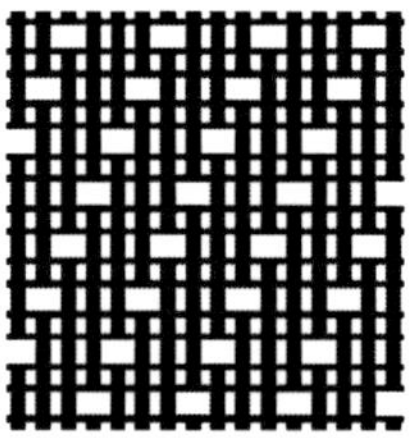

6.35 Warp-faced twill with more ends than picks (with an angle greater than 45°).

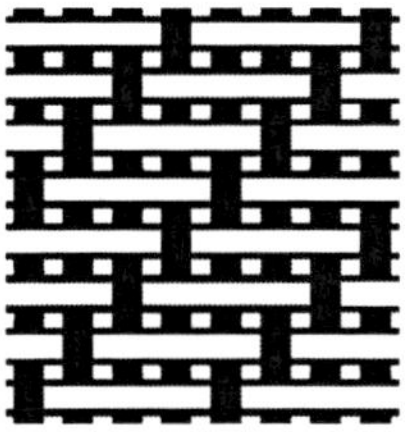

6.36 Weft-faced twill square sett (with an angle of 45°).

6.37 Weft-faced twill with more picks than ends (with an angle less than 45°).

Twilled hopsacks

Hopsacks can be arranged as twills (see Fig. 6.38).

6.6.5 Fancy diagonals

Different twills combined together are also known as fancy diagonals (see Fig. 6.39). Fancy diagonals can also use twilled hopsacks.

Colour and weave effects in twills

Just as with plain weave, the interlacings of twill weaves combined with colour patterns in the warp and weft can create interesting colour and weave effects, as shown in Figs 6.40–6.46.

6.6.6 Herringbone and pointed twills

Herringbone and pointed twills are produced by combining twills to the left and twills to the right. A herringbone twill is where the two twills join in what is termed a clean cut, that is the adjacent ends to the cutting point interlace in opposition – where one end goes over the weft then the adjacent end goes under the weft and vice versa (Fig. 6.47). The joins in pointed twills are just that – pointed (Fig. 6.48).

6.6.7 Russian twill

Russian twill is 2/2 herringbone – 2 ends 2/2 twill running right, 2 ends 2/2 twill running left, all clean cut. Note the weft runs alternately 2 and 2 and 1 and 1 (see Fig. 6.49).

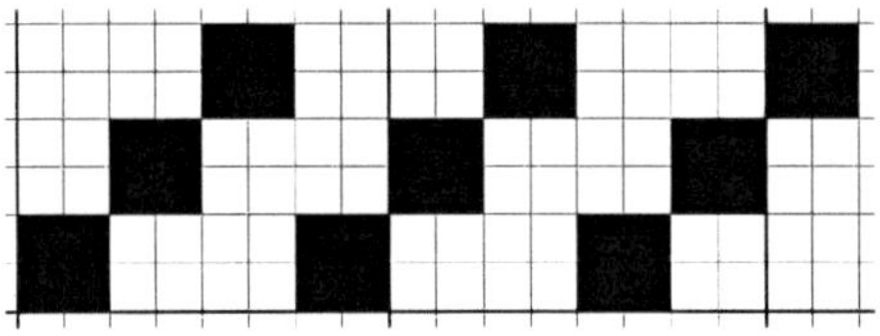

6.38 Twilled hopsack.

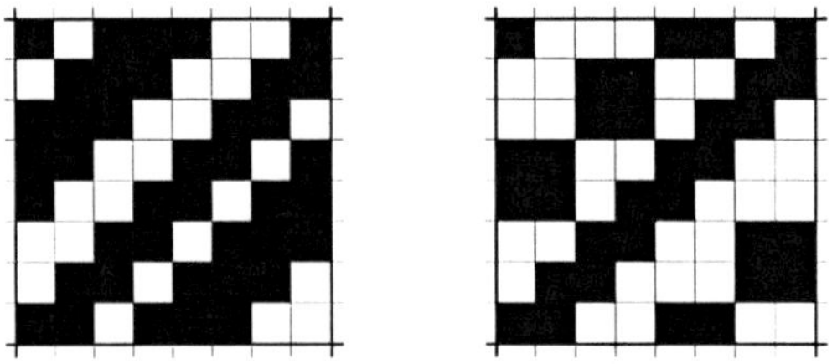

6.39 Fancy diagonals.

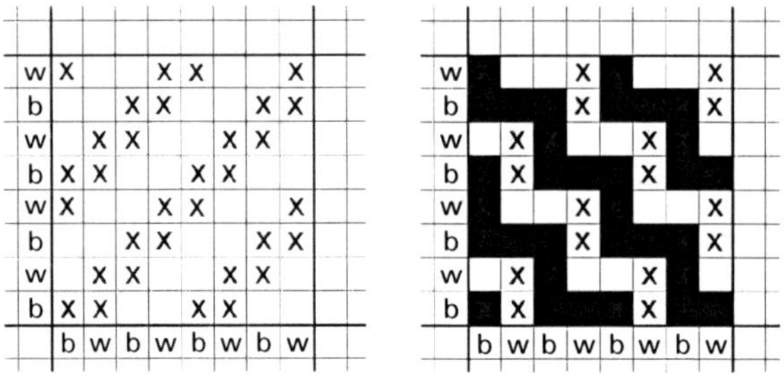

6.40 Step effect on 2/2 twill.

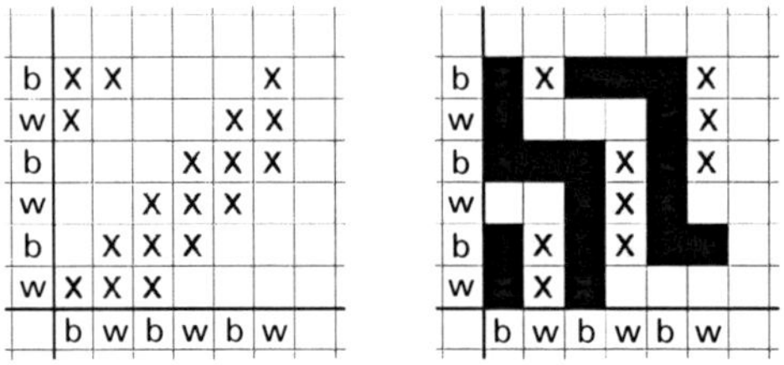

6.41 Step effect on 3/3 twill.

6.6.8 Bell Celtic

Bell Celtic is another standard weave based on 2×2 twill with both pointed and clean cut breaks. As in Russian twill, the warp threads run 2 and 2, this time, however, the weft threads run alternately 1 and 1 and 3 and 3 (see Fig. 6.50).

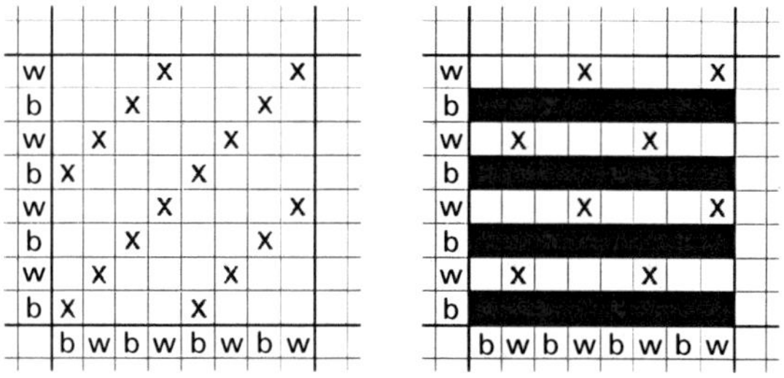

6.42 Weft hairline on 1/3 twill.

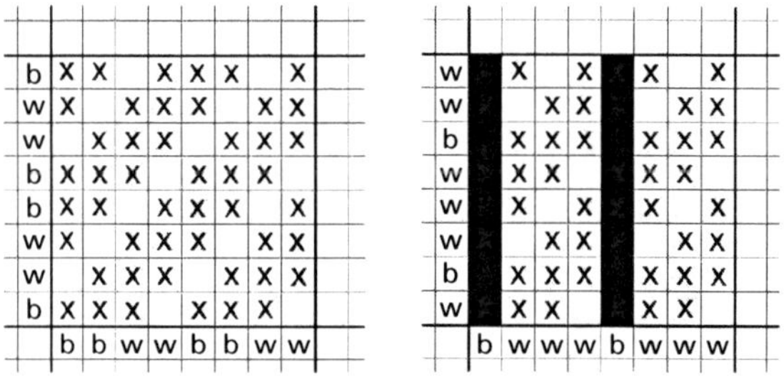

6.43 Warp hairline on 3/1 twill.

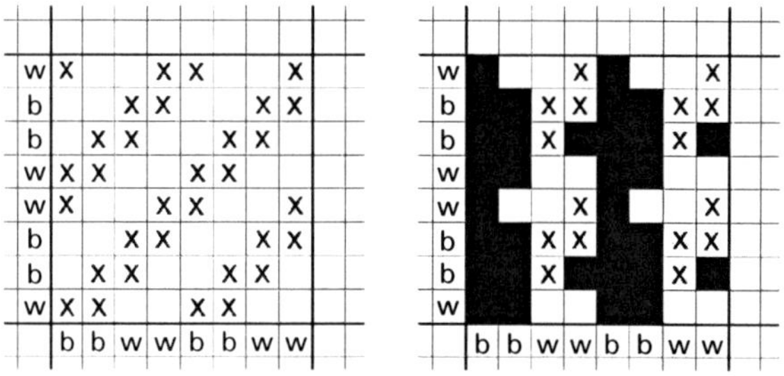

6.44 Warp-ways broken line effect on 2/2 twill.

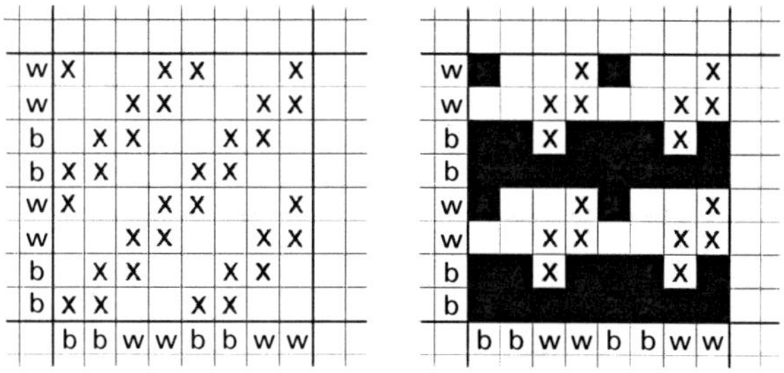

6.45 Weft-ways broken line effect on 2/2 twill.

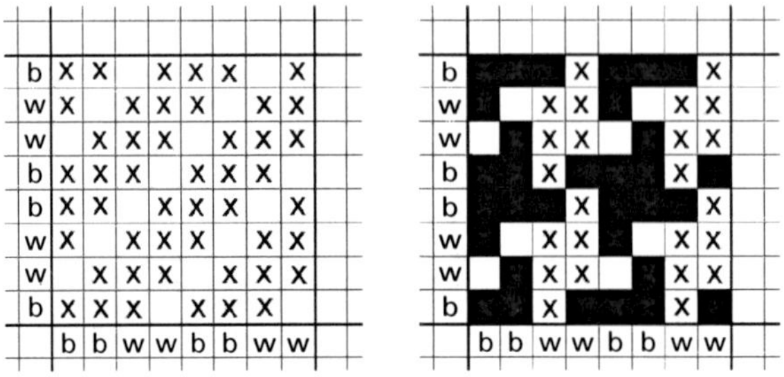

6.46 Four point star on 3/1 twill.

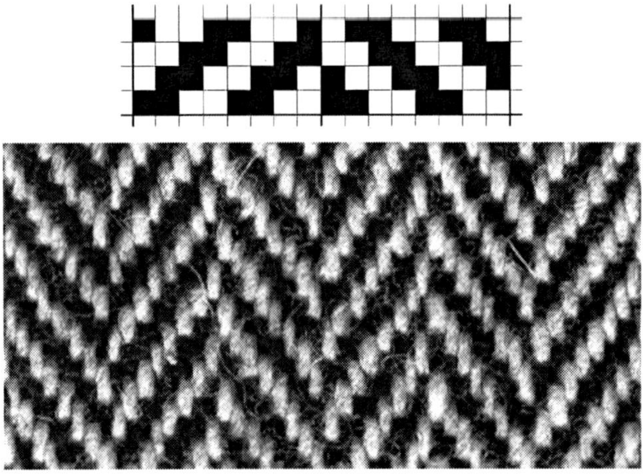

6.47 Herringbone twill – 2/2 twill 8 ends running right, 2 × 2 twill 8 ends running left, all clean cut (photograph by the author).

6.48 Pointed twill – 2/2 twill 8 ends running right, 2 × 2 twill 8 ends running left.

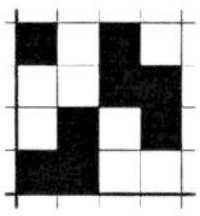

6.49 Russian twill.

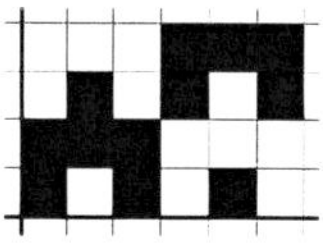

6.50 Bell Celtic (repeats on 6 ends × 4 picks).

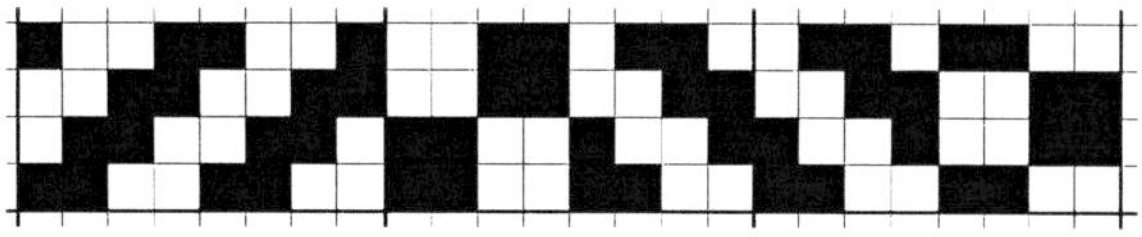

6.51 2/2 twill right for 8, 2/2 hopsack for 4, 2/2 twill left for 8, 2/2 hopsack for 4 – all clean cutting (repeats on 24 ends × 4 picks).

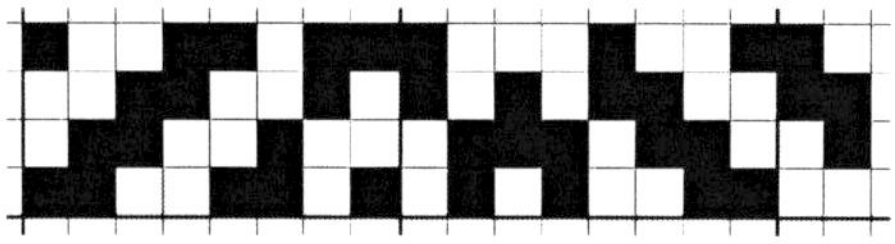

6.52 2/2 twill right for 6, bell Celtic for 6, 2/2 twill left for 6 – all clean cutting (repeats on 18 ends × 4 picks).

6.6.9 Stripe designs

Different stripe effects can be achieved by combining standard weaves (see Figs 6.51 and 6.52).

6.6.10 Warp and weft satins

Typically fabrics made from satins have a smooth and lustrous appearance. A warp satin is warp-faced, while a weft satin (sometimes referred to as a sateen) is weft-faced. These weaves are often used together to create patterns in the fabric. In warp- and weft-faced satins the intersections of warp with weft are distributed so that they are never adjacent. In a closely woven cloth the floats conceal the binding points. 5 and 8 end satin weaves are common.

The smallest number of ends and picks on which it is possible to construct a regular satin is 5. A satinette (broken twill) can be created on 4 ends and 4 picks. They are not true satins as there are adjacent binding points (see Figs 6.53 and 6.54).

Warp satins usually have a substantially higher number of ends than picks (see Figs 6.55 and 6.56). The warp yarn consists very often of fine silk or worsted while the weft may be an inferior yarn. Weft satins usually have a higher number of picks than ends (see Figs 6.57 and 6.58). The weft is of a superior quality and covers the warp. Production costs are high because of the high number of picks.

In any repeat of a warp satin, all the shafts are lifted except for one. This can cause problems with weft insertion as the cloth can tend to lift off the raceboard which is what supports the shuttle as it is inserted into the shed. For this reason the majority of warp satins are woven upside down. To achieve the correct warp satin, the weft satin weave must be a mirror image. It is advisable when producing 5 or 7 end satins to increase the number of shafts to an even number; for example, 5 end sateen is often woven on 10 shafts.

6.6.11 Honeycombs

These weaves get their name from their resemblance to the cellular structure of bee's honeycombs. Essentially honeycombs are constructed with alternating

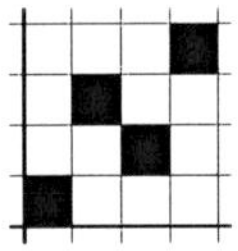

6.53 Weft satinette.

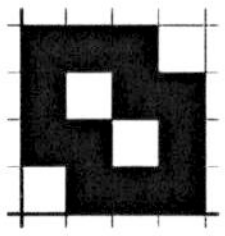

6.54 Warp satinette.

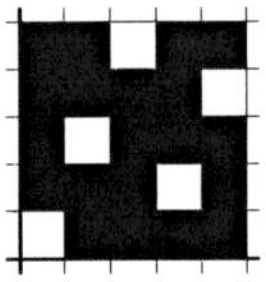

6.55 Regular 5 end warp satin.

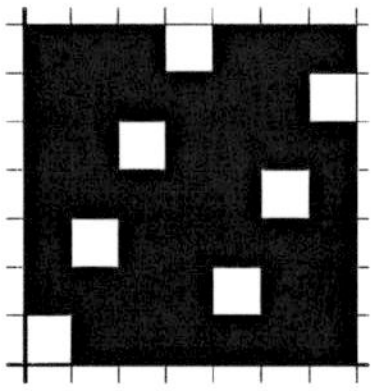

6.56 Regular 7 end warp satin.

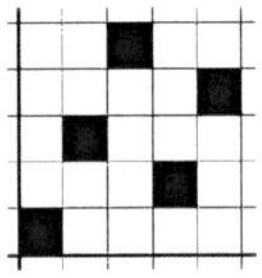

6.57 Regular 5 end weft satin.

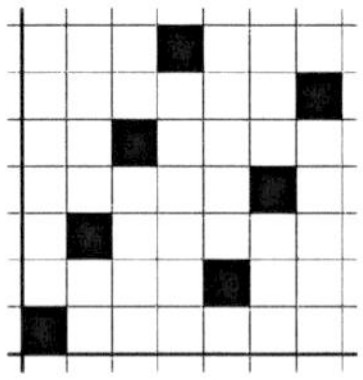

6.58 Regular 7 end weft satin.

diamonds of warp and weft floats bordered by diagonal lines working in plain weave (see Fig. 6.59). The effect of cells is created with differently intersecting ends – some ends and picks interlace tightly while other ends and picks are long floats. A variety of patterns can be created on the honeycomb principle. They can be used for blankets when made using coarser yarns and for towels when made using cotton yarns. The honeycombs trap air in the blankets to keep warmth in and in towels absorb moisture well.

Grecian honeycombs

These are a variation on regular honeycombs. They are constructed with sections of tightly interlacing plain weave and sections with long warp and weft floats arranged in a horizontal or vertical direction (Fig. 6.60).

Huckaback weaves

Huckabacks are hard-wearing and relatively thick with floats in the warp or weft direction adding texture to a plain weave base (Fig. 6.61). Their

6.59 Honeycombs.

6.60 Grecian honeycomb.

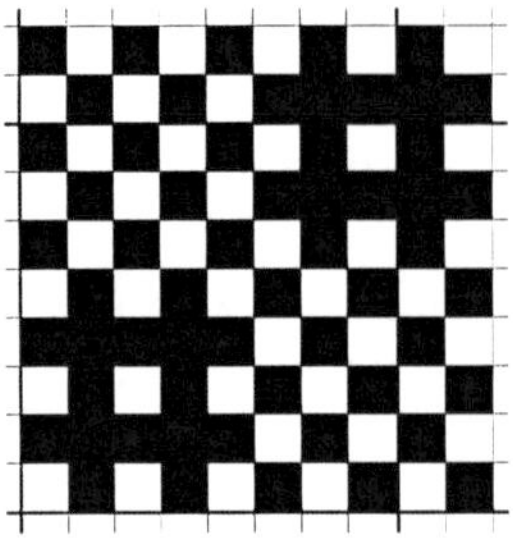

6.61 Huckaback.

structure also means that they are good at absorbing moisture and so they are often used for towels.

6.6.12 Mock leno

Leno weaves are open structures woven with special healds (see later in this chapter) to allow warp ends to cross one another. Mock leno structures give a similar effect on an ordinary loom. The weave is arranged in groups with threads working in plain weave alternating with threads floating on the face or back of the fabric (Fig. 6.62). This bunches the floating ends together causing a slight gap or opening in the fabric similar to a leno weave. Where possible, threads in each group should be drawn into the same dent of the reed.

6.7 Introduction of extra threads

Extra threads in either the warp or weft direction may be introduced into a woven fabric construction for adding weight or for decorative effect. When adding weight the extra threads usually work mainly on the back of the fabric with the weave chosen to hide these threads from the fabric face. When used for decorative effects, these extra threads are used to float on the surface of the ground weave for the parts where they are required for decorative effect. At other times these threads will be hidden, floating below the ground fabric or working in the ground weave.

6.7.1 Backed fabric

In general with regard to backed fabrics, a heavy single fabric can only be made by using heavier yarns and will be coarser. By using extra warp and/or weft a heavy weight cloth can be woven that has a fine face. Good quality yarns can be used on the face while the weight can be achieved on the back with cheaper, poorer quality yarns. Weaker yarns can be used for the backing in weft-backed cloths as these backing threads will be under less

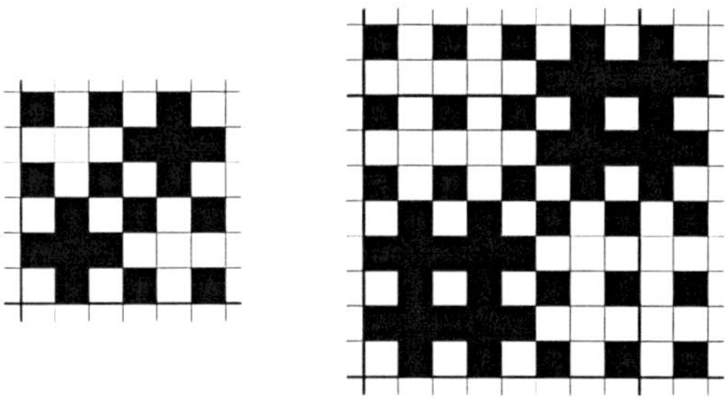

6.62 Mock lenos.

tension than if they were used for warp backing threads. However, warp-backed structures are faster to weave than weft-backed fabrics. Figure 6.63 shows how a warp-backed 2/2 twill is constructed: the twill is expanded to allow for the extra warp ends and these are stitched to the face so they are hidden between the warp floats of the 2/2 twill. Figure 6.64 shows how a weft-backed fabric is constructed.

6.7.2 Figured threads

Selected threads in either the warp or weft can be selected to figure on the surface of a fabric to create a design effect. This can be done by selecting some yarns to float on the surface at particular parts of a weave such as plain weave. A plain weave fabric can be changed so that there are floats which make a design effect as in the left-hand weave shown in Fig. 6.65. Where the yarns are not floating on the surface they are weaving plain weave. Alternatively, a base fabric of plain weave can have extra threads introduced in either the warp or weft. These can be used to create design effects by figuring on the face. When these threads are not figuring on the face, they float behind the plain weave ground as in the right-hand weave in Fig. 6.66.

If the figured ends which make the ornamental effect are in the warp these extra ends have to be wound on a separate warp beam to the ground fabric to enable both sets of warps to be woven at suitable tensions.

6.8 Double and treble cloths

Double cloths are woven fabrics in which two sets of warps and two sets of weft yarns are woven together to form a two-layered cloth. Complex patterns and surface textures can be created by interchanging the layers. Double cloth structures are also used to create heavier fabrics which have the appearance of finer fabrics. Filling threads can be used to lie between the layers to add weight and warmth. Treble cloths can also be woven with three sets of warps and wefts.

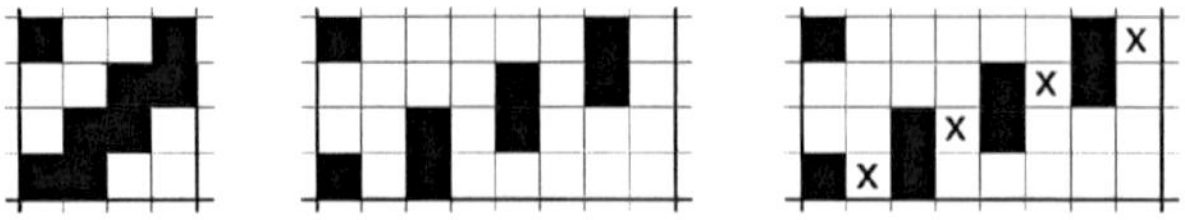

6.63 Warp-backed 2/2 twill.

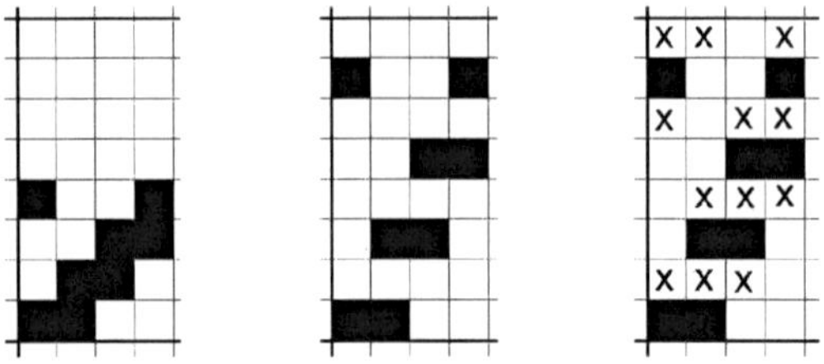

6.64 Weft-backed 2/2 twill.

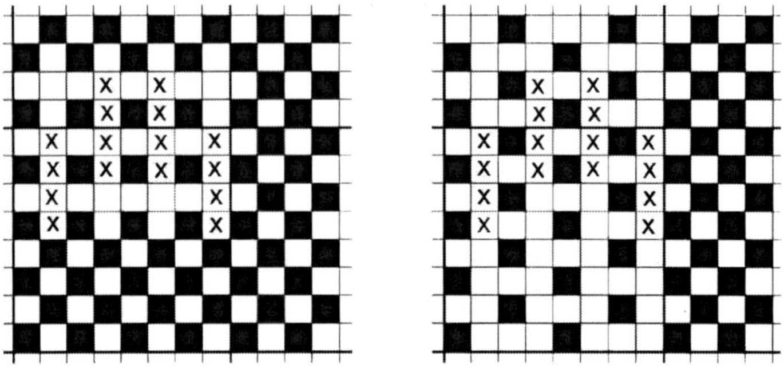

6.65 Warp-figured ends.

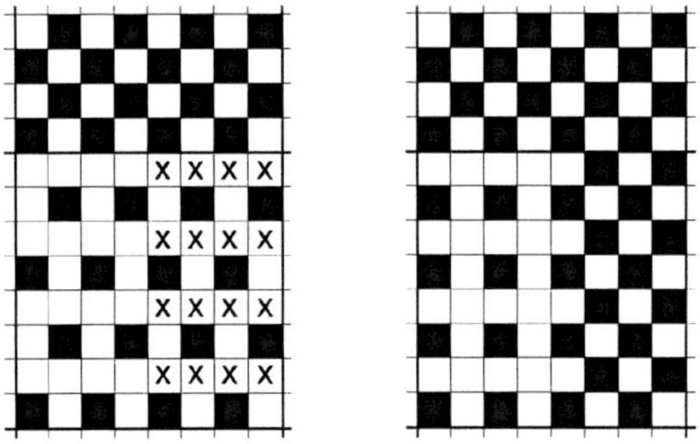

6.66 Weft-figured picks.

6.9 Repeating patterns

As has already been stated, woven fabric designs repeat not only in terms of the weave used but also the arrangement of yarns in the warp and weft. While the repeat size of the actual weave is limited by the number of shafts or the arrangement of jacquard hooks, the warp and weft patterns are only limited by practicalities in set up and weaving.

The representation of the check in Fig. 6.67 has been taken from a designer's initial sample blanket. Without any indication of how this would repeat, it might be assumed that the check shown is one repeat and the fabric would look like the representation shown in Fig. 6.68. A large number of checks and stripes, however, can be derived from this one check as shown in Figs 6.69–6.72.

6.67 Initial design idea.

6.68 Four repeats of initial ideas.

6.69 Derived check.

6.70 Derived symmetrical check.

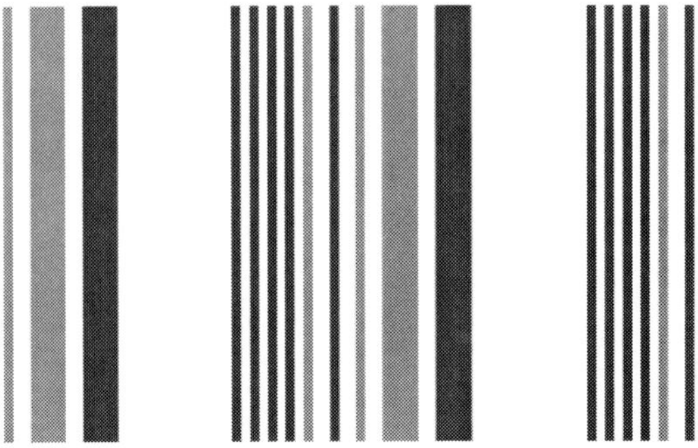

6.71 Derived stripe.

6.72 Derived symmetrical stripe.

Balanced designs are ones that are symmetrical about the vertical and horizontal axis, as shown in Fig. 6.73.

6.10 Centring

For most economic fabric usage it is better for fabrics to be symmetrical in both vertical and horizontal directions. This means that laying up plans for

6.73 Balanced check.

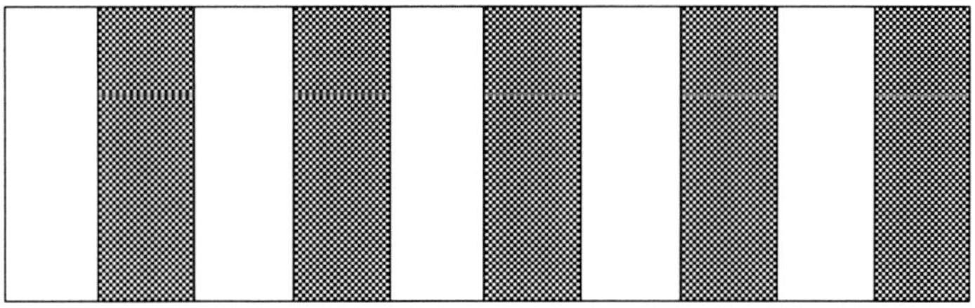

6.74 Five repeats.

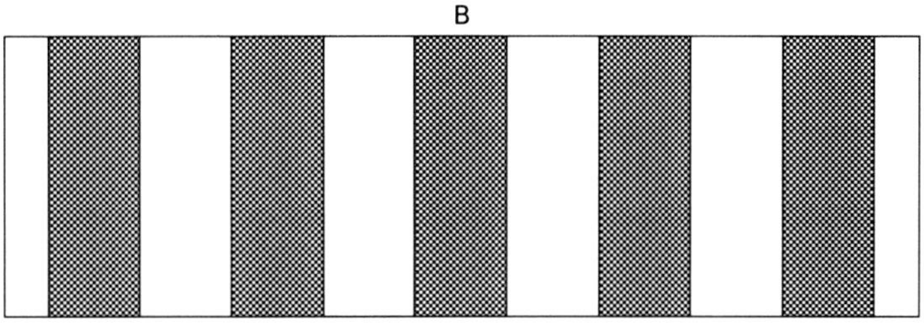

6.75 Five repeats centred.

cutting out pattern pieces can be more easily worked out. If a fabric has a pile (e.g. velvet), it is important that the pile lies the same way on all garment pieces. Centring is when a fabric design is organised in such a way that it is balanced about the middle line of a fabric in a vertical direction.

Take a colour woven fabric with a warp stripe of 30 white and 30 grey ends with sett at 10 ends per centimetre. To produce a fabric width of 30 cm would require a warp of 300 ends. This means that with a warp repeat of 60 ends there will be 5 repeats exactly across the width of the fabric. The fabric could be warped starting with 30 ends of white as shown in Fig. 6.74. However, at one edge there would be a white stripe while at the other side there would be a grey stripe. This is clearly not balanced and could cause problems with laying up and cutting. It is better to centre the warp about a stripe and this can be done by starting the warp with 15 white ends. The warp is now symmetrical in a horizontal direction (centred about the grey stripe B). At both selvedges there are now white stripes (Fig. 6.75).

6.11 Drafting and lifting

To create any weaves the warp ends need to be lifted appropriately to allow the weft to be interlaced in the required way. In the simple loom shown earlier, a system of shafts is used to lift the warp ends. The warp ends are threaded onto shafts via heald wires so that all the warp ends drawn onto one shaft will all lift when that shaft is lifted. The draft is the instruction as to when and what shafts the threads in the warp are drawn through. For plain weave at least two shafts are needed. The odd threads need to be drawn onto one shaft and the even threads onto another. The lifting plan tells which shafts to raise and when. In dobby weaving the lifting plan is called a peg plan; this is in reference to the pegs used to operate the dobby mechanism.

In Fig. 6.76, the first, third, fifth and seventh ends are drawn on the first shaft, the second, fourth, sixth and eighth ends are drawn on the second shaft. When the first shaft is lifted, the odd threads lift and the first pick is inserted. The second shaft is then lifted raising the even ends and the second pick inserted.

However, four shafts are used more commonly than two (see Fig. 6.77 for an example). This is to reduce friction between the warp ends and also to allow a variety of weaves in addition to plain weave to be woven (if so desired) on the same warp without having to redraw the threads. Figure 6.78 shows that the same straight draft on four shafts can also be used to produce 2/2 twill.

A wide range of weave structures can be created on a straight draft on four shafts. The only restriction is that the weave repeat widthways is restricted to four ends. Being restricted to four shafts limits the weaves you can create. However, it does not limit you to a weave that repeats on four ends. A weave can have a weave that repeats on more ends, but there can only be four differently interlacing ends (see Fig. 6.79).

6.11.1 Establishing a draft

To establish the draft required for any weave, a simple analysis technique is usually used. The first end in the repeat of the weave (left side) is allocated to shaft number 1. The lifting sequence of the second end in the weave is then compared with the first. If it is exactly the same, then it is allocated the same shaft. If it is in any way different, it has to be allocated a new shaft (shaft 2). All other remaining ends in the weave repeat are now compared

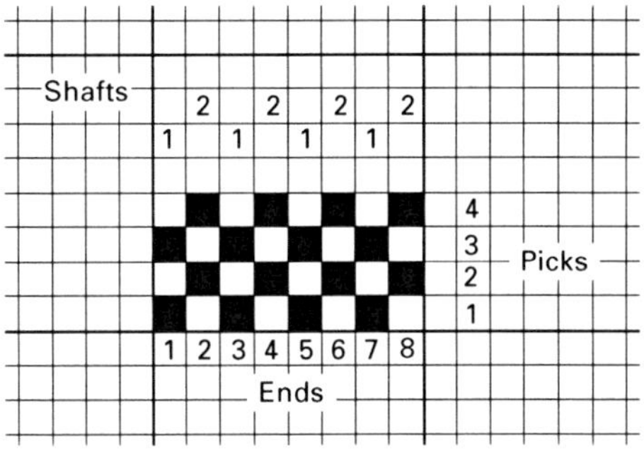

6.76 Draft for plain weave using two shafts.

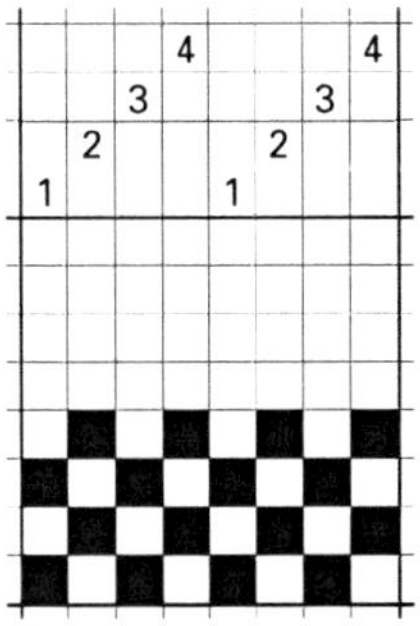

6.77 Draft for plain weave using four shafts.

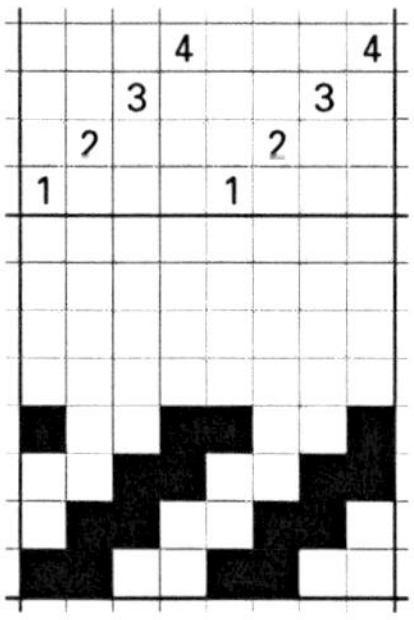

6.78 Draft for 2/2 twill using four shafts.

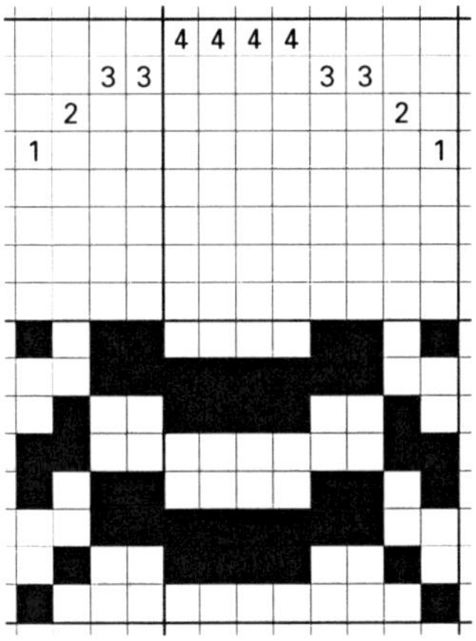

6.79 Weave repeating on 12 ends × 8 picks on four shafts.

with previous ends. Where it is found that they exactly compare with a previous end, the end is allocated the same shaft, where it is found that the end is different from previous ends, it has to be allocated a new shaft. At the end of the analysis, a drafting sequence for one repeat of the weave will have been created. This may not always be the best draft for the fabric being woven – see later in this chapter.

Figure 6.80 presents an example of a weave with a simple draft. The draft has been worked out as explained previously working from left to right.

6.11.2 Block drafting

With two weaves which can both be woven on four shafts, we can weave these simultaneously side by side if there are eight shafts available (see Fig. 6.81).

Block, grouped or unit drafts

Drawing the warp threads in blocks allows different weaves to be created on different sections. by using what is termed a block, grouped or unit draft (see Figs 6.82 and 6.83). The ability to weave blocks vastly extends patterning possibilities.

By changing the weave applied to each draft block, weaves with different block sections in the weft can be created (see Fig. 6.83).

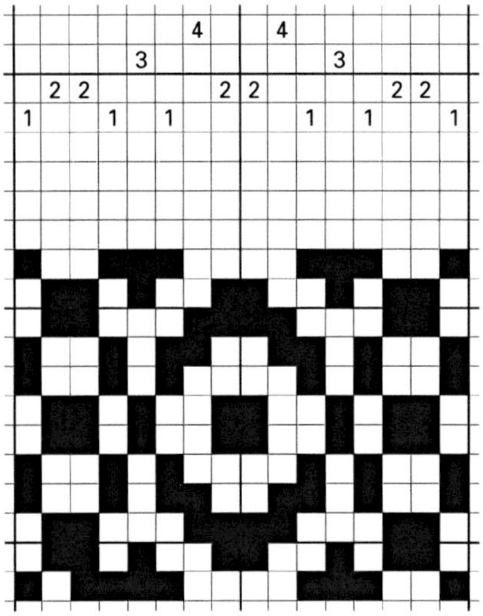

6.80 Weave with simple draft.

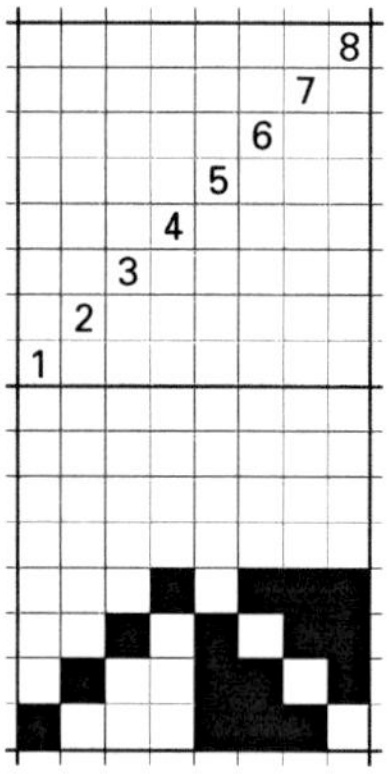

6.81 Two weaves drafted together on eight shafts.

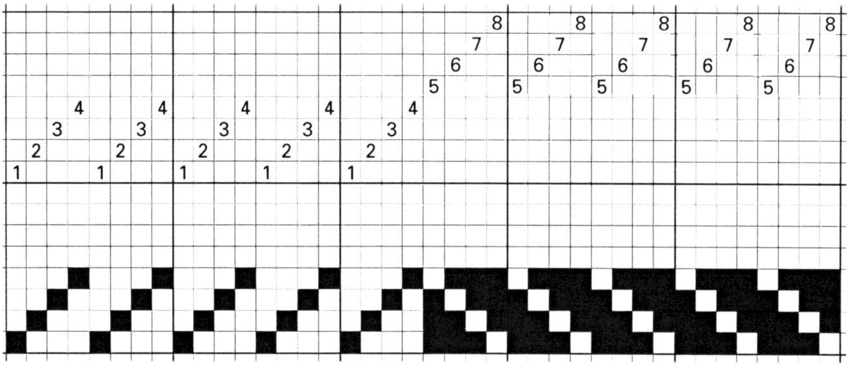

6.82 Block draft.

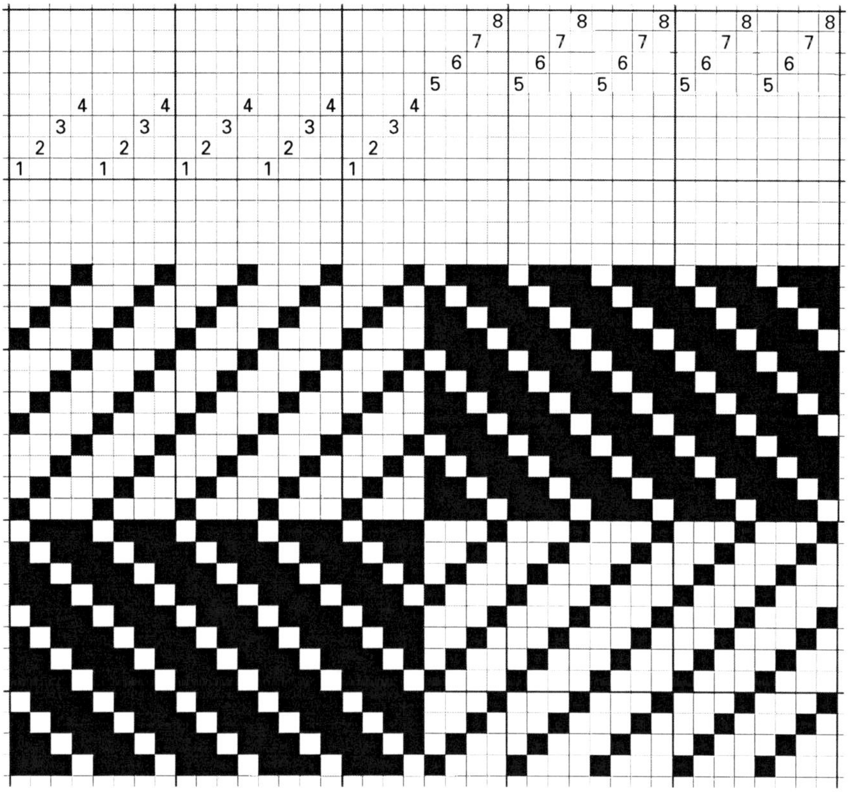

6.83 Block draft with different weaves in weft.

6.11.3 Lifting plans

The lifting plan (or peg-plan) is the instruction for the lifting of the shafts. Warp ends that interlace the same are drawn on the same shaft. In dobby shedding, the plan is used to peg a set of lags. The lifting plan (to the right in Fig. 6.84) gives instructions as to the order in which the shafts are lifted for plain weave. For the first pick the first shaft must be lifted, for the second pick the second shaft must be raised, for the third pick the first shaft must be raised, for the fourth pick the second shaft must be raised.

With four shafts as in Fig. 6.85, for plain weave, for the first pick the first and third shafts must be lifted, for the second pick the second and fourth shafts must be raised, for the third pick the first and third shafts must be raised, for the fourth pick the second and fourth shafts must be raised.

6.11.4 Straight drafts on eight shafts

In the example in Fig. 6.86, on the first pick the first, second, fifth and sixth shafts must be lifted, for the second pick the second, third, sixth and seventh shafts must be raised, for the third pick the third, fourth, seventh and eighth shafts must be raised, for the fourth pick the first, third, fourth and eighth shafts must be lifted.

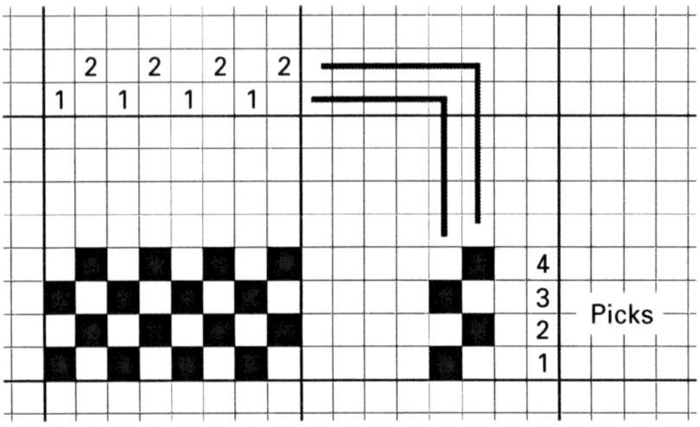

6.84 Lifting plan for plain weave on two shafts.

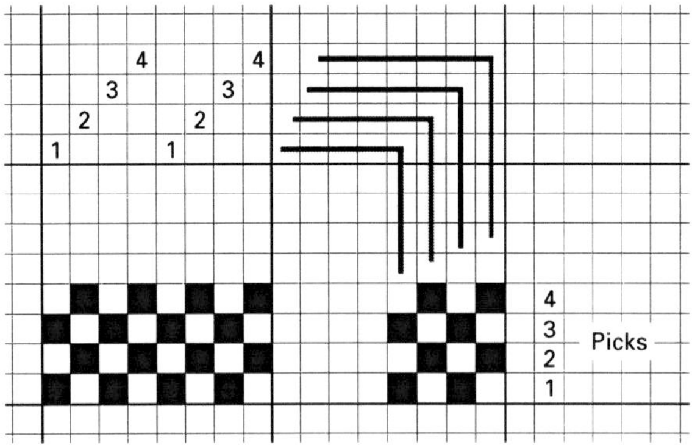

6.85 Lifting plan for plain weave on four shafts.

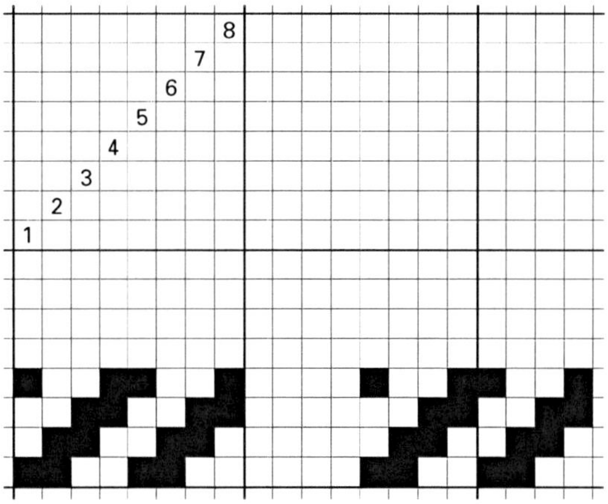

6.86 Straight draft 2/2 twill.

In the example in Fig. 6.87, on the first pick the first, second, fifth and sixth shafts must be lifted, for the second pick the second, third, fourth and eighth shafts must be raised, for the third pick the third, fourth, seventh and eighth shafts must be raised, for the fourth pick the first, fourth, sixth and seventh shafts must be lifted. When a straight draft is used, the lifting plan is a copy of the weave (Fig. 6.86).

6.11.5 More about drafts

One weave may be drafted in more than one way. Figure 6.88 shows two examples of a drafting plan for the same weave.

6.11.6 Factors that should be considered when drafting

These include:

- heavy shafts should be kept to the front;
- weak yarns, wherever possible, should be drawn on the front shafts;
- tight working ends should be drawn on the front shafts;
- the draft should be kept as simple as possible to follow;
- too many healds per centimetre cause friction and therefore breakages (this is the reason why for plain weave four shafts are more commonly used than two);
- the possibility of using the draft for other weaves.

6.11.7 Pointed drafts

This is where a straight draft is returned in the opposite direction at predetermined intervals. In combinations with appropriate weaves a pointed

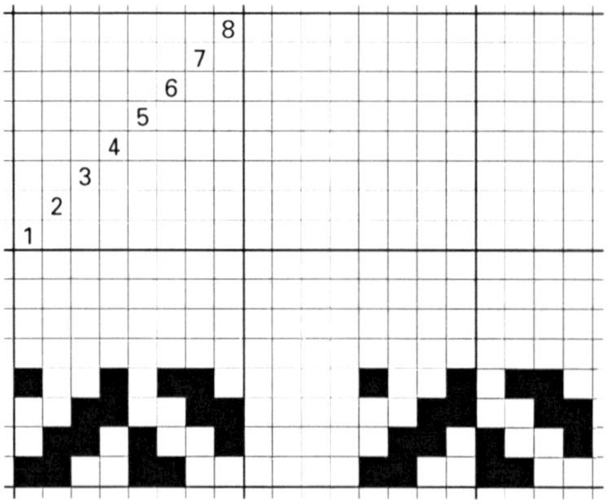

6.87 Straight draft 2/2 twill herringbone.

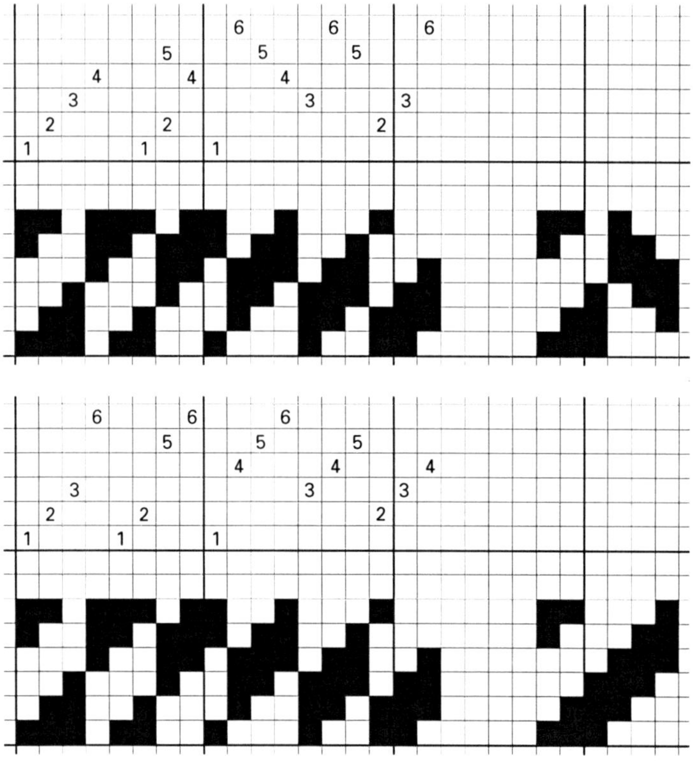

6.88 One weave with two different drafts.

draft will cause the weave to be mirrored, creating waved or diamond effects, as shown in Fig. 6.89.

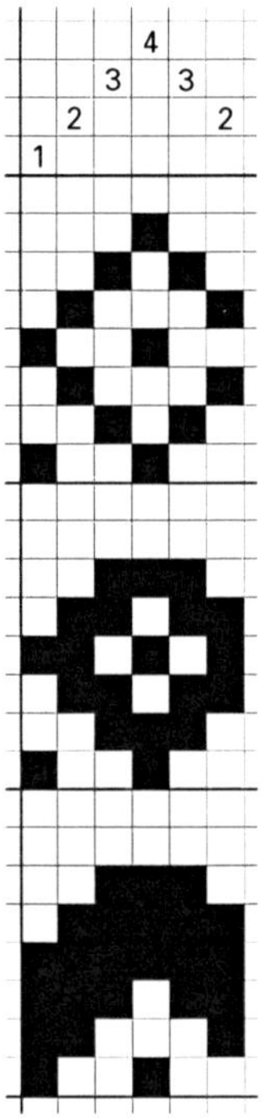

6.89 Pointed draft.

6.11.8 Skipped drafts

Skipped drafts are used in densely set warps as they reduce the friction between adjacent warp ends (see Fig. 6.90).

6.11.9 The relationship between weaves, drafts and lifting plans

A weave can be drafted in different ways and several weaves can be created using the same draft. The lifting plan is directly related to weave and draft.

6.12 Denting

Denting plays an important part in achieving a flawless fabric. The number of dents per centimetre and the way the ends are dented determine the number of ends per centimetre, and thus the density of the resulting fabric. It is necessary to state the number of dents per 10 cm of the reed to be used and the number of ends to be drawn into one dent of the reed. For densely set warps, four ends in a dent are preferred to two.

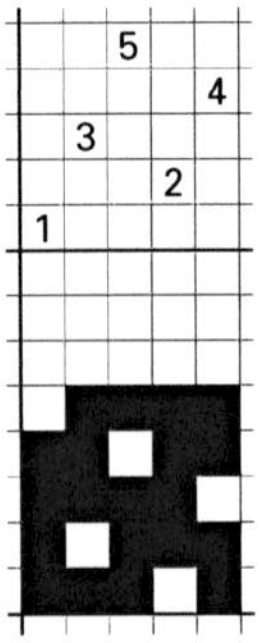

6.90 Skipped draft.

As previously stated, hopsacks require particular care when denting. Denting can prevent ends rolling over each other in the warp (see Fig. 6.91). In the example to the left in Fig. 6.91, above the denting plan indicates that the first and second end are together in the one dent in the reed as are the third and fourth, etc. The example to the right is preferred as it will stop threads which follow the same intersecting path rolling over each other; here the denting plan indicates that the second and third ends are both threaded through the same dent, the fourth and fifth together, etc.

6.13 Combining weaves

Different weaves can be combined to give a variety of effects. Warp- and weft-faced weaves such as satins can be used to create colour and texture patterns, while combinations of plain weave and ribs can create distorted thread effects. Many different effects can be introduced into fabrics by using figured threads in combination with different weaves and introducing different coloured or fancy yarns. Figure 6.92 shows weft-figured ends combined with a mock leno and a variety of fancy yarns.

6.13.1 Combining warp- and weft-faced weaves

Warp- and weft-faced structures (such as warp and weft satins) can be combined to make ornamental effects. Patterns can be made by using blocks of warp- and weft-faced structures in conjunction with contrasting yarns for the warp and the weft. The yarns may be contrasting in texture (shiny/matt) or in terms of colour. In Fig. 6.93 3/1 twill right is used with 1/3 twill right, and if the warp is dark and the weft is light, a patterned block effect is achieved.

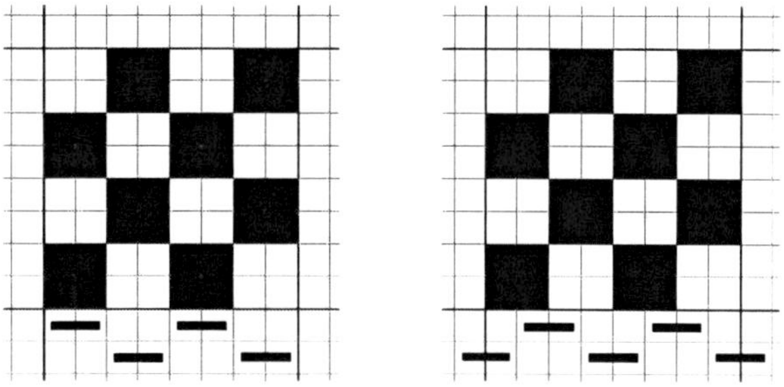

6.91 Denting

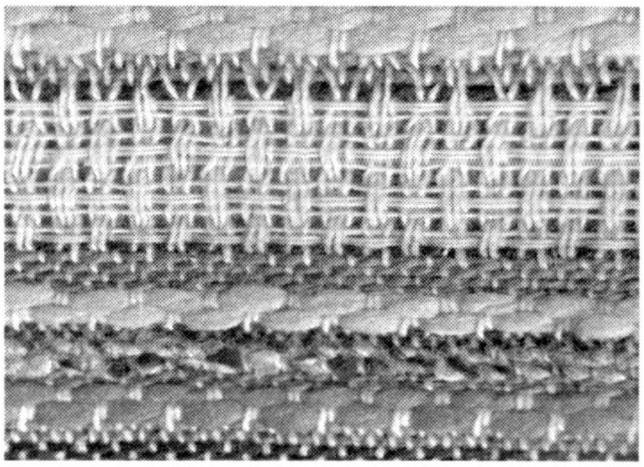

6.92 Combining mock leno with plain and other weaves (photograph by the author).

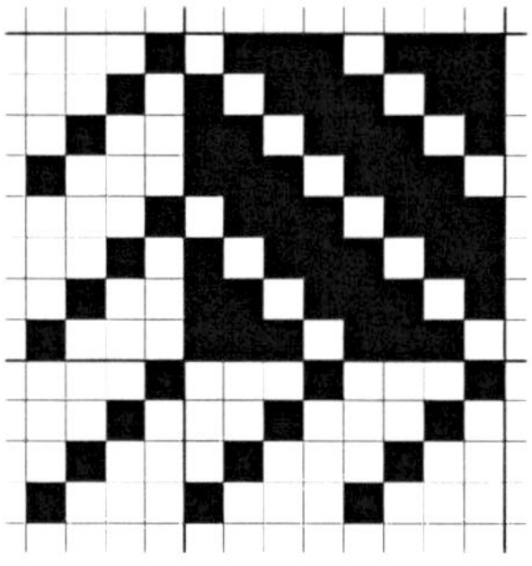

6.93 Block effect of warp- and weft-faced weaves.

6.13.2 Distorted thread effects

Distorted thread effects can be created by combining plain weave and rib sections. The plain weave sections hold their shape while the cord sections allow the picks to come together. Figure 6.94 shows the notation of such a fabric while Fig. 6.95 shows an actual fabric which combines plain weave and ribs. Distorted thread effects can be created in both warp and weft directions.

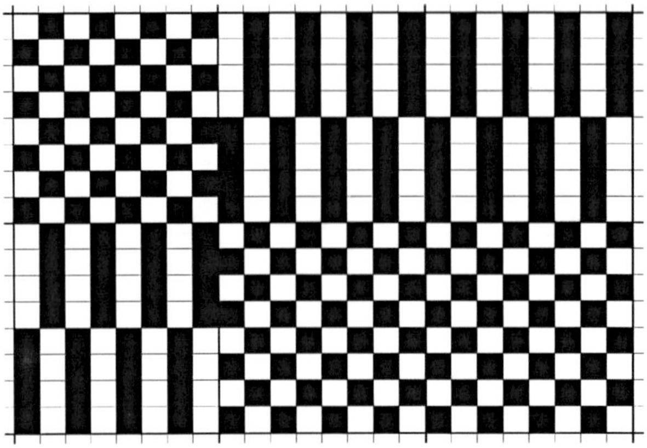

6.94 Distorted thread effect.

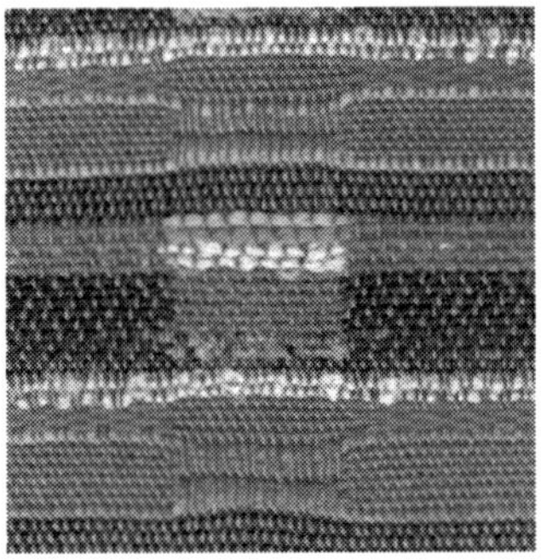

6.95 Distorted thread effect fabric (photograph by the author).

6.14 Fabric types

6.14.1 Dobby fabrics

A dobby is a type of mechanism that allows patterning of woven fabrics. The dobby controls the lifting of shafts that control the warp ends in a fabric. Different looms will have different capabilities when it comes to the number of shafts they can accommodate. The most shafts usually found on a dobby loom is 24 and more often 12 or 16. Patterned fabrics with a small repeat size are frequently termed dobby weaves.

6.14.2 Jacquard fabrics

A jacquard mechanism on a loom allows greater patterning possibilities than a dobby. Instead of the warp ends being controlled by shafts which raise up and down, they are controlled by hooks arranged in a jacquard harness.

Jacquard mechanisms allow larger designs to be produced. Jacquard fabrics are the patterned fabrics that can be woven on a jacquard loom.

6.14.3 Leno fabrics

Leno fabrics are light open weaves. Woven on special looms, the warp yarns work in pairs twisting around each other and are kept in place by the weft yarns.

6.14.4 Pile fabrics

Pile fabrics can be made in a number of different ways. Essentially weft pile fabrics are created with the yarn that is to form the pile being inserted as long floats which are then cut to form the pile. Warp piles were originally made by inserting rods into the weft. The warp ends used for the pile formed loops over the inserted rods ('loop pile'), or were cut to form 'cut pile'. Face to face weaving techniques can also be used to create pile fabrics. These involve the weaving of two layers of fabric on top of each other with the fabrics connected by pile ends which are cut after weaving.

6.15 Future trends

Woven fabrics are used for a variety of end uses; for industrial applications as well as apparel and interiors. Highly engineered performance weaves are being made stronger, lighter and smarter than ever before for a wide range of high-tech applications.

6.16 Sources of further information and advice

6.16.1 The Textile Institute

The Textile Institute is an international organisation which covers all sectors and all disciplines in textiles, clothing and footwear. The aim of the Institute is to 'facilitate learning, to recognise achievement, to reward excellence and to disseminate information' (http://www.texi.org/).

6.16.2 The Textile Society

The Textile Society promotes itself as forum for textile professionals and enthusiasts. It promotes 'the study of textile disciplines and celebrates the history and culture of textiles, both traditional and contemporary' (www.textilesociety.org.uk/).

6.17 Bibliography

Denton MJ and Daniels PN (eds) (2002), *Textile Terms and Definitions 11th Edition*, Woodhead Publishing, Cambridge.

Goerner D (1986), *Woven Structure and Design – Part 1: Single Cloth Construction*, Wira Technology Group, Leeds.

Goerner D (1989), *Woven Structure and Design – Part 2: Compound Structures*, British Textile Technology Group, Leeds.

Grosicki Z (ed.) (1975), *Watson's Textile Design and Colour: Elementary Weaves and Figured Fabrics*, (7th edn), Woodhead Publishing, Cambridge.

7
Computer aided design (CAD) systems for woven textile design

P. SINHA, University of Huddersfield, UK and University of Manchester, UK

Abstract: This chapter introduces some key issues regarding the use and role of CAD within the global textiles trading environment, such as cost, expertise and skills training, impact on the supply chain and development of new products and new market areas. The chapter concludes by discussing some of the current research being undertaken with CAD applications for woven textiles with a view towards future trends and sources of useful information for readers wishing to explore the issues raised in greater depth.

Key words: computer aided design (CAD), consumer woven fabrics, technical textiles, supply chain impact.

7.1 Introduction

Weaving is one of the oldest crafts known to man. Fragments of woven cloth have been dated to the Neolithic/Eneolothic period (Soffer, 2004). The woven fabric has been regarded as possessing mystical qualities: woven rags are offered as part of the prayers in some Buddhist shrines, form the requisite set of ceremonial accoutrements for many different religions' prayer settings and, as observed by Kramrisch (1968) 'in the Rig Veda and the Upanishads, the universe is envisioned as a fabric woven by the gods. The cosmos, the ordered universe, is one continuous fabric; the uncut fabric is a symbol of totality and integrity' (pp. 67–68). Knowledge and mastery of the skill has long been a matter of intellectual pride, as illustrated by the Greek myth of goddess Athena's punishment meted out to Arachnida of being turned into a spider to forever spin and weave as a result of her demonstration of pride in winning a weaving competition between them. To understand and be knowledgeable about woven design is still a matter of mastery of weaving technique and mathematics, and the competitive environment within which the textiles industry operates requires computer aided design (CAD) skills to develop new products and new textiles applications.

This chapter introduces the reader to some of the key issues regarding the use of CAD within woven textile design. Section 7.2 begins with an overview of the scope and standing of the textiles industry within the global trading environment and the role that CAD has to play. Sections 7.3–7.7 then discuss key issues when using CAD within the woven design process such as cost, expertise and skills training, impact on the supply chain and development of new products and new market areas. Textiles, as has been noted already, have consumer markets (such as clothes and interiors) as well as technical (industrial) applications. Section 7.8 concludes this chapter by discussing some of the current research being undertaken and briefly considers some of the aspects of CAD expertise required within the expanding area of technical textile area and presents sources of useful information for readers wishing to explore the issues raised in greater depth.

7.2 Computer aided design (CAD) and the global textiles industry

According to the Organisation for Economic Co-operation and Development (OECD), the biggest exporters of textile and clothing products are China (followed by Hong Kong China, Italy, Germany, United States and India). The latest figure regarding world trade in textiles and clothing was about US$610 billion, i.e. about 4.3% of the world's exports (Kowalski and Molnar, 2009).

The textiles sector is wide and diverse with applications as diverse as clothing, geology, architecture, auto industries, civil engineering, medicine, aerospace, etc. (Fig. 7.1; Allwood *et al.*, 2008). The textiles industry has traditionally been regarded as consisting of several discrete but related areas of production: fibres (natural or synthetic), yarns, fabrics, wet processes (such as dyeing or finishing) and the finished product (consumer goods, e.g. clothing, domestic interiors). Production of fabric is capital intensive, requiring technology and automation to produce requisite volumes of fabric. Although more labour intensive, the consumer end of the chain (import, distribution and retail) has become increasingly competitive and significant in the industry's value chain (OECD, 2004; Nordas, 2005) and new applications of textiles have led to expanding market potentials for technical textiles. New technology, research and development activities are regarded as essential factors to compete within the capital intensive countries (such as those within the OECD) as it helps to develop products that are differentiated from competing companies and are able to be traded within the saturated market conditions. CAD is an important factor in developing differentiated products and reacting to the changing nature of the supply chain.

7.3 Key issues in the use of computer aided design (CAD) for woven textile design

Before consideration of the issues raised by using CAD, a very brief reminder of the key operations regarding woven textile design is presented to illustrate how CAD initially became a vital part of the textiles industry.

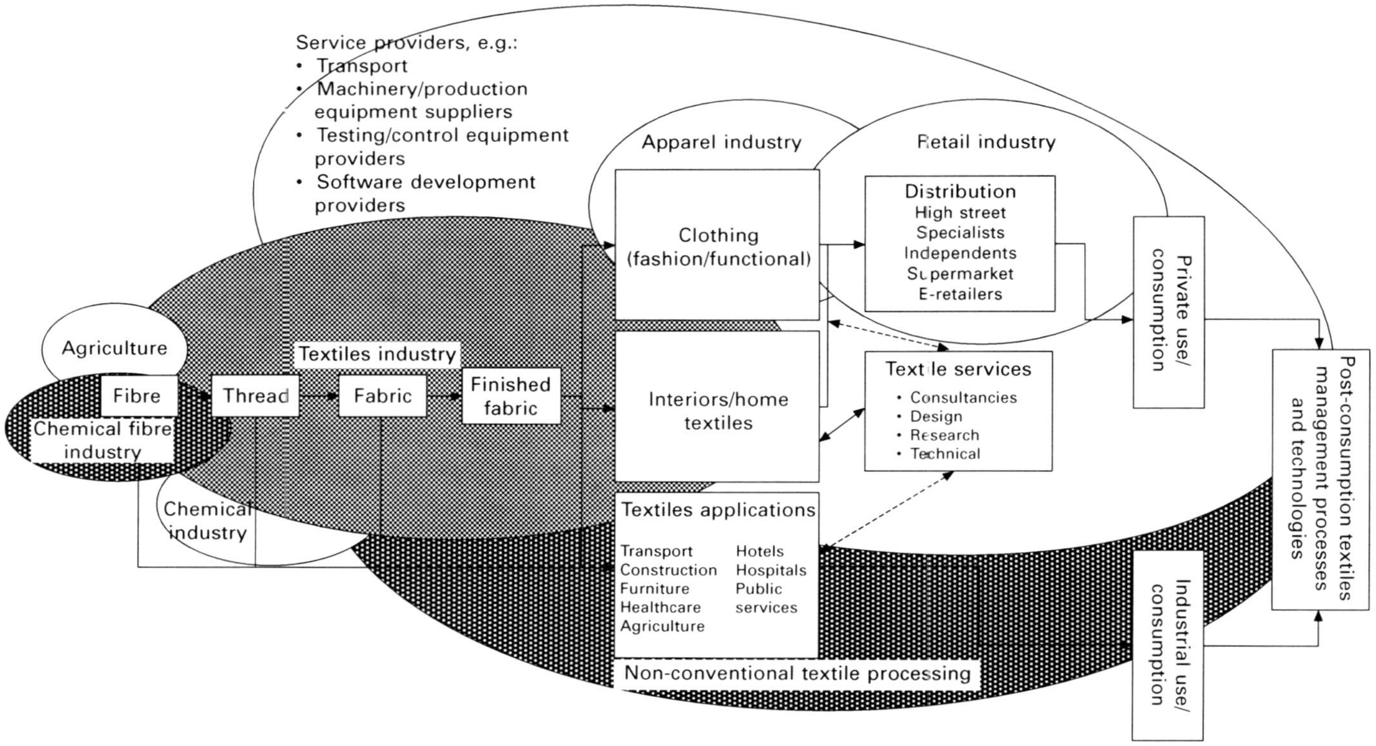

7.1 The textiles and clothing industry sector (Allwood *et al.*, 2008).

The first section demonstrates how basic weave knowledge of techniques and manufacture are required to work with CAD software for consumer textiles. The following section will appraise how knowledge of weave skills, techniques and manufacturing technologies have been coupled with CAD to further research and development of woven fabrics.

7.3.1 Basic textiles weave design features

Weaving has been described as 'the art of interlacing one element in and out of one another' (Gillow and Sentence, 2004, p. 68) and woven structures are formed by interlacing one set of yarns with another set oriented in the opposite direction. Yarns that form the length of the fabric are called the *warp*; the yarns that cross or interlace the warp to form the pattern are called the *weft* or filling yarns (Corbman, 1975). A weaving loom has the following components to enable fabric formation (see Fig. 7.2; Wingate, 1979):

- Warp beam: the warp yarns are wound onto this beam.
- Heddle: a wire with an 'eye' in the middle of it through which to feed a warp yarn.
- Harness: a frame to which a group of heddles are vertically attached in a series. The weave operation is achieved through the raising and lowering of the harness. A separate harness is used for each group of warp yarns arranged to be moved as a group to weave a desired pattern.
- Shuttle: carries the weft threads back and forth across the loom (this may be replaced by a variety of other methods, e.g. air, water etc.).
- Reed: maintains the warp spacing at the required spacing and presses or battens each filling yarn (weft) to form the fabric.
- Cloth beam: to collect the newly constructed fabric.

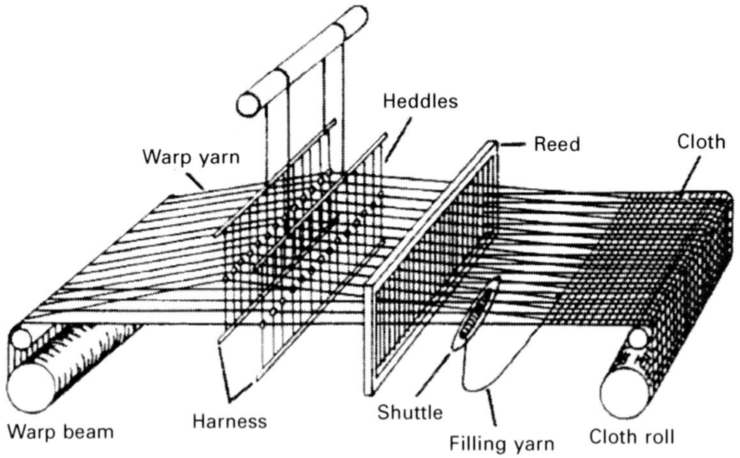

7.2 Components of a typical shuttle loom (Wingate, 1979).

Loom operation requires the following processes: warp let-off, shedding, filling insertion, beat-up and fabric take-up (Arora, 2008):

- *Warp let-off:* the mechanical or electronic mechanism that controls the rate of flow of warp yarn to be let off the beam with consistent and appropriate tension setting to achieve efficiency of the weaving process.
- *Shedding:* the action of some warp yarns raised and some yarns lowered to form a 'shed' through which the filling yarn passes. The warp yarns are grouped and passed through a harness. The grouping of warp yarns, the number of harnesses and the sequence in which the warp yarns are raised and lowered are determined by the design (weave structure).
- *Filling insertion:* the term for the filling yarn being inserted through the shed. The filling yarn is inserted by a carrier device such as shuttle or shuttle-less technologies such as projectile, air jet or water jet. As the filling yarn is inserted, the carrier device moves across from one side to the other of the shed, the edges of the fabric, selvage, are formed. A single movement across the shed is called a 'pick'.
- *Beating up:* an inserted filling yarn is not able to be placed directly into the correct position at the shed opening as it physically cannot fit into the acute angle of the shed opening. The filling yarn is pushed into its final position (the cloth fell) by a reed and this action is referred to as beating up. The cloth fell is an imaginary line where the cloth starts to be formed. Beating up takes place after the shuttle has moved through the shed. Beating up gives the fabric firm and compact construction. The speed with which a loom operates is quoted in 'picks per minute' and signifies how many times the weft yarn moves across the warp yarns and is beaten up within one minute.
- *Take up:* as the fabric is formed, it needs to be removed from the weave area and is wound onto a cloth beam. The rate at which this takes place is one of the elements that control the filling density (picks per centimetre or picks per inch), the other is the weaving machine speed (picks per minute).

All types of looms, in industry, may be computer controlled both at the design end through CAD and at the manufacture end through computer aided manufacturing (CAM). The interaction of CAD/CAM has helped to reduce the number of fabric faults and increased the production rate of fabrics. Weave patterns (structures) depend on the number of harnesses and the sequence of shedding, and it is important to understand the shedding motion. There are three different types of attachments for shedding motions:

- tappet
- dobby
- jacquard.

Tappet shedding is the simplest and least versatile shedding motion; it is a mechanical movement using tappets and cams to control the up/down movement of the harnesses. The mechanism is bulky and limits the loom to control up to 8 harnesses and the design is restricted to plain weave, simple twill and simple sateen or satin weave. Any design requiring more than 8 harnesses requires a dobby loom.

Dobby shedding is a compact, electronically guided shedding motion and capable of having up to 28 harnesses; it therefore is capable of more complex and versatile shedding motion with much greater weave repeat. Design may be woven with two or more basic weaves and their variation and fabrics woven using dobby shedding are often referred to as dobby cloths or dobby weave.

Jacquard shedding provides unlimited design width as there are no harnesses; instead, the individual warp yarns are independently controlled through cords attached to a jacquard head. Modern jacquard looms are electronic, and designs are input through CAD with the potential for figured, non-repeat or large and complex patterns.

7.3.2 Woven technical textiles

According to the Technitex group (www.technitex.org) the term 'technical textiles' was coined in the 1980s to follow on from the term 'industrial' textiles (which is still used in the United States), as its scope of applications was regarded as too narrow. 'Technical textiles' were initially regarded as 'textile materials and products manufactured primarily for their technical and performance properties rather than their aesthetic or decorative characteristics', but technical textiles are being used within sportswear apparel and the increasing use of design to attract consumer sales has rendered a purely functional approach somewhat narrow (Byrne, 2000). Knitted or woven fabric structures may be used for a range of technical textiles (Sherburn, 2007). Byrne (2000) provides some common features of technical textile products: they are the result of understanding and know-how about combinations of the techniques of manipulating fibres, fabrics and finishing and how they perform in different combinations and environments.

Woven fabrics may be single layer or multiple layers and the weave structure may be plain through to complex, depending on the final product requirements. Multilayer woven fabrics are also known as 3D weaves, as they have several layers of warp and weft yarns woven together in an almost infinite number of possible ways, but broadly categorised as orthogonal, through-thickness angle interlock and angle interlock (also known as layer-to-layer). Braided fabrics (created by interweaving two or more yarns in a diagonally overlapping pattern) are also commonly used to develop technical textiles. Two types of braided fabrics are widely available, biaxial (two sets of aligned yarns) and triaxial (three sets of aligned yarns). As with woven fabrics, multilayered (3D) braided fabrics are also manufactured. 3D weaving may result from specific loom technologies such as 3Tex and Biteam or through a conventional weaving loom (Chen, 2011). Use of a conventional loom has the advantage of readily available technology, flexibility in the types of 3D structures to be produced and lower costs. The disadvantage of using a conventional loom is that it can be complicated to design and manufacture 3D fabrics in this way (Chen, 2011).

Technical textiles are interdisciplinary in nature; 'cross-discipline collaboration between textiles and engineering' were noted to be increasingly commonplace within research and industry by woven textile designer and academic Julie Soden (2001). She also noted that, although these types of projects resulted in mutually beneficial outcomes, there were challenges regarding communication such as terminology and a basic understanding of specific technical requirements and limitations. To design technical textiles, the designer needs to have a very good understanding of the yarn and its properties, particularly because much of the 3D woven fabrics are used for applications such as textiles reinforced composite materials (TRCM), where both the materials used as the matrix and the reinforcing material (such as textiles) retain their individual properties at microscopic level. Textile composites may be textile structures within polymers, ceramics or metals (Ogin, 2000). Building upon this knowledge, the ability to conduct mechanical or geometrical modelling of the textiles is paramount. According to Sherburn (2007), geometric modelling of yarn as 'solid volumes representing the approximate bounds of the fibres contained within them' (Sherburn, 2007, p. 5) for application in composite textiles is important because it enables:

- determination of the mechanical properties of the fabric forming behaviour
- prediction of the permeability of fabrics for processing of composites
- modelling of the mechanical properties of composite parts and their adverse behaviour for use in engineering applications.

7.4 Necessary expertise and skills training for woven computer aided design (CAD) textile designers

Weave designers need to be multi-skilled and understand both the supply and cost issues, the specifications for production such as fabric analysis, drafting the pattern, tensions as well as computer aided design packages. Skillfast UK (2008) conducted a survey of fashion and textiles industry to identify the skills considered necessary for a variety of roles. Table 7.1 illustrates the skills regarded as necessary for a weave designer. Most textile companies

Table 7.1 National Database of Accredited Qualifications: Woven Fabric Design (reference no. R/502/2281)

Learning outcome		Assessment criteria	
1	Understand the process of woven fabric design and development from initial idea to finished product	1.1	Analyse the factors impacting on design development
		1.2	Describe the different types of repeat and their application
		1.3	Explain the processes of design from initial idea to finished product to include: • Research • Interpretation of design brief • Customer requirements • Costings • Design specification • Sample requirements • Sample production • Production specification
2	Understand the principles of technical drafting	2.1	Explain the principles of technical drafting for each of the following: plain fabric, twill, stain and sateen weaves
		2.2	Draw a technical draft for two designs to be woven on two different types of machine
3	Be able to use CAD techniques in woven fabric design processes	3.1	Describe CAD – computer aided design techniques and processes
		3.2	Produce a design for a woven fabric using CAD – computer aided design
4	Understand fabric analysis techniques	4.1	Describe sample analysis techniques for: fibre type, yarn type and count and type of fabric
		4.2	Analyse two woven fabrics of different construction and for each of the following: produce fabric notations, measure fabric parameters and carry out calculations
5	Understand the principles and processes of weaving	5.1	Describe the primary motions of weaving
		5.2	Explain the manufacturing processes and production cycle of woven fabrics
		5.3	Explain factors influencing handling characteristics of materials during processing

require new employees to have a working knowledge of CAD, even where they provide some in-house training from the software providers. Most new textile designers will learn their skills at universities or colleges of textiles. To become adept in the use of CAD, one must understand the principles of weaving. The basic principles tend to be learnt initially through hand weaving classes; at university or college, this often requires a year to master before the student undertakes designing through CAD. Proficiency in CAD for technical textiles requires not only fundamental understanding of fibre properties and CAD but also mathematical modelling techniques and skills that can be applied to leverage the fibre and fabric properties. To acquire the specialist knowledge and expertise of a technical textiles CAD weaver requires study and training at Masters Level or beyond, i.e. this is a highly specialised area that tends to be located in the R&D areas of companies or universities. There are a number of weave CAD software packages available to both education and industry. As noted earlier in this section, a credible weave designer needs to display a working knowledge, if not some degree of expertise, in using CAD. There are, however, two general, main issues with regard to using CAD in weaving: cost and weave application (i.e. commercial/domestic or industrial/technical use).

7.5 Costs incurred in using computer aided design (CAD)

A term most often associated with CAD is 'investment' and it is often the larger companies who are suppliers or part of a global network that spend the most and benefit from this. The costs of software are varied and cannot be estimated in a simple manner. The costs of buying a CAD package may vary from thousands to tens of thousands of pounds; factors that influence the actual and final cost include the following:

- The licence to use the software (and how many – each person requiring the use would need to have the licence).
- Whether 'floating' licences and 'keys' are required (to use *in situ* at the company offices or studio or on a laptop away from the office).
- Training to use the software, which includes:
  - a trainer's daily fee
  - the costs of travel for the trainer to come to the studio
  - any further expenses incurred by the trainer for overnight stay and sustenance.
- Computer platforms to support the use of the software (and what they are compatible with).
- Peripherals (e.g. printers, links to CAD/CAM at other sites, links to factories, etc.).

- Servers to store or access the information.
- Any annual subscription or payment for annual software update.

7.6 Computer aided design (CAD) software applications

As noted earlier, the scope of the textiles industry is vast and spans domestic and commercial applications to industrial and technical textiles (such as geo-engineering, medicine and heavy industry). The types of software systems and packages vary between the two types of industries, in terms of both the systems and use of the software.

7.6.1 Domestic and commercial applications

Windows® is the most common operating platform for the software but Apple Mac® and Linux are also used. The designs are generally able to be read across the different platforms to avoid, as far as possible, technology 'lock in' by the textile manufacturers. In general, weave software is built up as separate packages of a variety of functions (e.g., dobby, jacquard, presentation – to present mood boards and specifications of the designs to potential customers and clients, drape function – to visualise fabric designs on e.g. shaped sofas and bedcovers, etc.). All software packages should be able to print out loom cards or be linked to peripherals such as printers and looms such as sending the instructions directly to looms for manufacture (i.e. links to CAM processes, most often Bonas or Staubli for jacquard weaving). Four commonly used weave CAD packages (AVA, Scotweave, Textronic and Arahne) have been compared for the purpose of this chapter and they have the following similarities:

- Yarn consumption and cost implications are calculated as part of the design function.
- All have the following functionalities, either as a separate module or as part of the package:
  - draping functions (to visualise the woven fabric design applied to a product),
  - data storage, retrieval, sharing and management systems and presentation methods,
  - presentation – such as texture mapping, story-boarding.

Table 7.2 illustrates the differentiating features between the software packages, but may be briefly summarised as:

- **Scotweave** software has a package for face-to-face velvet and technical textiles (e.g. for the automotive industries, geo-textiles, medical textiles,

Table 7.2 Comparison of some consumer application CAD software

Software package	Platform/ compatibility	Special features	Weave types				Drape	Peripherals	Systems management	Sharing	Presentation
			Dobby	Jacquard	Velvet	Technical					
AVA Weave (UK-based company)	Platform – Apple Macintosh computer, supports Mac OS X; compatible with PowerPC and Intel Mac computers		Yes – plain weave to mixed construction blankets; cartridge can be linked to AVL loom	Yes – requires AVA Quick Separations; design exported to be read by either a Bonas or a Staubli jacquard loom			Grid and calibrated colour	Print out to EPROM Cartridge Loader	AVA QA Pro: central control (lead retailer controls if quality is appropriate)	Information about colour and design shared through the AVA server called Synchronise	AVA Materialise and AVA Colour

Table 7.2 Continued

Software package	Platform/ compatibility	Special features	Weave types				Drape	Peripherals	Systems management	Sharing	Presentation
			Dobby	Jacquard	Velvet	Technical					
Scotweave (UK-based company)	Microsoft Windows®, can export as TIFF, BMP, JPEG, etc. for any Windows®-compatible with colour printer, scanner or other hardware	Yarn Designer – create/ scan in fancy yarns. Weave 3D to move around weave structure in real time. Drape 3D – drape with real-time movement. Auto Drape to visualise design in a real setting	Yes. Production data printed (loomcard) or sent directly to electronic dobby looms or direct loom control	ScotWeave Jacquard Designer with an electronic jacquard loom for design concept to woven fabric	Face-to-face woven velvet dobby – seperate software to Jacquard Velvet Designer for velvet jacquard fabrics from artwork to loom disk. Includes full weave editing (hole finding) features, also terry towelling	Technical Weaver – technical textiles for industrial, commercial and geo-textiles – dobby or jacquard. Includes yarn design, cross section view/edit as well as 3D weave	3D visualisation tool to see ScotWeave fabrics realistically draped over images of garments and other real-life objects	Any Microsoft Windows, printer, etc.	Electronic dobby or jacquard loom controllers or looms		Presentation Program story-board layout – separate programme from others

Textronic (Indian–based company)	Windows-based	Yarn design facility; also carpet design module integrates various carpet production techniques – supports Van de Wiele – EP files, Staubli and Bonas. Also for hand knotted carpet weaving	Production data printed (loomcard) or sent directly to electronic dobby looms or direct loom control	Yes – also design exported to be read by either a Bonas, Staubli, Grosse, etc. jacquard loom		Yes		Archiving, data management system – storage and retrieval of design work, security, reports of design and productivity	Yes	Storyboarding program

Table 7.2 Continued

Software package	Platform/ compatibility	Special features	Weave types				Drape	Peripherals	Systems management	Sharing	Presentation
			Dobby	Jacquard	Velvet	Technical					
Arahne (Ljubljana, Slovenia) – CAD/CAM established 1992, from tight cooperation between local textile industry and universities. Used: China, Croatia, Czech Republic, Denmark, Germany, Egypt, Finland, France, India, Iran, Italy, Morocco, Macedonia, Mali	Linux platform, free Apache server. Email as GIF and JPEG or XML for larger production plants. Compatible with Windows, or Mac. Firewalled for security	Both dobby and jacquard are on one programme – ArahWeave. Yarn consumption calculation is in HTML	Yes	Yes			Yes	Internet-based communication	Retrieval system through Perl	Yes – internet but through firewall	

anti-ballistics etc.) and has a cross-sectional and 3D visualisation feature which enables the weave structure to be viewed from any desired angle and to identify any defects in the weave structure.

- **AVA** is highly popular within the printed textiles industry and so has many specific functions regarding colour matching of yarns to ensure that colour specifications for the designs are maintained throughout the production process.
- **Arahne** presents its software as the most cost effective package to be used by industry and educational institutes; it offers the dobby and jacquard software together in one package.
- **Textronic** software has a carpet weaving software package which is also capable of being linked to CAM operating looms such as Staubli.

7.6.2 Heavy industrial and technical textiles

The discussion so far has demonstrated the impact that CAD has had on the textiles industry in areas that we are commonly familiar with (clothing, automotives, interiors, etc). The use of CAD has also developed and propelled the technical textiles sector of the industry into prominence. CAD has been developed for the technical textiles sector on a commercial basis by software development companies but as the research is continuing and developing, academic institutes are also developing software applications from within textiles-related or materials departments. The following discussion is merely an overview of the main features; it is difficult to provide an in-depth analysis of each of the types of software applications available because each application has been developed in response to a particular product requirement and is outside the scope of this chapter.

Various commercial software packages have been produced for designers within textiles composites industries. For example, as described in the previous section, ScotCad Textiles Ltd has been providing CAD software for weaving since 1982. Used by academia to replicate industry environments for students, the ScotWeave package is used to help develop textile designs largely for the commercial arena. According to Sherburn (2007), many of the functions in the Scotweave package (yarn costing data, scanning feature, image edit tools, library of over 700 weaves, float checking, auto-drape, fabric finishing, import/export weave data, output instructions directly to the looms, etc.) are of limited value to the technical textile researcher. Scotweave's Technical Weaver module, however, models technical textiles at the mesoscopic scale providing some of the functionalities required by the technical textiles designer. Textiles may be examined at macroscopic, microscopic or mesoscopic levels: the microscopic scale examines the yarn, the macroscopic scale examines the woven structure, and an intermediate (mesoscopic) scale examines a few intertwined yarns, which 'define the unit cell reproducing the whole structure by a periodic translation' (Boubaker *et al.*, 2010). A unit cell has been defined as 'the smallest unit of textile that, when tiled, will recreate the full scale textile' (Sherburn, 2007, p. 5). The unit cell needs to be geometrically modelled in order to determine how the yarns are arranged, what the strains on them are and therefore develop predictions regarding their behaviour (shape forming, porosity, breaking, impact damage, etc). In the ScotWeave 'Technical Weaver' package, yarn cross-sectional shape and weave pattern can be specified to create a 3D geometrical model. However, according to Sherburn (2007), the ScotWeave 'Technical Weaver' software did not enable calculation of mechanical properties and was limited to modelling orthogonal woven fabrics.

The necessity to be able to mechanically or geometrically calculate and model yarns has led companies and universities to develop software systems of their own. For example, CAD modelling systems to provide predictive functions for fabric permeability and composite mechanical properties have been developed: TexGen, WiseTex and TexEng Software Ltd (TexGen; Sherburn, 2007; Robitaille *et al.*, 2004; Robitaille and Gauvin 1998; WiseTex; Lomov and Verpoest, 2000). TexGen, a free and open source licence software (General Public Licensing), operates on Windows and Linux. TexEng Software Ltd is a spin out company developed from research on technical textiles at the University of Manchester (www.texeng.co.uk). TexEng Software Ltd developed Weave Engineer® to support speedy and accurate design and manufacture of 3D solid fabric of multilayer as well as single-layer and Hollow CAD®, to support design of 3D hollow woven architectures with uneven or flat surfaces when opened up, and can be used in the investigation of lightweight textile composites (Chen, 2011).

7.7 The impact of computer aided design (CAD) on the supply chain

Fabric production has three broad phases: yarn formation, winding and fabric formation (knitting or weaving) followed by a wet processing such as dyeing or printing. Computer aided design (CAD) may design yarn compositions or combination of yarns, as well as develop weave structures for woven designs. Printed designs may be undertaken by CAD systems as well, but this is usually undertaken once the fabric is woven. The final stages of finishing and volume production of textiles may be completed through computer aided manufacture (CAM). Although the techniques of knitting and weaving share a common heritage of looping/interlacing of yarns to create fabric, they have remained distinct and there is little overlap in the technology used. Within the textile industry there are numerous types of companies that produce or service textile-related processing fabric manufacture (fibre, yarn, fabric) for fashion, interiors or industrial goods. Figure 7.3 illustrates the process of

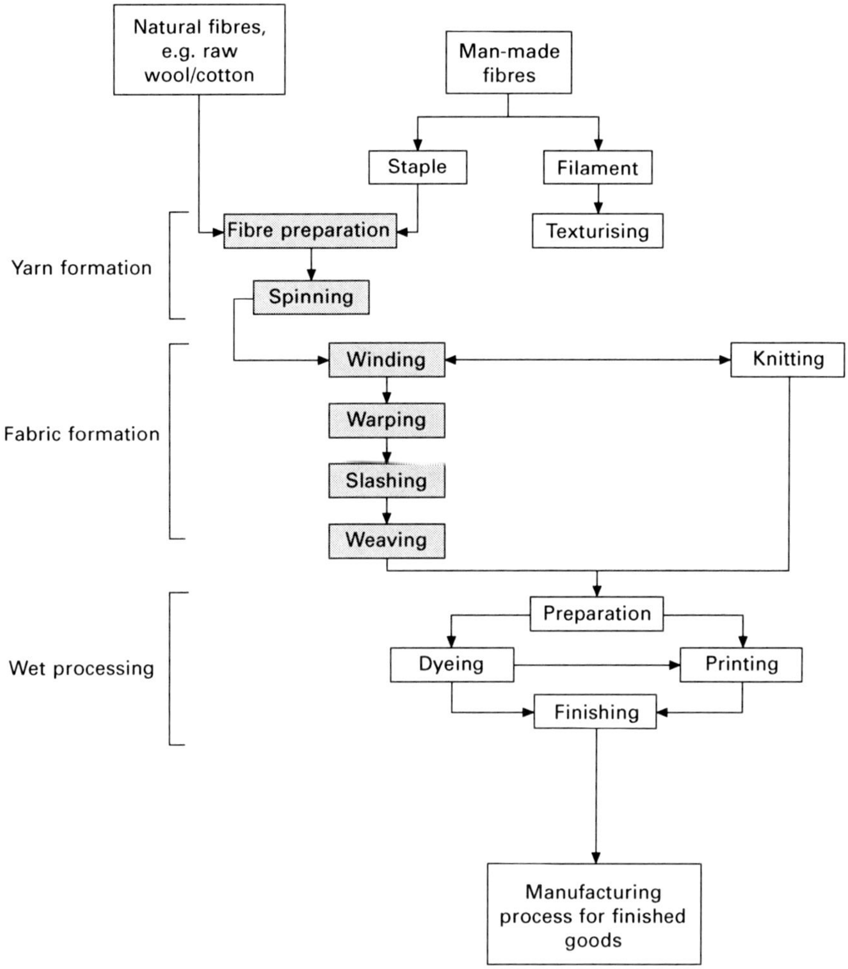

7.3 Fabric formation (after EPA, 1997).

textile manufacture by weaving and knitting (EPA, 1997) and the shaded boxes are the areas that woven CAD design is concerned with.

It is well known that the use of CAD reduces time spent on product development. Indeed, in some areas of weaving CAD is essential (e.g. in jacquard weaving). According to Price *et al.* (1999), CAD came to be used in the large textile manufacturing firms in the United States since the 1980s. The weave design process in the United States would often take weeks or months to complete as the weave design was first conceived of and drawn onto graph or point paper by the fabric designer and then copied onto board by a painter to the exact and precise colours of the design. This board would be sent to the mill for a sample to be woven for approval by the client. Any amendments would be noted and fed back to the studio by the mill. With the advent of CAD, the designer is able to design or decide which yarns will be employed, draw out the weave design in the computer, either print or email (as a JPEG or another file format) to the customer and make any amendments as required. Computer screen and printer calibration technology has enabled a 'real-life/virtual' sampling technique, although further approvals may still be required before final production commences. Figure 7.4 illustrates the changes in timing and the closer relationship between customer/client and designer enabling quicker communications and more creative decision making regarding the final product.

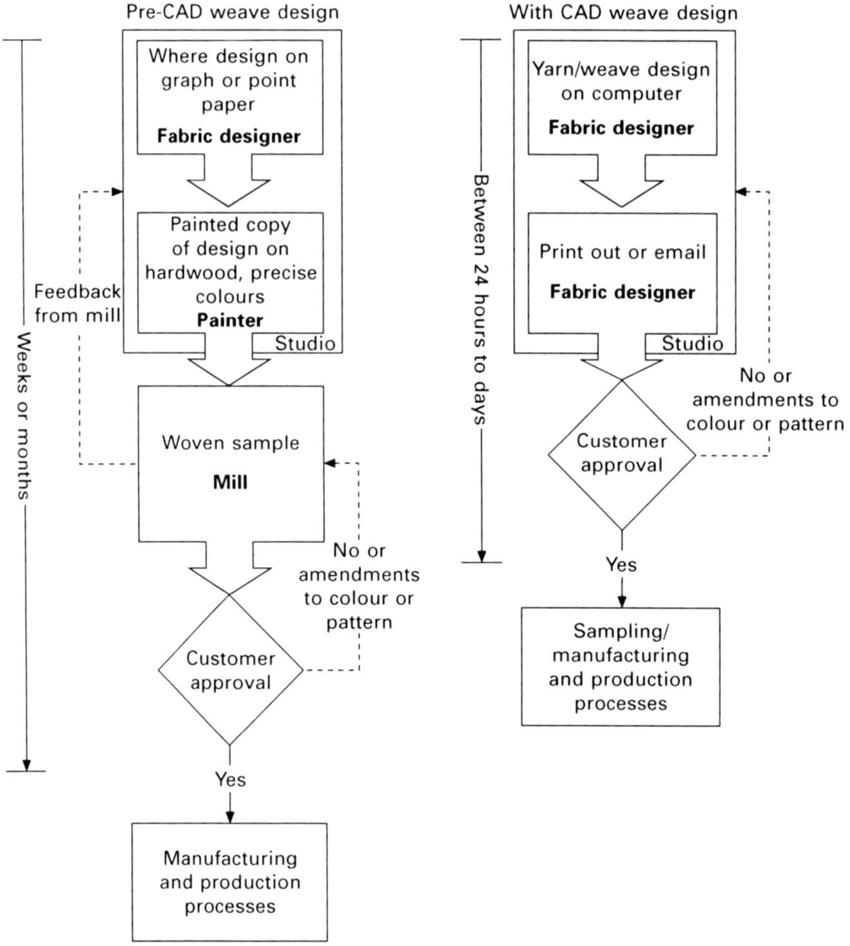

7.4 Weave design before and after introduction of CAD technology (after Price *et al.*, 1999).

CAD therefore, in the first instance, enabled the following: value added merchandise to be designed, shorter production cycles and direct communication between buyer and designer. As CAD systems commonly now have draping and texture mapping functions, it is also possible to visualise the end products (be they interiors, garments, etc.). It is also possible to request a particular design from the designer as they can scan in a design and reinterpret it through the software into a weave design for the client.

The use of CAD is of increasing strategic importance in the supply chain (Wang and Chan, 2010). The nature of the textiles supply chain has changed from a linear sequential or serial system to a network system. Globalisation and the effects of globalisation have profoundly changed sourcing and manufacturing as well as trading and the marketplace. To compete, a company needs to have greater flexibility and quicker response. As supply chains have dispersed geographically they have become increasingly dynamic in nature, requiring ever greater communication and networking. Intra-organisational networking has been commonplace as communication within a company is vital for it to progress. Information technology has made inter-organisational networking increasingly common as different companies begin to form virtually integrated operations to cope with competitive pressures.

Virtual integration differs from vertical integration as it enables flexibility and responsiveness (Wang and Chan, 2010). CAD is a form of IT and this has helped partnering companies in product development to achieve greater and more effective co-operation, communication and higher productivity with lower costs. Therefore, use of the same IT and CAD systems helps companies to understand each other better (e.g. product specifications are in the same 'language') but raises the following issues that have to be addressed in order for the network to continue to flourish:

- trust between members within a network regarding technology, intellectual property, time to market of new products,
- shared resources and competencies,
- agreement on size of network (for the market size and its constraints),
- optimisation of the global supply and, in some cases,
- possible technology 'lock-in' whereby the use of a particular platform is necessary to enter the virtually integrated network.

7.8 New products and markets and future trends through the use of computer aided design (CAD)

While the textiles industry has gradually become ever more mechanised and reliant on the use of CAD/CAM architecture of its production plants, the area of technical textiles has been growing and is now a means of competitive advantage for saturated textiles producing economies. This sector has been described as a growth area, and, indeed, a report in 2006 stated that the technical textiles sector was a dynamic and buoyant one with an annual growth of 5% (Chen *et al.*, 2006; Miozzo *et al.*, 2005); however, because of the problems in defining the sector, measuring the size of the market is a difficult one. Kowalski and Molnar (2009) described some of the efforts to try to estimate the size of this sector which include dividing the market into (i) traditional, (ii) technical and (iii) mixed textile product categories.

- Traditional textiles have been described as being: silk, cotton, wool and other vegetable textile fibres.
- Technical textiles have been described as man-made filaments made of, or containing, nylon, polyamide, polyester, polypropylene, viscose; man-made staple fibres and impregnated, coated or laminated textile fabric (e.g. various man-made tyre cord fabrics).
- Mixed textiles products are those manufactured from both traditional and man-made fibres such as: wadding, felt, nonwovens, yarns, twine, cordage, carpets, knitted or crocheted fabric.

Despite these problems of categorisation, Kowalski and Molnar (2009) noted that exports from China and India are increasingly technical textiles. According to Byrne (2000) a major, influential textiles industrial exhibition, Techtextil, was launched in 1985 which provided a taxonomy that has been used to describe the various industry and market sectors that technical textiles are applied in:

- **Agrotech** – agriculture, horticulture, forestry and aquaculture textiles
- **Buildtech** – building and construction textiles
- **Clothtech** – technical components of shoes and clothing, e.g. linings
- **Geotech** – geotexiles and civil engineering materials
- **Hometech** – technical components of furniture, household textiles and floor coverings
- **Indutech** – textiles for industrial applications, e.g. filtration, conveying, cleaning, etc.
- **Medtech** – hygiene and medical products
- **Mobiltech** – automobiles, shipping, railways and aerospace
- **Oekotech** – environmental protection
- **Packtech** – packaging materials
- **Protech** – personal and property protection
- **Sportech** – sport and leisure.

Notwithstanding the actual size and share of the textiles market, CAD knowledge and expertise has had a powerful influence on the technical textiles sector. Of increasing interest are composite textiles (textile reinforced composite materials) which have been described as ‘multiphase materials within which the phase distribution and geometry have been deliberately tailored to optimise one or more properties' (Ogin, 2000, p. 264) and as such that 'they do not dissolve or merge completely into each other' (Wiggers, 2007, p. 1). Ogin (2000) goes on to describe textile composites such as fibre reinforced plastics (FRP). FRP use fibrous materials, such as carbon fibres, glass fibres, boron, ceramic, aramid or textiles, to mechanically enhance the strength and elasticity of a plastics matrix (which may be polymers, such as epoxide, polyester, nylon, or metals). FRP is used where the following are necessary: weight savings, precision engineering, finite tolerances, and the simplification of parts in both production and operation for low costs and high volume production while maintaining or even improving on performance using more traditional materials such as steel or aluminium. They have been used in the automotive and aerospace industries. Textiles have been used in industrial composites for over a century (Wiggers, 2007) and therefore it is a mature application. However, the use of CAD, advances in computational modelling and mathematical analysis techniques have seen an increasing replacement of other fibres with textiles as opposed to carbon or glass fibres.

Textiles have the advantage of cost, enhanced by processability and adequate or improved mechanical properties. Woven textile composites have been used very successfully in aerospace because they are light, have good drapeability, allow complex shapes to be formed without gaps and lower costs as biaxial fabrics have replaced two layers of non-woven fabrics. Woven textiles also have the advantage of increased resistance to damage from impact and compression strength but the disadvantage of lower stiffness and strength than non-woven textiles and so are less resistant to deformation (Wiggers, 2007).

Technical textiles require high input of R&D efforts as well as new technology. This subject area has been advanced through the development of user-friendly software systems for designing woven structures based on mechanical and geometrical modelling with the aim of creating close-to-reality geometrical models for analysing mechanical, fluid, and thermal properties. This type of modelling is essential in systems such as predicting fluid filtration performance through filtration textiles, 3D hollow composites for energy absorption, shock absorption or anti-ballistics performance, etc. (Chen *et al.*, 2006; Chen, 2011).

Efforts in textiles woven CAD applications are continuing in the areas of colour, drape and yarn modelling, which can have applications in either the commercial or technical areas. Linked with other digital applications, there are possibilities for visualising, characterising, modelling and predicting textiles properties with a view towards manufacture. Traditional jacquard designing, reliant on the skills of craftsmanship of freehand design for copying through the mechanical weave loom, is assisted through the application of

CAD systems. Digital jacquard technology incorporates CAD and digital production technology (electronic jacquard machine and new generation weaving looms). This enables total digital control of design and production, and all data from design to weaving are processed, controlled and transmitted via the computer. Digital jacquard technology has enabled jacquard fabrics with photo-realistic effects with a mega-level colour number on the face of the fabric, raising the potential for security applications of this research (Ng and Zhou, 2009). Advances in modelling dry drape of textiles have their applications in the development of predictive modelling for 3D woven technical textiles (composites) as well as garment design and manufacture and reverse engineering of weave structures from samples of fabric, useful in security applications (Xin *et al.*, 2009).

7.9 Sources of further information and advice

The reader is directed to visit the following websites for further information and advice:

- www.UCAS.com – UCAS an organisation that manages applications to higher education courses in the UK, in order to help students make informed choices about their applications, there is a datbase of courses at all universities around the UK. This website will provide the initial information regarding any details about courses at universities to attend regarding textiles and CAD.
- http://www.skillset.org/fashion_and_textiles/ – Skillset, an industry body which supports skills and training for people and businesses in the UK creative industries: they provide news and information about several creative sectors, such as advertising, animation, computer games, facilities, fashion and textiles, film, interactive media, photo imaging, publishing, radio, TV.
- https://ktn.innovateuk.org – set up by the government and in collaboration with industry, the Technology Strategy Board has developed a Knowledge Transfer Network which is a tool to help collaborate, disseminate and share knowledge and seek funding for projects. There are a number of groups: environmental sustainability, materials, technical textiles, creative industries, etc.

Further information regarding the software programmes reviewed under the consumer applications is available at the following websites:

- www.ScotWeave.com
- www.arahne.si
- www.textronic.com
- www.avacadcam.com

7.10 References

Allwood, J.M., Laursen, S.E., Russell, S.N., Malvido de Rodríguez, C. and Bocken, N.M.P. (2008), 'An approach to scenario analysis of the sustainability of an industrial sector applied to clothing and textiles in the UK', *Journal of Cleaner Production* 16, 1234–1246.

Arora, J.K. (2008), *Modern Weaving Technology*, Global Media, Chandigarh, India.

Boubaker, B., Assidi, B.M. and Ganghoffer, J.F. (2010), 'Evaluation of poisson's ratio of textiles from mesoscopic models', *International Journal of Material Forming* 3, 81–84.

Byrne, C. (2000), 'Technical Textiles Market – an Overview', in Horrocks, A.R. and Anand, S., *Handbook of Technical Textiles*, Textile Institute, Manchester, pp. 1–23.

Chen, X. (2011), 'Mathematical modelling of 3D woven fabrics for CAD/CAM software', *Textile Research Journal*, 81(1), 42–50.

Chen, X., Lawrence, C.A. and Stylios, G.K. (2006), 'Engineering the Performance and Functional Properties of Technical Textiles', EPSRC Final Report, GR/S09227/01, GR/S09210/01 and GR/S09203/01.

Corbman, B.P. (1975), *Textiles: Fiber to Fabric*, 5th edn, McGraw-Hill, New York.

EPA (1997), EPA Office of Compliance Sector Notebook Project: Profile of the Textile Industry, US Environmental Protection Agency, Washington, DC, EPA/310-R-97-009.

Gillow, J. and Sentence, B. (2004), *World Textiles: a Visual Guide to Traditional Techniques*, Thames and Hudson, London.

Kowalski, P. and Molnar, M. (2009), 'Economic Impacts of the Phase-out in 2005 of Quantitative Restrictions Under the Agreement on Textiles and Clothing', OECD Trade Policy Working Paper No. 90, TAD/TC/WP(2007)14/FINAL.

Kramrisch, S. (1968), *Unknown India: Ritual Art in Tribe and Village*, Falcon Press, New York.

Lomov, S.V. and Verpoest, I. (2000), 'Compression of woven reinforcements: a mathematical model', *Journal of Reinforced Plastics and Composites* 19, 1329–1350.

Miozzo, M., Dewick, P. and Green, K. (2005), 'Globalisation and the environment: the long-term effects of technology on the international division of labour and energy demand', *Futures* 37, 521–546.

Ng, M.C.F. and Zhou, J. (2009), 'Innovative layered-combination mode for digital jacquard fabric design', *Textile Research Journal* 79, 737–743.

Nordas, H.K. (2005), 'The Global Textile and Clothing Industry post the Agreement on Textiles and Clothing', Discussion Paper No. 5, World Trade Organization, Geneva, Switzerland.

OECD (2004), 'A New World Map in Textiles and Clothing: Adjusting to Change', OECD, Paris, http://www.oecd.org/dataoecd/43/14/33824605.pdf.

Ogin, S.L. (2000), 'Textile reinforced composite materials', in Horrocks, A.R. and Anand, S., *Handbook of Technical Textiles*, Textile Institute, Manchester.

Price, A., Cohen, A.C. and Johnson, I. (eds) (1999), *J. J. Pizzuto's Fabric Science*, 7th edn, Fairchild Publications, New York.

Robitaille, F. and Gauvin, R. (1998), 'Compaction of textile reinforcements for composites manufacturing. ii: Compaction and relaxation of dry and H_2O-saturated woven reinforcements', *Polymer Composites*, 19, 543–557.

Robitaille, F., Long, A., Sherburn, M., Wong, C. and Rudd, C. (2004), 'Predictive modelling and performance properties of textile composite unit cells: current status and perspectives', in *11th European Conference on Composite Materials, Spn. ESCM: European Society for Composite Materials*, Rhodes, Greece, May.

Sherburn, M. (2007), 'Geometric and Mechanical Modelling of Textiles', Unpublished PhD thesis, University of Nottingham.

Skillfast UK (2008), 'Sector Qualifications Strategy', at http://www.ukces.org.uk/assets/ukces/docs/publications/sector-qualifications-strategy-skillfast-uk.pdf

Soden, J.A. (2001), 'Design and CAD innovation in woven textile research', *Point*, 9, http://www.point.ac.uk/articles/jsoden.htm

Soffer, O. (2004), 'Recovering perishable technologies through use wear on tools: preliminary evidence for upper paleolithic weaving and net making', *Current Anthropology* 45, 407–424.

Wang, W.Y.C. and Chan, H.K. (2010), 'Virtual organization for supply chain integration: Two cases in the textile and fashion retailing industry', *International Journal of Production Economics* 127, 333–342.

Wiggers, J. (2007), 'Analysis of Textile Deformation during Preforming for Liquid Composite Moulding', Unpublished PhD Thesis, University of Nottingham.

Wingate, I.B. (1979), *Fairchild's Dictionary of Textiles*, 6th edn, Fairchild Publications, New York.

Xin, B., Hu, J., Baciu, G. and Yu, X. (2009), 'Investigation on the classification of weave pattern based on an active grid model', *Textile Research Journal* 79, 1123–1134.

8
Modelling the structure of woven fabrics

B. K. BEHERA, Indian Institute of Technology Delhi, India

Abstract: The chapter starts with a discussion of the fundamentals of woven structure. This is followed by a section on design methods and sections on deterministic and non-deterministic modelling methodologies. The latter part of the chapter discusses authentication and testing, and reverse engineering.

Key words: woven fabric, deterministic and non-deterministic modelling, reverse engineering.

8.1 Introduction: fundamentals of woven structure

A woven fabric is produced by interlacing two sets of yarns called the warp and the weft at right angles to each other in the plane of the fabric. The warp is along the length and the weft along the width of the fabric. Individual warp and weft yarns are called ends and picks. The interlacement of ends and picks with each other produces a coherent and stable structure. The yarns are held in place due to the inter-yarn friction [1, 2]. The repeating unit of interlacement is called the weave. In this fabric formation process, there is a great scope in choosing fibres with particular properties, arranging fibres in the yarn in several ways and organizing in multiple ways interlaced yarns within the fabric. This gives the textile designer great freedom and variation for controlling and modifying the fabric structure. Figure 8.1 shows plain weave in plan view and in cross-section along warp and weft. The weave representation is shown by a grid in which vertical lines represent warp and horizontal lines represent weft. Each square represents the crossing of an end and a pick. A mark in a square indicates that the end is over the pick at the corresponding place in the fabric that is warp up. A blank square indicates that the pick is over the end that is weft up. One repeat of the weave is indicated by filled squares and the rest by crosses.

The properties of the fabric depend on the fabric structure. The formal structure of a woven fabric is defined by weave, thread density, crimp and yarn count. The interrelation between fabric parameters can be obtained by considering a geometrical model of the fabric. The geometrical model is concerned mainly with the shape taken up by the yarn in the warp or weft cross-section of the fabric. Mathematical deductions obtained from simple

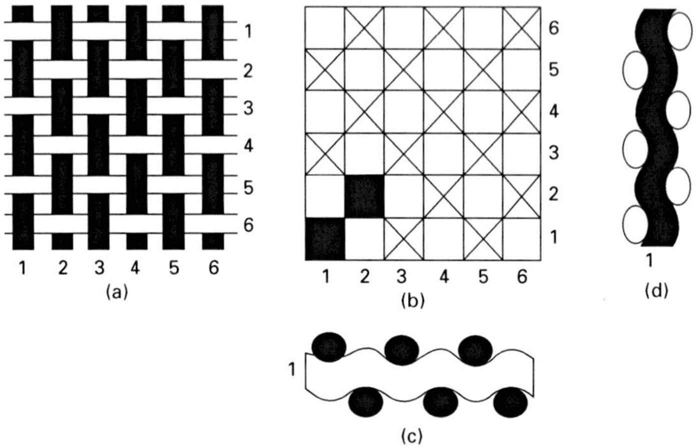

8.1 Plan (a), weave representation (b), cross-sectional view along weft (c) and cross-sectional view along warp (d) of a plain woven structure.

geometrical form and physical characteristics of yarn combined together help in understanding various phenomena in fabrics. The basic model of Peirce's [3] analysis is shown in Fig. 8.2. It represents a unit cell interlacement in which the yarns are considered inextensible and flexible. The yarns have circular cross-section and consist of straight and curved segments. The main advantages in considering this simple geometry are:

- It helps to establish the relationship between various geometrical parameters.
- It enables calculation of the resistance of the cloth to mechanical deformation such as initial extension, bending and shear in terms of the resistance to deformation of individual fibres.
- It provides information on the relative resistance of the cloth to the passage of air, water or light.
- It provides a guide to the maximum density of yarn packing possible in the cloth.

From the two-dimensional unit cell of a plain woven fabric, geometrical parameters such as thread spacing, weave angle, crimp and fabric thickness are related by deriving a set of equations. The symbols used to denote these parameters are:

d = diameter of thread
p = thread spacing
h = maximum displacement of thread axis normal to the plane of cloth (crimp height)
θ = angle of thread axis to the plane of cloth (weave angle in radians)

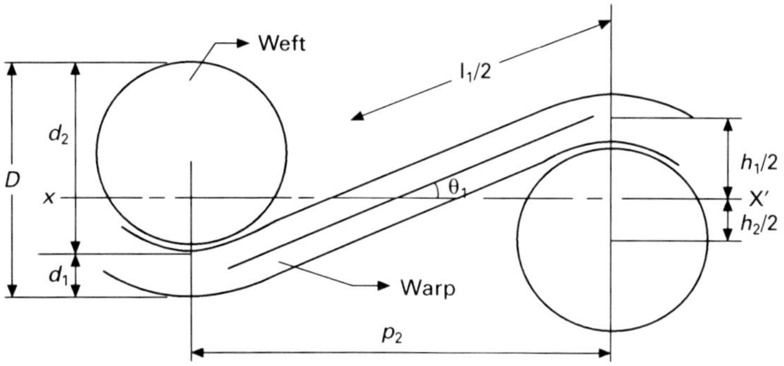

8.2 Peirce model of plain weave.

l = length of thread axis between the planes through the axes of consecutive cross-threads (modular length)
c = crimp (fractional)
$D = d_1 + d_2$

Suffixes 1 and 2 to the above parameters represent warp and weft threads, respectively in Fig. 8.2. The projection of yarn axis parallel and normal to the cloth plane gives the following equations:

$$c_1 = \frac{l_1}{p_2} - 1 \qquad [8.1]$$

$$p_2 = (l_1 - D\theta_1)\cos\theta_1 + D\sin\theta_1 \qquad [8.2]$$

$$h_1 = (l_1 - D\theta_1)\sin\theta_1 + D(1 - \cos\theta_1) \qquad [8.3]$$

Three similar equations are obtained for the weft direction by interchanging suffixes 1 and 2 or vice versa as follows:

$$c_2 = \frac{l_2}{p_1} - 1 \qquad [8.4]$$

$$p_1 = (l_2 - D\theta_2)\cos\theta_2 + D\sin\theta_2 \qquad [8.5]$$

$$h_2 = (l_2 - D\theta_2)\sin\theta_2 + D(1 - \cos\theta_2) \qquad [8.6]$$

Also, $d_1 + d_2 = h_1 + h_2 = D$ [8.7]

In all there are seven equations connecting eleven variables. If any four variables are known, then the equations can be solved and the remaining variables can be determined. Unfortunately, these equations are difficult to solve. Researchers have tried to solve these equations using various mathematical means to find new relationships and also some useful simplified equations.

8.2 Fundamentals of design engineering

Designing is the process of delineating a product to meet the functional and the aesthetic performance critera with efficient use of resources. Engineering is an applied science dealing with relationships between the raw material, the fabric and the finished product. Therefore, the design engineer has to engineer fabric construction for pre-determined properties that fit specific applications with quick response. Designing of textile-related products has been carried out for thousands of years in a traditional and intuitive way. However, in the last few decades the importance of design has increased due to the growing needs of the industry and the consumer for complex and specific requirement of functionality. This has called for a system of engineered manufacturing of textile product in place of traditional experience-oriented manufacturing. The design process itself may be subjected to certain restrictions such as time, manpower, cost, etc. Nevertheless, no standard procedure or logic is available or followed as far as textile product design is concerned. The design is usually carried out manually, based on experience and trial and error. In recent years, inexpensive computers have made the design of textile products more popular. Also, simulation and other technologies are brought into use to design and develop numerous quality products.

8.2.1 Traditional designing

In the traditional designing method [4], as shown in Fig. 8.3, the market expert formulates the rough concept of fabric requirements; thereafter, the fabric designer, in consultation with the experts from various departments such as marketing, production, quality assurance, costing and others, prepares a detailed description of the fabric design. With this derived design, a sample fabric is woven, tested, and undergoes the whole process until the customer

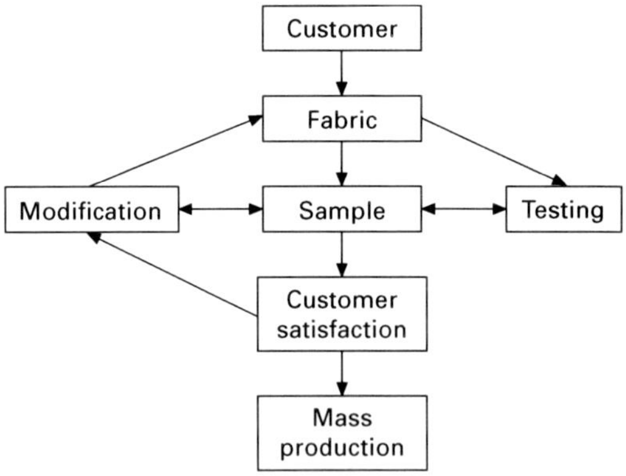

8.3 Traditional fabric design cycle.

is satisfied; then the fabric is taken for mass production. This fabric design cycle is the result of experience, heuristic reasoning and intuition. The basic problem here is the amount of expert knowledge available and its consistency, and also the fact that no single expert is usually available for a group of domains. A further development introduced in the fabric design process is known as manual design. A typical manual design procedure [5] for industrial fabrics is illustrated in Fig. 8.4. This is a rough procedure for synthesizing a fabric structure through several sub-tasks from top-down resolution of the overall design task. At first, for a set of property constraints, the designer chooses the fibre type for interlacing yarns governed by its cost, physical properties, etc., along with the type of yarn and its strength. The yarn linear density with strength of individual yarn components is then determined. At this point, the fabric tensile strength is specified, and fine adjustment of yarn properties is made using yarn twist followed by consideration of yarn density. The fabric weave structure is then selected, depending on the tearing strength, thickness, fluid permeability, etc. The manual design procedure for other types of textile materials, such as technical textiles, is almost the same with the only difference in properties requirement dependent on end use applications.

8.2.2 Traditional designing with structural mechanics approach

The stages of the traditional approach based on mathematical and physical principle are shown in Fig. 8.5 [6]. It is strongly mathematical with computations brought on at the end in order to solve equations. Fibre properties are specified by the parameters of the linear or non-linear equations. Geometry can be defined by algebraic and trigonometric equations. Mechanics can be analysed by differential or integral calculus.

8.3 Designing of textile products

Overall, the major areas of designing textile products can be broadly classified into colour or pattern and material design. Table 8.1 presents the difference between an artistic and engineering design approach. The colour or pattern design deals with the visual appearance and fashion of the products that

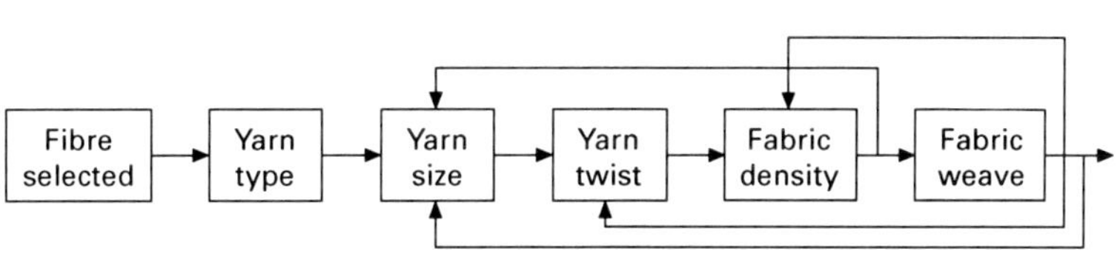

8.4 Manual design procedure for industrial fabrics.

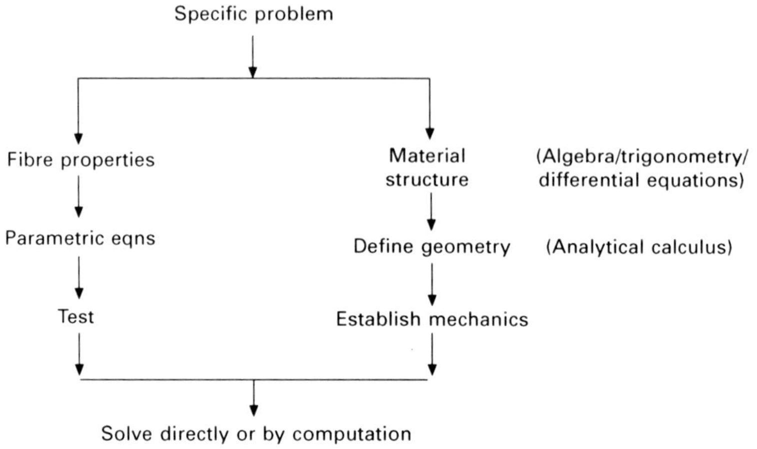

8.5 Mechanics of textile structure: traditional route.

Table 8.1 Artistic design versus engineering design

CAD for artistic design	CAD for engineering design
• Colour	• Thread count
• Pattern	• Thread density
• Weave	• Weave
• Motif	• Areal density, etc.

includes the motif [7–10], colour [11–17], patterns preparation [18–30], etc. This involves creating infinite combinations of motifs and arriving at an arrangement that may be aesthetically pleasing to the consumer.

In artistic design, traditionally the logic and main process is achieved through 'pencil-paper' sketching. To begin with, the creative idea is expressed by freehand sketching. If the design is thought to have a commercial potential, it is then processed by the designer or design technicians from the sketch onto a special design paper. At this stage, the design is modified according to the process for which it is intended, that is knitting, weaving or printing, etc. [31–33]. Then, this design paper becomes an enhanced sketch with additional information. With the advent of the computer, the creative idea of designers is translated to the computer environment for design purposes. A large amount of innovative work has been carried out in the area of colour and pattern designing [34] by using computer aided design (CAD). In fact, the application of colour graphics technology for colour pattern designing is becoming ever more popular because of its effectiveness.

Material design can be defined broadly as design that deals with the structural, aesthetic and functional behaviour of the material based on product type [35]. The logic of material designing involves mapping a specified function onto a realizable structure [36]. A correct material design is one whose basic structure correctly sets up the specified function or effect. In this system, material behaviour mediates function and structure. On the other hand, the colour or pattern and material design has a common region that is the surface or texture appearance caused by the material effect. Both will overlap effectively in producing a product. Although a considerable amount of research has been carried out on colour or pattern designing related to fabric appearance and fashion, the material design technology for textile products is still at the development stage. Compared to other fields of engineering, textiles may have taken the longest to realize the potential of CAD [37] for innovation and creativity in product design. This is primarily attributed to the constraint of material characteristics.

8.4 Design engineering using theoretical modelling

8.4.1 Model

A model is a description or an analogy which helps to visualize something that cannot be directly observed. It can be a system of postulates, data and inferences presented as a mathematical description of an entity or state of affairs. Modelling is an activity about which we think and make models to describe how the objects of interest behave. There are several ways in which objects and their behaviour can be described. We can use words, drawings, physical models, computer programs or mathematical formula. In designing an engineered product, normally, mathematical equations and/or logical concepts are used to simulate and predict real events and processes while making models. Mathematical models represent the behaviour of objects in mathematical terms. In modelling, it is important to know how to generate and validate mathematical representations (equations). It is also important to know how model equations are used and their limitations. Before analysing these issues, it is worthwhile to understand why we do mathematical modelling.

8.4.2 Theoretical modelling

Theoretical models describe a type of object or system by attributing to it what might be called an internal structure, a composition or mechanism that, when taken as a reference, will explain various properties of that object or system. A theoretical model therefore analyses a phenomenon that exhibits certain known regularities by reducing it to more basic components, and not simply by expressing those regularities in quantitative terms or by relating the known properties to those of different objects or systems. A theoretical model is considered as an approximation that is useful for certain purposes. Models can be used for the purpose of explanation, prediction, calculation, systematization, derivation of principles, etc. The most important characteristic of a theoretical model is that it consists of a set of assumptions about the concept or system. The models provide explanations; but these explanations are based on assumptions that may be simplified, and this condition must be borne in mind when one compares them with theories. It is also important that a theoretical model is often formulated and developed and sometimes even named on the basis of an analogy between the object or system that it describes. The model simulates the behaviour or activity of systems, processes or phenomena. They include the use of mathematical equations, computers and other electronic equipment.

8.4.3 The need for theoretical modelling

Modelling methodologies for predicting fabric properties are essential to design fabrics according to the specifications desired by the customer. If the relationships between different parameters that determine the specific fabric property are known, they can be handled to optimize that particular property for different end-use applications. Predictive modelling methodologies can also be used to identify different levels of combinations of process parameters and material variables that yield the desired fabric property. From this, a specific combination of process and material variables resulting in maximum savings in cost and time can be selected.

8.4.4 Mathematical modelling and the scientific method

Scientific methods help to identify the real world and the conceptual world. In the real world we observe various phenomena and behaviours which could either be natural or produced by artifacts. The conceptual world is the world of the mind. The conceptual world has three stages: observation, modelling and prediction. In the observation part of the scientific method we measure what is happening in the real world. Here we gather empirical evidence and facts on the ground. Modelling is concerned with analysing the above observations for at least one of the three reasons. The model describes the behaviour or results observed and allows us to predict future behaviours that are yet unseen or unmeasured. In the prediction of a scientific method, we use models to tell us what will happen in a future experiment or in an anticipated set of events in the real world. These predictions are followed by observations that serve either to validate the model or to suggest reasons for an inadequate model. The scientific method and engineering design have much in common, but there are differences in motivation and approach. In the practices of science and engineering design, models are often applied to predict what will happen in a future situation. Mathematical modelling is formulated using certain principles. The basic philosophical approach to formulate a model is shown in Fig. 8.6.

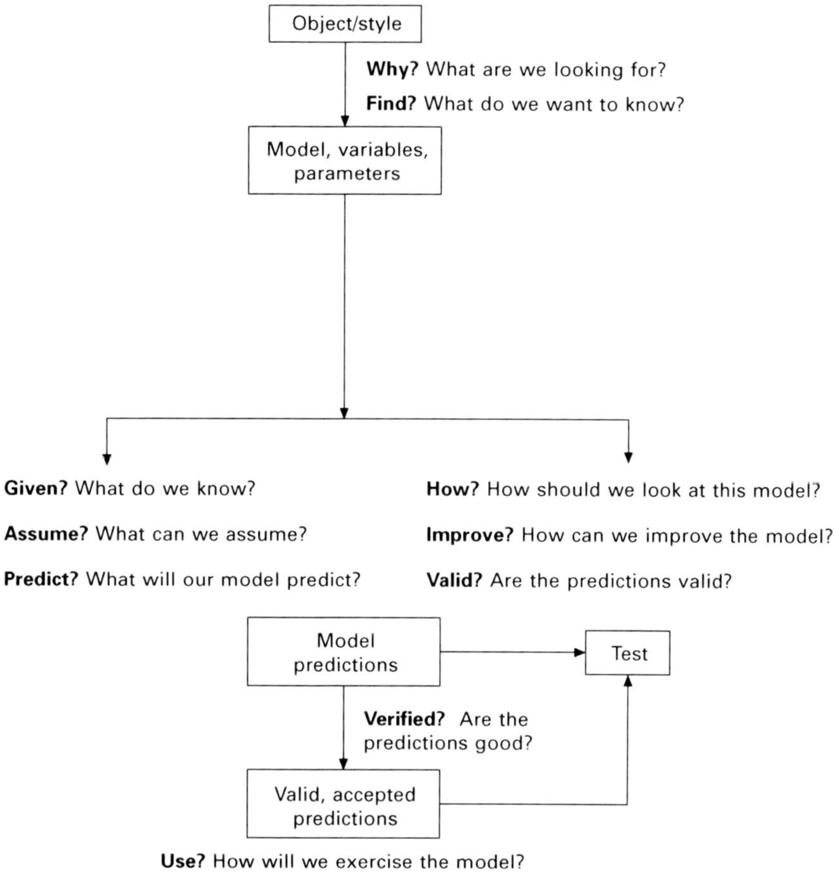

8.6 The philosophy of mathematical modelling.

8.5 Modelling methodologies: deterministic models

Over the years, many attempts have been made to develop predictive models for textile properties using different modelling methodologies. These are essentially of two kinds: deterministic and non-deterministic. Mathematical models, empirical models, computer simulation models such as finite element analysis are deterministic models, whereas models based on genetic methods, neural networks, chaos theory and soft logic are non-deterministic. Each method has its own merits and demerits. Nowadays, more and more processes and systems are modelled and optimized using non-deterministic approaches. This is due to the high degree of complexity of systems, and consequently, the inability to study them efficiently with conventional methods only. In non-deterministic approaches, there are no precise, strict mathematical rules. No assumptions regarding the form, size and complexity of models are made in advance. Hence this approach offers a flexible means to provide solutions to a wide variety of textile problems with reasonable prediction accuracy [38]. A brief overview of various computing tools is given in this chapter. For a detailed explanation on these modelling methodologies and other soft computing tools, one can refer to the standard texts available [39–41].

8.5.1 Pure geometrical model

Fabric properties are greatly affected by the choice of fabric parameters. The choice of fabric parameters influences the structure. The behaviour and relationship between the fabric parameters are a precursor to the optimal solution for fabric engineering problems. Many features of the cloth are essentially dependent on the geometrical relationships. The geometrical model of fabric provides some simplified formulae to facilitate calculations and specific constants which are of value for fabric engineering, problems of structure and mechanical properties [38]. These fabric parameters are a tool for an innovative fabric designer in creating fabrics for diverse applications. The theoretical relationship between the fabric parameters enables the fabric designer to play with different fibres, yarn linear density, thread density and weave to vary texture and fabric properties.

It is possible to predict the fabric parameters and their effect on the fabric properties by simple computing. This information is helpful in taking a decision regarding specific buyers' need. A simplified algorithm is used to solve fabric geometry equations and relationships are obtained between useful fabric parameters such as thread spacing and crimp, fabric cover and crimp, warp and weft cover, etc. Such relationships help in guiding the direction for moderating fabric parameters. For example, the parameters of a fabric requiring certain crimp in warp and weft even in a jammed state can be calculated. It may be desirable to make a non-jammed fabric with a certain amount of warp and weft. Even fabrics with maximum crimp in either warp or weft direction can be estimated. In this chapter only a few relations are graphically demonstrated to optimize engineering attributes of a plain weave fabric.

Needless to say, the pioneering work of Peirce on the geometrical model provides a basic platform for fabric engineering. Many basic fabric dimensional parameters can be worked out from the geometrical model by use of modern computing methods. Fabric parameters for any relationship between warp and weft cover can be evaluated. Last but not the least, the relationship between fabric mass and fabric cover can also be evaluated. This is the strength of the computing method, as the same results cannot be achieved by any other means like graphical or nomogram methods. The basic equations derived from the geometrical model given in Eqs [8.1]–[8.7] are taken as the basis for working out many derived relationships. For example, the solution of p_2/D

and h_1/D is obtained for different values of θ_1 (warp crimp angle) ranging from 0.1–$\pi/2$ radians. Such a relationship is shown in Fig. 8.7. It is a very useful relationship between fabric parameters for engineering desired fabric constructions. One can see its utility for jammed structures, non-jammed fabrics and even for some special cases in which cross threads are straight. For jammed structures, Fig. 8.7 shows a non-linear relationship between the two fabric parameters p and h on the extreme left. In fact, this curve is for jamming in the warp direction. It can be seen that the jamming curve shows different values of p_2/D for increasing crimp. Interestingly, this curve is a part of circle and its equation is:

$$\left(\frac{p_2}{D}\right)^2 + \left(\frac{h_1}{D} - 1\right)^2 = 1 \qquad [8.8]$$

with centre at (0, 1) and radius equals to 1. For jamming in the warp direction of the fabric the parameters p_2/D and corresponding h_1/D can be obtained

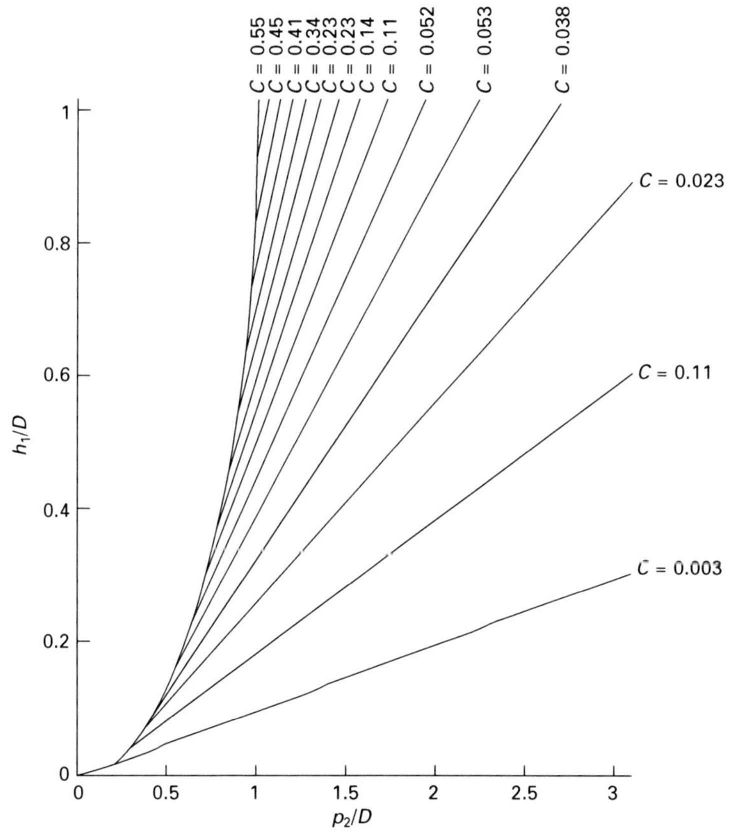

8.7 Relation between thread spacing and crimp height.

either from Fig. 8.7 or from the above equation. The relationship between the fabric parameters over the whole domain of the structure being jammed in both directions can be obtained by developing a suitable algorithm for computation using the following derived equations:

$$p_2 = D\sin\theta_1 = D\sin\frac{l_1}{D}$$

$$h_1 = D(1 - \cos\theta_1) = D\left(1 - \cos\frac{l_1}{D}\right) \qquad [8.9]$$

$$\frac{h_2}{D} = 1 - \frac{h_1}{D}$$

$$p_1 = D\sin\theta_2 = D\sin\frac{l_2}{D}$$

$$h_2 = D(1 - \cos\theta_2) = D\left(1 - \cos\frac{l_2}{D}\right) \qquad [8.10]$$

From the computation the relationships between different useful fabric parameters are obtained and are given in Figs 8.8–8.12. Figure 8.8 gives the relationship between p_2/D and p_1/D. This figure shows that the relationship between these parameters is less sensitive at the two extreme ends. The relationship is sensitive in a p/D range closer to 1. In fact this sensitive range corresponds to maximum crimp in one direction only. The scope for change

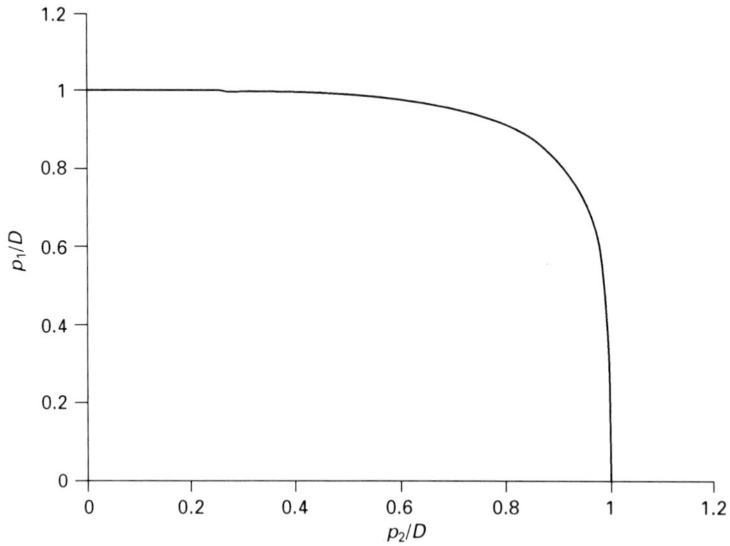

8.8 Relation between warp and weft thread spacing for jammed fabric.

in both p_2 and p_1 lies in the range 0.6 to 0.9. Beyond this range change in thread spacing is possible only in one direction.

A useful relationship between the crimps in the two directions is shown in Fig. 8.9. It indicates an inverse non-linear relationship between c_1 and c_2. The intercepts on the axis give maximum crimp values with zero crimp in the cross direction. This is a fabric configuration in which cross-threads are straight and all the bending is done by the intersecting threads. It may be concluded here that at either end a small change of crimp in one direction can permit a large change in crimp in the other direction.

Figure 8.10 shows the relation between h_1/p_2 and h_2/p_1. This shows linearity between them except at the two extremes. The behaviour is in fact the relationship between the square root of crimp in two directions and the constant is fixed for any jammed fabric. Other practical relations are obtained between warp and weft cover factor and between cloth cover factor and fabric mass (gsm).

Figure 8.11 gives the relation between warp and weft cover factor for different ratios of weft to warp yarn diameters (β). The relation between the cover factors in two directions is sensitive only in a narrow range for all values of β. The relations between the cover factors in two directions are interdependent for jammed structures. Maximum threads in the warp/weft direction depend on yarn count and weave. Maximum threads in one direction of the fabric will give unique maximum threads in the cross-directions. The change in the value of β causes a distinct shift in the curve. A comparatively

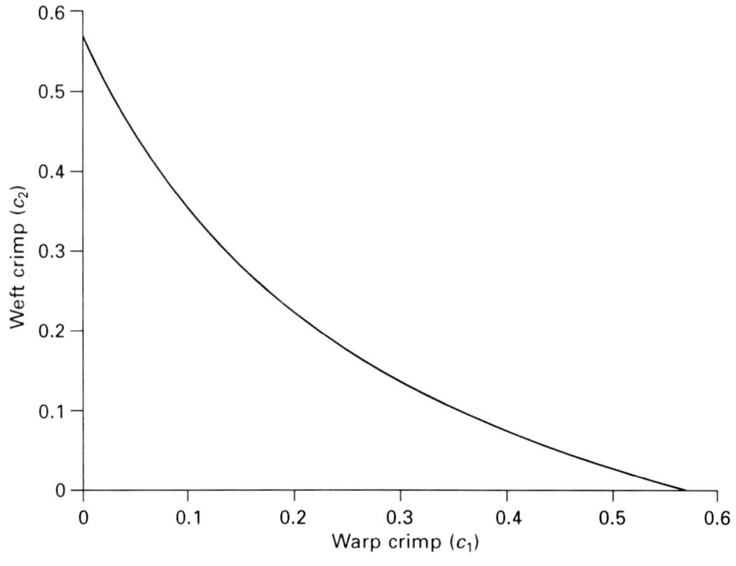

8.9 Relation between fraction warp and weft crimp for jammed fabric.

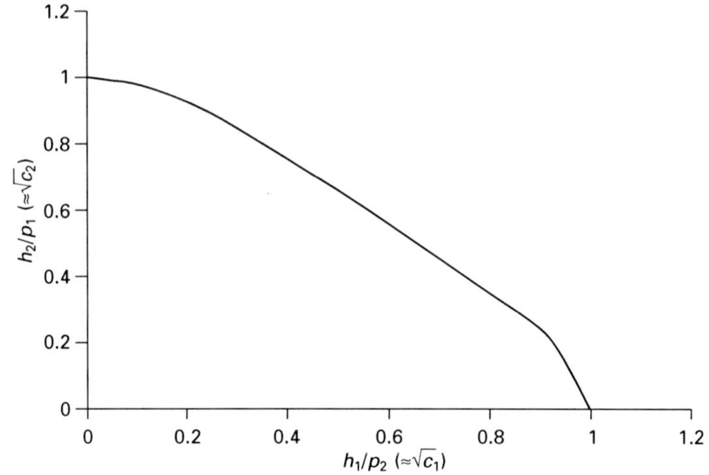

8.10 Relation between $\sqrt{c_1}$ and $\sqrt{c_2}$ for jammed fabric.

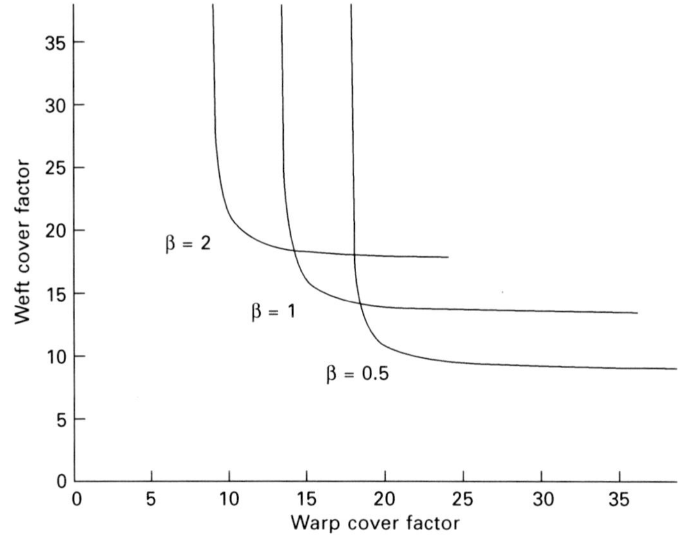

8.11 Relation between warp and weft cover factor for different β for jammed fabric.

coarse yarn in one direction with respect to the other direction helps in increasing the cover factor. For $\beta = 0.5$, the warp yarn is coarser than the weft, which causes an increase in warp cover factor and a decrease in weft cover factor. A similar effect can be noticed for $\beta = 2$, in which the weft yarn is coarser than the warp yarn.

The relation between fabric mass in gsm with the cloth cover ($K_1 + K_2$) is positively linear as shown in Fig. 8.12. The trend may appear to be self-

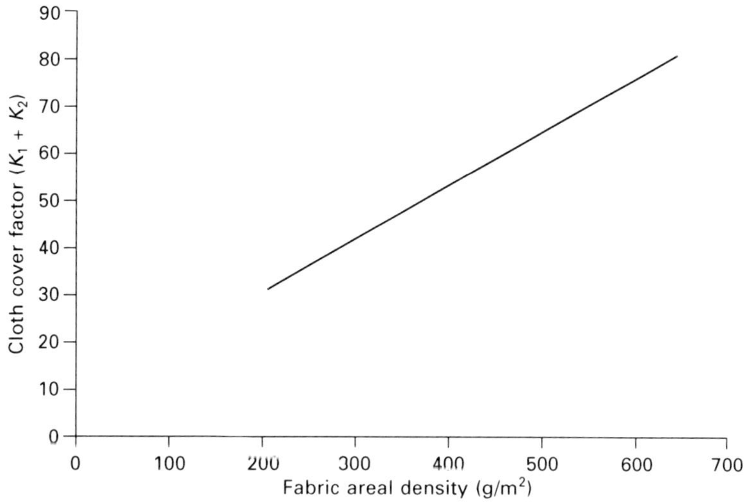

8.12 Relation between fabric cover factor and fabric mass for jammed structure.

explanatory. In practice, an increase in fabric mass and cloth cover factor for jammed fabrics can be achieved in several ways such as zero crimp in warp direction and maximum crimp in the weft direction, zero crimp in the weft direction and maximum crimp in the warp direction, equal crimp in the both directions and dissimilar crimp in both directions. This explanation can be understood by referring to the non-linear part of the curve in Fig. 8.7.

Similarly for a non-jammed structure, it can be seen that the relation between p_2/D and corresponding h_1/D is linear for different values of crimp. This relationship is useful for engineering non-jammed structures for a range of values of crimp. The fabric parameters can be calculated from the above non-jammed linear relation between p_2/D and h_1/D for any desired value of warp crimp. Then h_2/D can be obtained from $(1 - h_1/D)$ and for this value of h_2/D one can obtain the value of p_1/D for the desired value of weft crimp. Thus all fabric parameters can be obtained for desired value of p_2/D, picks/cm, warp and weft yarn tex, warp and weft crimp. One can choose any other four parameters to get all fabric parameters. The algorithm mentioned in Fig. 8.1 can be used to get several solutions for the non-jammed fabric.

Considering a fabric in which cross-threads are straight, it may further be realized from Fig. 8.7 that the intersection of horizontal line corresponding to $h_1/D = 1$ gives all possible structures ranging from relatively open to jammed configuration. In this case $h_2 = 0$, $h_1 = D$. The fabric designer gets the option to choose the possible fabric constructions. These options include jamming and other loose constructions. Using the above logic, it is also possible to get fabric parameters for a fabric jammed in both directions, a fabric with maximum crimp in one direction and cross-threads being straight and a fabric which is neither jammed nor has zero crimp in the cross-threads.

Applications of geometrical models

In the weaving and processing industry, computer and mathematical models of fabric can facilitate rapid development of required products without the need to produce many fabric samples. Geometrical modelling can be used as a tool to achieve desired functionality for industrial fabric production. The model and analysis system would make it possible to realistically model the fabric structure and predict its performance before it is used or even manufactured, thereby saving both time and energy. Examples of some properties for specific industrial applications are discussed below.

Weavability and maximum sett

This is helpful in realizing maximum areal density of the fabric and the suitability of the fabric for applications like waterproof fabrics and airproof fabrics. A very accurate value of end and pick densities can be determined, thereby reducing the extent of experimentation.

Fabric design and engineering

The fabric can be engineered and designed for specific applications by using the structural parameters such as crimp, thickness, cover factor, etc. An example is filter fabrics, where thickness of the fabric plays an important role.

Manipulation of the fabric shape and structure

This is possible in some specific applications like stentering and calendering (the geometrical model is very much applicable here because of the type of forces in action) in which, under the application of force, the width or length of the fabric is changed. The final length and width of the fabric can be predicted using the computation method embedded in the algorithm. It may be mentioned here that the geometrical model can predict only the structural parameters of the fabric and therefore it is of very little use for prediction of mechanical properties of the fabric.

8.5.2 Mathematical models

Mathematical models are derived from first principles. These models are very appealing because they have their basis in applied physics. They can be used to explain the reasons that determine structure–property relationships.

The main purpose for carrying out these structural analyses is to provide tools to enable the designer to design textile structures to meet various end use specifications.

Limitations of mathematical model

Mathematical modelling has certain limitations. The development of theory is cumbersome and requires many years to yield results. The models are normally problem specific and any change in the system requires a new analysis and new programs to solve equations. It often produces large prediction error and the procedures are not user friendly.

8.5.3 Empirical modelling

The most common approach has been to develop predictive models for fabric properties and performance through experimental investigations. In this case, a large number of experiments are made under controlled conditions and statistical techniques are used to derive empirical models. Data generated from experimental results are used to develop regression models to predict desired property parameters. The empirical equations are based on linear multiple regression. In the statistical method of modelling, multilinear regression equations between fabric property and constructional parameters are applied to the same data, to find the predictability of the statistical method. The coefficient of multiple determination (R^2) which defines the fraction of variability in the dependent variable explained by the regression model is taken as a guiding parameter. If the model's R^2 value is high, it suggests that the empirical model fits the data reasonably well.

Limitations of empirical modelling

Empirical models can have limited applications for two reasons. First, because the size of the experiments is generally limited due to cost and time factors, the selection of materials and processes can be considered in a narrow range only. Secondly, the existing tools and techniques are inadequate for accurately modelling and optimizing complex non-linear processes like woven fabric manufacturing. Hence, there is a need for models that can accurately predict process and product design for woven fabrics.

8.5.4 Finite element modelling (FEM)

The finite element method is a powerful tool for the numerical solution of a wide range of engineering problems. Application of FEM ranges from deformation and stress analysis of automotive, aircraft, building and bridge structures to field analysis of heat flux, fluid flow, magnetic flux and other flow problems. In recent years, finite element analysis has also been applied to flexible materials like textile fabric for modelling and simulation. In this method of analysis, a complex region defining a continuum is discretized into simple geometric shapes called finite elements. The material properties and the governing relationships are considered over these elements and expressed in terms of unknown values at element corners. An assembly process, duly considering the loading and constraints, results in a set of equations. Solution of these equations gives the approximate behaviour of the continuum.

FEM is considered useful for textiles because the complexity of the structure, the anisotropic properties of yarns and fabrics, and the interaction phenomena between yarns at the contact areas preclude the use of analytical methods. With FEM-based calculation models, an in-depth understanding of the physical phenomena can be obtained, and it is possible to change important parameters in a short time in order to improve textile-based constructions and also to generate new products [42]. From the structural point of view, woven fabrics are discontinuous in microstructure and therefore they do not satisfy the continuity requirement in solid mechanics. However, discontinuity in the fabric is small in comparison to the finite element mesh size, and therefore fabric continuity is a reasonable assumption to make. Several research works have been reported in recent years on the application of FEM analysis for prediction of fabric mechanical properties which can be used for engineering of woven construction. Lloyd [43] has applied FEM to study fabric in-plane deformation using membrane elements with no bending resistance. The fabric deformation considering shell plate elements has been studied by Gan and Ly [44] using nonlinear finite elements. In this study, fabrics are assumed to be linearly elastic and orthotropic. Constitutive relation of fabrics was determined and used in shell elements. Two-dimensional bending of fabrics was studied and the results were compared with the experimental values obtained by cantilever bending test. Bending hysteresis was not accounted for by the finite element analysis.

Jeong and Kang [45] developed a computer model for analysing the compressional behaviour of a woven fabric using FEM. A three-dimensional unit cell was defined to describe the compressional deformation of fabric in the three-dimensional viewpoint. The geometry of unit cell is determined using the equilibrium requirements of planar-elastica-boundary value problems. The yarns in the fabric were assumed to be elastic and isotropic. The contact conditions at the yarn crossover point were determined by the equilibrium equations. Two kinds of second-order isoparametric solid elements were used to describe warp and weft yarns. The total number of elements and nodes for the modelling of the three-dimensional geometry were 876 and 5,426 respectively. The analysis was carried out with the aid of ABAQUS FEM package. The result showed that the compressional resistance of fabric is influenced by the geometrical structure of the fabric unit cell as well as the yarn property. Also, the compressional resistance increased with Poisson's ratio [46].

8.6 Modelling methodologies: non-deterministic models

Models based on genetic methods, neural networks, chaos theory and soft logic are non-deterministic models and known as soft computing methods according to convention for predicting properties and performance. Soft computing is a collection of methodologies, which aim to exploit tolerance for imprecision, uncertainty, partial truth and approximation to achieve tractability, robustness and low solution cost. It differs from conventional (hard) computing in that, unlike hard computing, it is tolerant of imprecision, uncertainty, partial truth and approximation. The soul of soft computing is to make the computer as soft as the human brain, and is capable of carrying out both quantitative and qualitative computing. Along this direction, fuzzy logic has been developed to handle qualitative information, and in this sense it is called fuzzy computing. A neural network is a kind of soft computing technology as it provides a relatively easy way to acquire information about a system through learning. The inclusion of neural computing and genetic computing in soft computing came at a later stage. There are a number of publications in this field of computing covering a wide range of application domains including that of textiles. Notable ones are a book on soft computing in textile sciences by Sztandera and Pastore [47], a textile institute monograph on artificial neural network applications in textiles by Chattopadhyay and Guha [48] and research work carried out by Behera and Muttagi [77–83] on development of ANN embedded expert system for woven fabric engineering.

8.6.1 Fuzzy logic

Knowledge representation and processing are the keys to any intelligent system. In logic, knowledge is represented by propositions and is processed through reasoning by the application of various laws of logic, including an appropriate rule of inference. Fuzzy logic uses the fuzzy set theory and approximate reasoning to deal with imprecision and ambiguity in decision making. A crisp set is defined by the characteristic function that can assume only the two values {0, 1}, whereas a fuzzy set is defined by a 'membership function' that can assume an infinite number of values; any real number in the closed interval [0, 1]. The idea of fuzzy logic was introduced by Lofti A. Zadeh in 1965 in his paper on fuzzy sets [49].

Consider a universe of discourse X with x representing its generic element. A fuzzy set $\underset{\sim}{A}$ in X has the membership function $\mu_{\underset{\sim}{A}}(x)$ which maps the elements of the universe onto numerical values in the interval [0, 1]:

$$\mu_{\underset{\sim}{A}}(x): X \rightarrow [0, 1] \qquad [8.11]$$

Every element x in X has a membership function $\mu_{\underset{\sim}{A}}(x) \in [0, 1]$. $\underset{\sim}{A}$ is then defined by the set of ordered pairs:

$$\underset{\sim}{A} = \{(x, \mu_{\underset{\sim}{A}}(x)) \mid x \in X, \mu_{\underset{\sim}{A}}(x) \in [0, 1]\} \qquad [8.12]$$

A membership value of zero implies that the corresponding element is definitely not an element of the fuzzy set $\underset{\sim}{A}$. A membership value of unity means that the corresponding element is definitely an element of fuzzy set $\underset{\sim}{A}$. A grade of membership greater than zero and less than unity corresponds to a non-crisp (or fuzzy) membership of the fuzzy set $\underset{\sim}{A}$. Classical sets can be considered as special cases of fuzzy sets with all membership grades equal to unity. Membership functions characterize the fuzziness in a fuzzy set [50].

Models designed based on fuzzy logic, usually consist of a number of fuzzy if-then rules expressing the relationship between inputs and desired output. For instance, in the case of two-input single-output systems, it is expressed as:

$$R_i\text{: IF } x \text{ is } A_i \text{ and } y \text{ is } B_i \text{ THEN } z \text{ is } C_i \qquad [8.13]$$

where R_i is a fuzzy relation representing the ith fuzzy rule; x, y, z are linguistic variables representing two inputs and the output; and A_i, B_i, C_i are linguistic values of x, y, z, respectively. In these models, inputs are fuzzified, membership functions are created, association between inputs and outputs are defined in a fuzzy rule base, and fuzzy outputs are restated as crisp values.

Fuzzy modelling can be categorized into two groups: subjective modelling, and objective modelling. In the subjective modelling approach, it is assumed that *a priori* knowledge about the system is available and that this knowledge can be directly solicited from experts. By contrast, in objective modelling, it is assumed that either there is no priori knowledge about the system, or the expert's knowledge is not trustworthy enough. Therefore, instead of any *a priori* interpretation of the system, raw input and output data is used to augment human knowledge or even generate new knowledge about the system. This approach was initially proposed by Takagi-Sugeno-Kang (1985) and called TSK fuzzy modeling.

Applications of fuzzy logic

Fuzzy logic has been used in several areas of textiles which include colour grading of cotton into different classes, prediction of tensile strength and yarn count of melt spun fibres, an intelligent diagnosis system for fabric inspection and automatic recognition of fabric weave pattern. Raheel and Liu [51] used a fuzzy comprehensive evaluation technique to calculate fabric handle of lightweight dress fabrics. Thickness, weight, flexural rigidity, wrinkle recovery and 45° filling elongation were used to describe the hand of lightweight fabrics. In order to obtain the fuzzy transformation *R*, five membership functions corresponding to the five properties were selected. A decreasing half Cauchy distribution was used to describe the membership degrees of fabric weight, fabric thickness, fabric flexural rigidity and 45° filling elongation. For wrinkle recovery, a linear membership function was used. From a survey of judges, the importance of each property selected was ascertained and expressed as a weighted vector. By using the weighted vector and the fuzzy transformation matrix, fabric handle was calculated. The same approach was followed by Park and Hwang [52] for predicting total hand value from selected mechanical properties of double weft knitted fabrics and by Chen *et al.* [53] for grading softness of 100% cotton and cotton/polyester blended fabrics. Huang and Yu [54] investigated the use of fuzzy logic controller for controlling concentration, pH and temperature in the dyeing process.

8.6.2 Neural networks

The term neural network is used to describe a number of different models intended to imitate some of the functions of the human brain, using certain of its basic structures. The development and use of neural networks are part of an area of multidisciplinary study that is commonly called neural computing, but is also known as connectionism, parallel distributed processing and computational neuroscience. ANN is a powerful data modelling tool that is able to capture and represent each kind of input–output relationship. A neural network is composed of simple elements called neuron or processing element operating parallel, which are inspired by biological neuronal systems. As in nature, the network function is determined largely by weighted connections between the processing elements. The weights of the connections contain the knowledge of the network.

There are different types of neural networks. An important distinction can be made between supervised and unsupervised learning. In supervised learning, the network is presented with pairs of input and output data. For each set of input values there is a matched set of output data. The key point is that if the desired output of the network is known for each set of input values, then the weights of the network can be modified accordingly. Supervised neural networks are typically used to solve what is known as function approximation, using examples of the function in the form of input–output pairs. In unsupervised learning, the output is not known; the network is simply presented with input data.

Neural networks that are based on the multi-layer perceptron (MLP) account for approximately 80% of all practical applications. In an MLP, the units are arranged in distinct layers and each unit receives weighted input from each unit in the previous layer. A neural network is usually adjusted or trained so that a particular input leads to a specific output. The process of training is adjusting these weight values and sliding down the error surface. Among the various kinds of algorithms for training neural networks, back propagation is the most widely used.

Artificial neural networks produce the least error, as well as lower spread in the error, compared with mathematical and regression methods of modelling. ANN also has the excellent property of approximating any functional relationship between large numbers of input–output (independent variables–dependent variables) parameters. No prior assumptions are required to be made on the statistical nature of the variables of the data, since ANN are non-parametric in nature. ANN requires a much smaller dataset than that required for conventional regression analysis for capturing the non-linear relationships between the input and output parameters. Even with a small training dataset, the network generalizes the functional relationships very well. In the industry, where large data is continuously available, ANN can be expected to perform significantly better. The major advantage of ANN is that there is no restriction on the levels of interaction between the variables; therefore, it can capture the dynamics of the real world very well.

Back propagation algorithm

The input vectors X for the input layer are expressed in the vector form as $X = (x_1, x_2 \ldots . x_n)$. Predicted parameters (network outputs) are denoted as Y. As shown in Eq. [8.14], the ith component of the input signal x_i comes out from the unit i and is transferred to the unit j of the model through the synapse weight W_j, where b_j is the bias term connected to the jth unit.

$$u_j = \left(\sum_{i=1}^{n} W_{ji} x_i + b_j \right) \qquad [8.14]$$

The unit j non-linearly transforms the total input u_j (Eq. [8.14]) by means of a transfer function (hyperbolic tangent transfer function used here), which is propagated forward to the unit of the next layer as the input signal y_i (Eq. [8.15]):

$$y_j = \frac{2}{1 + e^{(-2u_j)}} - 1 \qquad [8.15]$$

The difference in the output y_i from the target output t_j is used to adjust the synapse weights according to the calculated mean squared error (MSE), as shown in Eq. [8.16]:

$$MSE = \frac{\sum_{j=0}^{p} \sum_{i=0}^{N} (t_{ij} - y_{ij})^2}{NP} \qquad [8.16]$$

were y_{ij} is the network output for data set i at neuron j, t_{ij} is the target network output for data set i at neuron j, P indicates the number of output neurons, and N refers to the number of datasets. The ANN is trained by updating the weights with a back propagation rule. The change in synapse weight w_{ji} is based on the gradient descent rule according to Eq. [8.17]:

$$\Delta w_{ji} = -\eta \frac{\partial(MSE)}{\partial w_{ji}} \qquad [8.17]$$

where η is the learning rate.

Applications of ANN

Image processing analysis and neural networks have been widely used for fabric defect detection. The basic principles underlying this technique along with numerous applications are detailed by Behera [84]. Lin [55] used feed-forward back propagation neural nets to find the relationships between the shrinkage of yarns and the cover factors of yarns and fabrics. A typical multilayer feed-forward network is shown in Fig. 8.13. Beltran *et al.* [56] also studied the use of MLP-BP neural networks to model the multi-linear relationships between fibre, yarn and fabric properties and their effect on the pilling propensity of pure wool knitted fabrics. Behera and Muttagi [57]

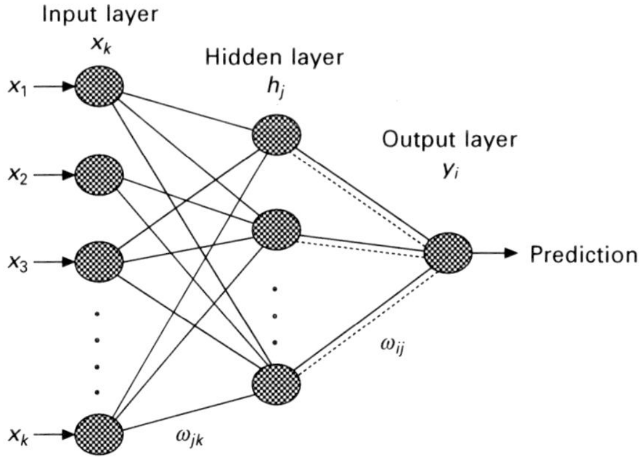

8.13 Multilayer feed-forward network.

predicted the low-stress mechanical, dimensional, and tensile properties of woven suiting fabrics using back propagation network (BPN) and radial basis function neural network (RBFN). Fibre, yarn and fabric constructional parameters of wool and wool-polyester blended fabrics were given as input variables. Radial basis function neural networks were found to have better predictability and are faster to train and easier to design than back propagation neural networks. A reverse engineering approach is also reported for prediction of constructional particulars from the fabric properties.

Hui *et al.* [58] predicted sensory fabric hand from fabric properties using a resilient back propagation neural network (RBP). Shyr *et al.* [59, 60] studied the use of neural networks for discriminating generic hand of cotton, linen, wool and silk woven fabrics. They established translational equations for the total hand value of fabrics using back propagation nets. Wong *et al.* [61] investigated the predictability of clothing sensory comfort from psychological perceptions by using a feed-forward back propagation network. ANN based prediction of fabric appearance index by Behera and Mishra [85, 86] can be used as an objective method of fabric engineering to achieve desired aesthetic performance. This work provides measurement of an integrated fabric appearance index given in Eq. [8.18] using image processing and neural network computation method. Fabric appearance index is termed as FAI and given by:

$$FAI = \sum_{i=1}^{n} A_i W_i \qquad [8.18]$$

where n is total number of properties, A_i is grade of the ith property obtained by digital image processing and W_i is weighting of the ith property. The properties such as drape, texture, wrinkle and pilling are used to access the aesthetic appearance of an apparel fabric.

8.6.3 Genetic algorithms

A relatively new area of study in artificial intelligence is that of genetic algorithms (GAs). GAs are a powerful set of stochastic global search techniques that have been shown to produce very good results for a wide class of problems. GAs can find good solutions to non-linear problems by simultaneously exploring multiple regions of the solution space and exponentially exploiting promising areas through mutation, crossover and selection operations [62]. GAs are programs that attempt to find optimal solutions to problems when one can specify the criteria that can be used to evaluate the optimal solution. They are useful when a problem has multiple solutions, some of which are better than others. Unlike deterministic, linear and non-linear optimization models, GAs test a variety of solutions and, through an evolving process, attempt to find the best solution through processes that parallel the metaphors of survival of the fittest, genetic crossover, mutation and natural selection. Evolutionary algorithms differ substantially from more traditional search and optimization methods. The most significant differences are:

- GAs work with a coding of the parameter set, not the parameter themselves.
- GAs search from a population of designs, not a single design.
- GAs use objective function information, not derivatives or other auxiliary knowledge.
- GAs use probabilistic transition from design to design; they do not use deterministic rules.
- Evolutionary algorithms can provide a number of potential solutions to a given problem. The final choice is left to the user.

In building a genetic algorithm, six fundamental issues that affect the performance of the GA must be addressed: chromosome representation, genetic operators, selection strategy, initialization of the population, termination criteria and evaluation measures. The initial population is randomly generated, which is the most common method, while the GA is run for a specified number of generations as its termination criterion. For any GA, a chromosome representation is needed to describe each individual in the population of interest. The representation scheme determines how the problem is structured in the GA and also determines the genetic operators that are used. The operators are used to create new solutions based on existing solutions in the population. There are two basic types of operators: crossover (recombination) and mutation. Mutation operators tend to make small random changes in one parent to form one child in an attempt to explore all regions of the state space. Mutation serves the crucial role of preventing the system from being stuck in the local optimum. Crossover operators combine information from two parents to form two offspring such that the two children contain a 'likeness' (a set of building blocks) from each parent. The application of these two basic types of operators and their derivatives depends on the chromosome representation used. The selection of individuals to produce successive generations plays an extremely important role in a genetic algorithm. A probabilistic selection is performed based upon the individual's fitness such that the better individuals have an increased chance of being selected. However, all of the individuals in the population have a chance of being selected to reproduce into the next generation. Evaluation functions of many forms can be used in a GA, subject to the minimal requirement that the function can map the population into a totally ordered set. The evaluation function is independent of the GA (i.e. stochastic decision rules) [63].

The basic procedure of genetic algorithms can be explained as follows:

Let P(g) and C(g) be parents and offspring respectively in the existing generation g;
Procedure
for
g: = 0;
initialize population P(g);
evaluate P(g);
for
recombine P(g) to generate C(g);
evaluate C(g);
select P(g + 1) from P(g) and C(g);
g: = g + 1
end

Applications of genetic algorithms

Genetic algorithms are being used to solve a wide variety of problems in textiles right from production of fibres to apparel design and manufacturing. Amin *et al.* [64] reported detection of the spinning fault source from spectrograms by using genetic algorithms. Blaga and Draghici [65] reported the application of genetic algorithms in knitting technology. Lin [66] investigated the use of GA for searching weaving parameters for woven fabrics. A search mechanism was developed to find the best combinations of warp and weft counts and yarn densities for fabric manufacturing, taking into account costs. This helps the designer to select appropriate combinations of these parameters to achieve the required weight of fabric at a pre-controlled cost. Grundler and Rolich [67] developed an evolutionary algorithm based software for creating different weave patterns. Only the weave and yarn colour were considered as attributes for fabric appearance and different patterns can be created by various combination of weave and colour of warp and weft threads. Jasper *et al.* [63] investigated fabric defect detection using a genetic algorithm tuned wavelet filter. Hui *et al.* [68] studied the application of GA on the roll planning of fabric spreading in apparel manufacturing. It was demonstrated that the use of GAs to optimize roll planning will result in reduced wastage in cutting and hence can reduce cost of apparel production. Chan *et al.* [69] investigated the problem of handling the assembly line balancing in the clothing industry. Inui [70] presented a computer aided system using GA applied to apparel design. In this study, a computer technique that helps consumers to take part in apparel design was presented. An interactive computer-aided system for designing was constructed on the basis of the GA search method.

8.6.4 Hybrid modelling

A trend that is growing in visibility relates to the use of fuzzy logic in combination with neurocomputing and genetic algorithms as shown in Fig. 8.14. The marriage of fuzzy logic with computational neural networks has a sound technical basis, because these two approaches generally attack the design of intelligent systems from quite different angles. Neural networks are essentially low level computational algorithms that offer good performance in dealing with large quantities of data often required in pattern recognition and control. Fuzzy methods often deal with issues such as reasoning on a higher (i.e., on a semantic or linguistic) level than do neural networks. Consequently, the two technologies often complement each other: neural networks supply the brute force necessary to accommodate and interpret large amounts of data and fuzzy logic provides a structural framework that utilizes and exploits these low level results. Neural networks (NNs) are known for their ability to perform complex, non-linear mapping of input–output data. However, it is difficult to decide which input data, network structure and learning parameters to utilize. Genetic algorithms can be applied as an optimization search to determine the optimal in neural network structure design including input data combination optimization, network structure optimization, learning rate and momentum optimization. In this way, computational complexity and the time required to design the NN are reduced [71, 72].

Applications of hybrid modeling

Hybrid modelling has been used in several cases of prediction of fibre, yarn and fabric properties. The prediction accuracy of the neuro-fuzzy system has been found superior to that of a conventional multiple regression model and

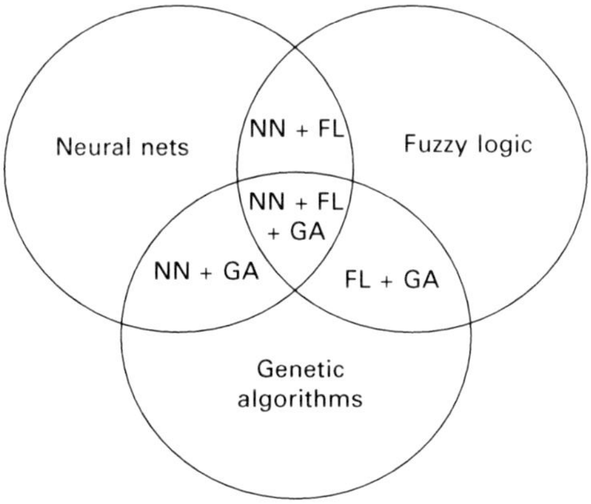

8.14 Hybrid models using soft computing tools.

comparable with an artificial neural network model. Wong *et al.* [73] predicted clothing sensory comfort from fabric physical properties by building eight different hybrid models combining statistics, fuzzy logic and neural network methodologies. Results showed that the TS-TS-NN-FL model has the highest ability to predict overall comfort performance followed by the TS-TS-NN-NN model (TS, NN and FL refer to statistical, neural network, and fuzzy logic, respectively). The data reduction and information summation ability of statistics, self-learning ability of neural nets and fuzzy reasoning ability of fuzzy logic were exploited to develop these hybrid models.

8.7 Authentication and testing of models

When building mathematical models, one has to use numbers derived from experimental or empirical data, or from analytical or computer-based calculations. Errors are thus produced due to data reading or data manipulation. Normally researchers use statistics to deal with error. Error is defined as the difference between a measured or calculated value and its true or exact value. Error is unavoidable but how much error is present depends on how skillfully the data are read or manipulated. Therefore, error analysis is an integral part of the modelling process. There are two types of error: systematic error and random error. Systematic error occurs when an observed or calculated value deviates from the true value in a consistent way. This error occurs in experiments when instruments are improperly calibrated. Random errors do not occur due to chance. They arise mainly in experimental work because unpredictable things happen due to ignorance or by accident. There is one absolute error which is defined as the difference between the true or experimental value and the measured value. The true value may be known or it may have an expected value based on a calculation or some other data source. For example, in case of a neural network modelling of woven fabric engineering, the ability of the model to predict the fabric structure–property relationships can be judged from percentage error in the prediction of fabric properties for a given set of input fibre, yarn and fabric constructional parameters. In such cases, the network is trained on fabric constructional parameters such as weave float, thread density, thread resultant tex, yarn crimp, fabric mass and fabric cover as inputs to the networks, and fabric shrinkage, hygral expansion, extension, bending rigidity, formability, shear rigidity, thickness, breaking strength and elongation as network outputs. Table 8.2 shows the prediction error in percentage as the average of test dataset for each fabric property predicted by the neural network. The overall absolute prediction error is 13.02%.

$$\text{Error \%} = \left(\frac{P - E}{E}\right) \times 100$$

Table 8.2 Prediction performance of the radial basis function neural network

Fabric properties	Absolute prediction error (%)
Warp extension (%)	12.3
Weft extension (%)	13.2
Warp bending rigidity (μN.m)	14.8
Weft bending rigidity (μN.m)	13.9
Warp formability (mm^2)	14.9
Weft formability (mm^2)	14.3
Shear rigidity (N/m)	12.6
Warp-way breaking strength (N)	10.3
Weft-way breaking strength (N)	4.1
Warp-way breaking elongation (%)	16.2
Weft-way breaking elongation (%)	16.5
Average prediction error (%)	13.02

where Error % = Absolute error %, P = predicted value by the network and E = experimental value.

The ability of the network to model the fabric structure–property relationships can best be judged from the network's response to the influence of constructional parameters on fabric properties. This can be further authenticated by plotting the relationship between actual values and predicted values of fabric properties. Figure 8.15 shows the correlation between actual and predicted values by the network which demonstrates a very good correlation between the actual and the predicted values ($r = 0.985$) in this example.

8.8 Reverse engineering

Reverse engineering is the process of discovering the technological principles of a device, object or system through analysis of its structure, function and operation. It often involves taking something apart and analysing its workings in detail; used in maintenance or making a new device or program that does the same thing without copying anything from the original [4]. The purpose is to deduce design decisions from end products with little or no additional knowledge about the procedures involved in the original production. Reverse engineering is carried out under several circumstances like lost documentation (the documentation of a particular device has been lost or was never written and the person who built it is no longer available), product analysis (to examine how a product works, what components it consists of, estimate costs, identify potential), security auditing, removal of copy protection, circumvention of access restrictions, creation of unlicensed and unapproved duplicates and also for academic-learning purposes. Reverse engineering is taking apart an object to see how it works in order to duplicate or enhance the object. Someone doing reverse engineering on software may use several

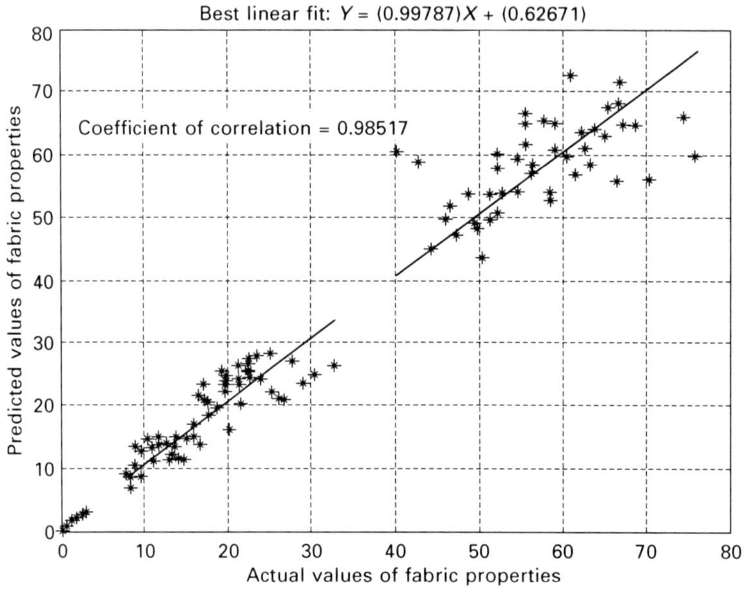

8.15 Actual values of fabric properties vs values predicted by neural network.

tools to disassemble a program. The term forward engineering is sometimes used in contrast to reverse engineering in various literatures.

8.9 Future trends in non-conventional methods of design engineering

8.9.1 Knowledge-based systems

The engineering design of woven textile structures is a complex task and makes extensive use of empirical knowledge accumulated over time. Only few, highly experienced experts in the industry know rule-of-thumb principles and methods of complex structural synthesis under performance constraints. Therefore, if the knowledge and expertise of the experts in the field and from the literature is used to develop a scientific database, then computer software can be prepared with such acquired knowledge, which in turn will provide expert advice on the problem. Such a knowledge-based system is called an expert system. It is a system that has been engineered to simulate a human expert. They are well suited to database manipulations, symbolic reasoning and decision making. They have conflict resolution principles so that mutually antagonist requirements are satisfied to the greatest extent possible and with minimal compromise. Expert systems permit the use of different design knowledge representational modes such as heuristic, procedural and factual knowledge and the reasoning methods working with each of these representational modes.

The technology of artificial intelligence (AI) and expert system (ES) enables computers to be applied to less deterministic design tasks, which require symbolic manipulation and reasoning, instead of routine number processing. ES can be used to diagnose, repair, monitor, analyse, interpret, consult, plan, design, instruct, explain, learn and conceptualize problems. ES is a powerful tool for solving problems like design, which is used for making decisions. Designing is typically a decision-making process, so the application of knowledge engineering and ES is suitable for it. ES is very suitable for database manipulation and decision making. The modularity of ES enables the changes or modifications to be accommodated easily by changing or adding merely the facts and rules contained in the knowledge [74–76].

In order to construct a successful ES, the problem domain must be well defined, and there must be at least one human expert acknowledged to perform well within the application area; however, additional knowledge relevant to the problem domain can be obtained from other sources. Experts must have special knowledge, judgement and experience. Also, the expert must be able to explain and justify the special knowledge and experience and the methods used to apply them to particular problems. Further, there must also be adequate programming tools, ideally a set of ES software building tools; specific high-level programming languages such as LISP and PROLOG are often considered to be languages of ES development.

8.9.2 Basic structure of expert system

The basic structure of an expert system is shown in the Fig. 8.16. The various components of ES are as follows.

1. The knowledge base consists of the domain facts and heuristics associated with the problem area, which must be well bounded and narrow.
2. The inference engine is the control structure or strategy for utilizing the knowledge base in the solution to the problem.
3. The dynamic and global database is the working memory for keeping track of the problem status, the input data for the particular problem and the relevant history of what has been done.
4. A user friendly interface to facilitate interaction of the system with the

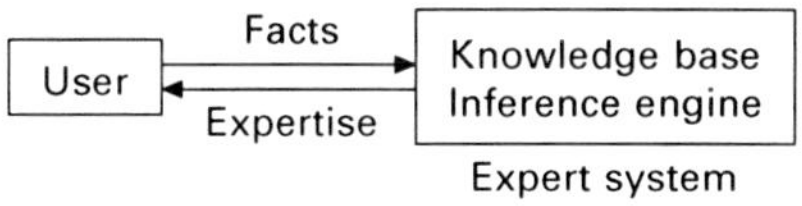

8.16 Basic structure of expert system.

user and provide a human window to its operations, preferably a natural language framework; additionally, an explanation module should be included to allow the user to raise queries and to help understand the reasoning process underlying the system's answers.

5. The knowledge acquisition module facilitates the transfer and transformation of problem-solving expertise from the knowledge source to the knowledge base.

Since knowledge is central to intelligence, the performance of ES is primarily a function of the size and quality of the knowledge base it possesses. The facts constitute a body of information that is widely shared, publicly available, and generally agreed upon by experts in the field. The heuristics are mostly private, little discussed rules of good judgement or guesswork that characterize expert-level decision making in the field. The knowledge must be represented or organized into a suitable form so that the inference engine can readily access it. In order to create a knowledge base for a given problem, the expert's knowledge must be formulated according to the knowledge representation scheme employed by the ES shell. The latter has a debugging facility and represents the knowledge acquisition module. The difference between ES and conventional computer programming can be understood from the following representation:

$$\text{Data} + \text{Algorithm} \rightarrow \text{Program}$$

$$\text{Knowledge} + \text{Inference} \rightarrow \text{ES.}$$

8.10 Conclusions

With globalization, there is increased need to reduce product lead times. Activities must be performed in parallel (i.e., integrated product development), to ensure that sufficient attention is paid to market needs and manufacturing technologies during the design process for successful product development. Designing of textile products is still based on traditional techniques, experience and intuition. This leads to (a) a limited number of experts and their expertise; (b) difficulty in finding a systematic approach for an optimum solution; and (c) increased time and costs. Compared to modelling from first principles and other techniques, computation methods can be a powerful tool to model the non-linearities and complexities involved in predictions of fabric properties. The system has to be developed to provide scientific databases, overall structure–function relationship, optimization procedures, suitable computer algorithm and standardization of these algorithms. The successful applications of soft computing in various applications indicate that the impact of soft computing will be felt increasingly in coming years. The employment of soft computing techniques leads to systems which have high MIQ (machine intelligence quotient).

8.11 References

1. Newton A (1993), *Fabric Manufacture: A Handbook*, Intermediate Technology Publications, London.
2. Robinson A T C and Marks R (1973), *Woven Cloth Construction*, The Textile Institute, Manchester.
3. Peirce F T (1937), The geometry of cloth structure. *J Text Inst*, 28 (3).
4. Muttagi S (2002), ANN embedded expert system for design of woven fabrics, Ph.D. Thesis, IIT Delhi.
5. Dastoor P H, Hersh H P, Batra S K and Rasdorf W J (1994), Computer-assisted structural design of industrial woven fabrics, Part I: Need, scope, background, and system architecture, *J Text Inst* 85 (2), 85–109.
6. Hearle J W S (1994), Fabric mechanics as design tool, *Ind J Fibre Text Res*, 19, 107–113.
7. Yoshida N *et al.* (1993), Technical Report of Fukui Prefecture Technological Center, Japan, No. 9, 55.
8. Maejima H (1995), Technical Report of Yamanashi Prefecture Technological Center, Japan, 39.
9. Ogami M and Nakamura S (1991), *Text Manufac*, 43 (2), 74.
10. Sakurai T, Kanemaru K and Kono M (1991), *Text Manufac*, 43 (2), 77.
11. Miller L (1984), in *Proc Text Inst Annual World Conf*, Hong Kong, 635.
12. Kokonoki T and Tanaka T (1985), Technical Report of Tochigi Prefecture Text. Ind. Lab., Japan, 75.
13. Takaoka T, Kugiya T and Kosago F (1993), Technical Report of Yamaguchi Prefecture Technological Center, Japan, No. 5, 55.
14. Della A V and Schettini R (1993), *Can Text J*, 110 (3), 27.
15. Rich D C (1986), *Text Chem Color*, 18 (6), 16.
16. Rich D C (1985), in *Proc Natl Tech Conf of Am Assoc Text Chem Color*, USA, 36.
17. Matsuda H, Nagata Y and Arizono Y (1986), *Fuji Elect J*, 59 (9), 626.
18. Hann M A and Thomson G M (1992), *Text Prog*, 22 (1).
19. Osek M and Koskova B (1984), in *Proc Text Inst Annual World Conf*, Hong Kong, 47.
20. Murai M *et al.* (1986), Technical Report of Tokyo Text. Ind. Lab., Japan, No. 34, 27.
21. Kawarai M *et al.* (1984), Technical Report of Tokyo Text. Ind. Lab., Japan, No. 33, 43.
22. Mukai T (1987), Technical Report of Yamagata Prefecture Technological Center, Japan, No. 19, 45.
23. Mukai T (1990), Technical Report of Yamagata Prefecture Technological Center, Japan, No. 22, 37.
24. Shouzhen Cao (1984), *J CTEA China*, 5 (3), 29.
25. Taylor B (1989), *Text Asia*, 20 (11), 85.
26. Shouzhen Cao (1984), in *Proc Text Inst Annual World Conf*, Hong Kong, 709.
27. Hihuchi Y *et al.* (1987), *Jap Women's Univ J*, 34 (34), 71.
28. Ohta K, Ishii T and Nakata A (1994), *Text Mach Soc Japan*, 40 (5), T103.
29. Kurihara K and Kato K (1990), Technical Report of Tochigi Prefecture Text. Ind. Lab., Japan, 248.
30. Kaunis R (1986), 'Pattern processing systems for dobby weaving – a review'. *Int Text Bull, Fabric Forming*, 32, 4/86, 4.

31. Nakano K *et al.* (1989), Technical Report of Fukui Prefecture Technological Center, Japan, No. 5, 48.
32. Nakano K *et al.* (1993), Technical Report of Fukui Prefecture Technological Center, Japan, No. 3, 108.
33. Nakano K *et al.* (1991), Technical Report of Fukui Prefecture Technological Center, Japan, No. 7, 28.
34. Aldrich W (1994), *Matric Pattern Cutting*, 3rd edn, Blackwell Scientific Publications, London.
35. Suresh M N (1995), in *Conf Proc 24th Text Res Symp*, Mt. Fuji, Japan, 127.
36. Matsuo T (1993), *J Text Mach Soc Japan* (English edn), 39 (4), 73.
37. Hearle J W S (1993), *Text Horiz*, 13 (5), TH15.
38. Dubrovski P D and Brezocnik M (2002), *Text Res J*, 72, 187.
39. Haykin S (1994), *Neural Networks: A Comprehensive Foundation*, Macmillan Publishing, New York.
40. Jang J S R, Sun C T and Mizutani E (1997), *Neuro-Fuzzy and Soft Computing*, Prentice Hall of India, New Delhi.
41. Goldberg D E (1989), *Genetic Algorithms in Search, Optimization and Machine Learning*, Addison-Wesley, Boston, MA.
42. Somodi Z, Hursa A and Rogale D (2003), Numerical simulation of textile flexibility testing, *Int J Cloth Sci Technol*, 15 (3/4), 276–283.
43. Lloyd D W (1980), 'The analysis of complex fabric deformations', in Hearle J W, Thwaites J J and Amirbayat J (eds), *Mechanics of Flexible Fibre Assemblies*, Springer, New York.
44. Gan L and Ly N G (1995), A study of fabric deformation using nonlinear finite elements, *Textile Res J*, 65(11), 660–668.
45. Jeong Y J and Kang T J (2001), Analysis of compressional deformation of woven fabric using finite element method, *J Text Inst*, 1–14.
46. Tarfaoui M and Akesbi S (2001), Numerical study of the mechanical behavior of textile structures, *Int J Clothing Sci Technol*, 13 (3/4), 166–175.
47. Sztandera L M and Pastore C (2003), *Soft Computing in Textile Sciences*, Springer, New York.
48. Chattopadhyay R and Guha A (2004), Artificial Neural Networks: applications to textiles, *Text Prog*, 35, 1–46.
49. Zadeh L A (1965), Fuzzy sets, *Information and Control*, 8, 338–353.
50. Gopal M (2004), *Digital Control and State Variable Methods*, Tata McGraw-Hill, New Delhi.
51. Raheel M and Liu J (1991), *Text Res J*, 61, 31.
52. Park S W and Hwang Y G (1999), *Text Res J*, 69, 19.
53. Chen Y, Collier B, Hu P and Quebedeaux D (2000), *Text Res J*, 70, 443.
54. Huang C C and Yu W H (1999), *Text Res J*, 69, 914.
55. Lin J J (2007), *Text Res J*, 77, 336.
56. Beltran R, Wang L and Wang X (2005), *Text Res J*, 75, 557.
57. Behera B K and Muttagi S B (2004), *J Text Inst*, 95, 283.
58. Hui C L, Lau T W, Ng S F and Chan K C C (2004), *Text Res J*, 74, 375.
59. Shyr T W, Lin J Y and Lai S S (2004), *Text Res J*, 74, 354.
60. Shyr T W, Lai S S and Lin J Y (2004), *Text Res J*, 74, 528.
61. Wong A S W, Li Y, Yeung P K W and Lee P W H (2003), *Text Res J*, 73, 31.
62. Michalewicz Z (1996), *Genetic Algorithms + Data Structures = Evolution Programs*, 3rd edn, Springer-Verlag, New York.

63. Jasper W, Joines J and Brenzovich J (2005), *J Text Inst*, 96, 43.
64. Amin A E, El-Gehani A S, El-Hawary I A and El-Beali R A (2007), *Autex Res J*, 7, 80.
65. Blaga M and Draghici M (2005), *J Text Inst*, 96, 175.
66. Lin J J (2003), *Text Res J*, 73, 105.
67. Grundler D and Rolich T (2003), *Text Res J*, 73, 1033.
68. Hui P C L, Ng F S F and Chan K C C (2000), *Int J Clothing Sci Technol*, 12, 50.
69. Chan K C C, Hui Patrick C L, Yeung K W and Ng F S F (1998), *Int J Clothing Sci Technol*, 10, 21.
70. Inui S (1994), *Sen-i-Gakkaishi*, 50, 593.
71. Liang Y H (2008), *Int J Quality Reliability Management*, 25, 201.
72. Ozturk N (2003), *Eng Computation*, 20, 979.
73. Wong A S W, Li Y and Yeung P K W (2004), *Text Res J*, 74, 13.
74. Waterman D A (1986), *A Guide to Expert System*, Addison-Wesley, London.
75. Hayes-Roth F, Waterman D A and Lenat D A (1983), *Building Expert System*, Addison-Wesley, London.
76. Forsyth R (1984), *Expert System*, Chapman and Hall, London.
77. Behera B K, Muttagi S B (2004), 'Performance of error back propagation vis-à-vis radial basis function neural network. – Part I: Prediction of properties for design engineering of woven suiting fabrics', *J Text Inst*, 95 (1–6), 283–300.
78. Behera B K and Muttagi S B (2004), 'Performance of error back propagation vis-à-vis radial basis function neural network. Part II: reverse engineering of woven fabrics', *J Text Inst*, 95 (1–6), 301–317.
79. Behera B K and Muttagi S B (2005), 'Comparative analysis of modeling methods for predicting woven fabric properties', *Journal of Textile Engineering*, 51 (1).
80. Behera B K and Muttagi S B (2006), 'Engineering design of polyester-viscose blended suiting fabrics using radial basis function network: Part I – Prediction of fabric low stress mechanical properties', *IJFTR*, 31, September.
81. Behera B K and Muttagi S B (2006), 'Engineering design of polyester-viscose blended suiting fabrics using radial basis function network: Part II – Prediction of fabric constructional parameters from its properties', *IJFTR*, 31, December.
82. Behera B K and Muttagi S B (2006), 'Artificial neural network embedded expert system for design of woven suiting fabrics, Part I', *Industria Textila*, both in English and Romanian, 21 (2).
83. Behera B K and Muttagi S B (2006), 'Artificial neural network embedded expert system for design of woven suiting fabrics', Part II', *Industria Textila*, both in English and Romanian, 21 (2).
84. Behera B K, 'Image Processing Applications in Textiles', *Text Prog*, 35 (2/3/4).
85. Behera B K and Mishra R (2006), 'Objective measurement of fabric appearance using digital image processing', *J Text Inst*, 97 (2), 147–153.
86. Behera B K and Mishra R (2007), 'Artificial neural network based prediction of aesthetic and functional properties of worsted suiting fabrics', *IJCST*, 19 (5).

9 3D woven structures and methods of manufacture

M. AMIRUL ISLAM, Bally Ribbon Mills, USA

Abstract: This chapter discusses various three-dimensional (3D) fabrics and structures, such as, T, tapered T, X, cruciform, Pi, tapered Pi, intersecting Pi, single and double blade joints, I-beam, H, truss core, polar structures, contour structures, fillets (noodles), sine wave, panels, and curved structures. Various 3D weaves are also discussed as are the different manufacturing technologies used to produce 3D structures.

Key words: 3D woven structures, 3D weaves, 3D weaving technologies, near net shape woven preforms, polar and contour weaving, woven composites.

9.1 Introduction: 3D woven structures, applications and advantages

Three-dimensional (3D) structures have been used in various industries, such as aerospace, for some time. Putting several plies of fabrics on top of one another, cutting them into shapes, and stitching them together were the original methods of forming 3D structures. These techniques are labor intensive, time consuming, and expensive. Moreover, the fibers are damaged due to stitching and the joints are weak because no continuous fibers run through the intersections. Current 3D structures are generally created by weaving, using a specially built loom, braiding, knitting, filament winding and pultrusion, among other methods. In this chapter only those 3D structures created using a loom, loom mechanisms and specially built machines are discussed.

The term 'yarn' is commonly used to describe both warp and fill (weft) in a regular woven fabric. In a 3D woven composite structure it is preferable to use the fiber tows without any twist. The term 'fiber tows' is used to refer to the fiber throughout this chapter. The warp yarns or fiber tows are referred to as '0°' or 'X direction fibers' or 'machine directional fibers'. Warp yarns or fiber tows numbers are expressed in ends per inch. The weft yarns or fibers are called 'fill fibers'. The fill yarns or fiber tows are referred to as 'Y direction fibers' or '90° fibers'. Fill yarns or fiber tows numbers are expressed in picks per inch. The yarns or fiber tows through-the-thickness are referred to as 'Z direction fibers' in this chapter.

9.1.1 Definition of 3D fabrics and structures

There are various definitions of 3D fabrics and structures because researchers and scientists differ in their opinions [1, 2, 3, and 4]. In this chapter any fabrics which take the shape of three dimensions are considered to be 3D fabrics. Woven structures with substantial thicknesses, thick structures, orthogonal structures or, which have been made using standard loom or special loom mechanisms and any woven three-dimensional shaped structures are considered to be 3D woven structures in this chapter. Layers are also referred to as 'ply' in the composite industries.

9.1.2 Various 3D structures

Various new applications have been found for three-dimensional woven fabrics and structures and composites made from 3D fabrics, panels and preforms are being increasingly used in aerospace, medical, automotive, armored vehicles, marine, boat and shipbuilding, construction, wind energy and industrial applications.

3D fabrics and structures

There are many 3D fabrics and structures, a few of which are shown in this chapter. Figure 9.1 [5] shows a 3D fabric made from carbon fiber. This fabric was made up of six layers. Each layer was connected with the next by a layer-to-layer interlock weave based on satin weave as shown later in Fig. 9.40. The fabric was drapeable and formed three dimensions without wrinkles. A rapier loom was used to weave this fabric, as shown in Fig. 9.1.

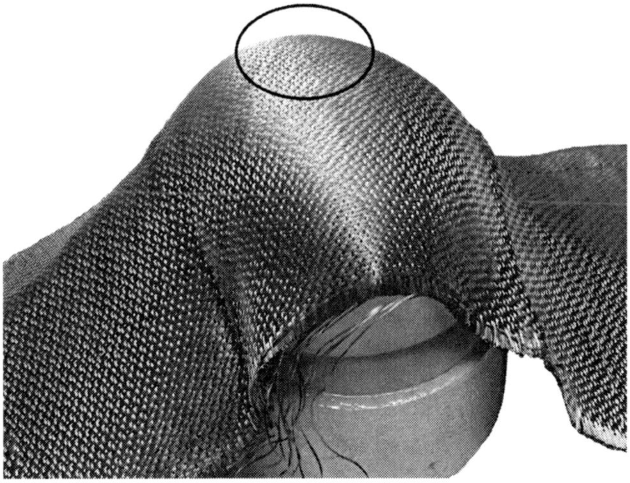

9.1 3D fabric [5].

Contour fabric

A contour fabric is shown in Fig. 9.2 [5]. This fabric, made of carbon fibers and woven on an Iwer (rapier) loom, was made into a composite for a jet engine. Contour weaving technology was used for the circumferential frame applications. Contour fabrics take the shape of the contours they are based on, meaning that they fit well, not only on cylindrical surfaces with different diameters, but also on flanges and complex curvatures. Regular fabrics do not fit so well on the contour surfaces and tend to wrinkle. When infused with resin, contour fabrics also make good composite parts. When used in this way it is important that the warp and fill fibers align close to 0°/90° without much angle variation.

Contour weaving is based on the same principles as standard weaving except that the take-up mechanism is modified to accommodate the different lengths of the warp fibers based on the configuration of the contour fabric. The take-up rate of the warp fibers remains the same along the whole width when weaving a regular fabric, which is why they will not conform to curvatures without cutting or darting them. When weaving a contour fabric, the length of every single warp fiber is calculated based on the specified configuration and the take up is adjusted accordingly to achieve the correct fiber length and orientation. Contour fabrics can therefore, easily conform to curvatures without wrinkling.

Contour weaving technology can be used for stiffener and aircraft engine fan blade containment. Aramid fibers are usually used for the latter. The benefits of contour weaving technology are as follows:

- reduced manufacturing costs since there is no cutting or darting
- wrinkles are eliminated
- warp and fill are aligned close to 0°/90° without much angle variation

9.2 Contour fabric [5].

- continuous and consistent parts are produced
- both convex and concave profiles can be created
- structures can be made in small scale without a full scale configuration.

Various processes, such as RTM (resin transfer molding), VARTM (vacuum assisted resin transfer molding), and RFI (resin film infusion) can be used to make composite parts.

T structure

Figure 9.3 [5] shows a T structure made from carbon fibers. Each flap consists of four layers and these layers are connected using layer-to-layer interlock weave based on a plain weave [2, 13, 14] as shown later in Fig. 9.37. IM7-12K carbon fibers were used and a tapered T was woven with tapering in all three sides. This tapering was created by dropping one layer at a time. A narrow fabric shuttle loom was used to weave the T structure.

A 3D loom [6] was used to weave the T-bar profile in Fig. 9.4. The mechanism of this machine is discussed later.

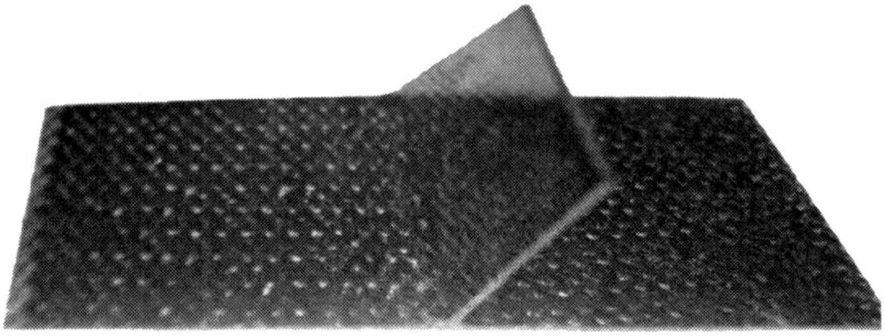

9.3 T structure [5].

9.4 T-bar profile [6].

Single and double blade joints

Figures 9.5 [5] and 9.6 [5] show a single blade joint and double blade joint. In both cases IM7-12K carbon fibers were used and the layers were connected using a layer-to-layer interlock weave based on the plain weave shown later in Fig. 9.37. A narrow fabric shuttle loom was used to weave both structures.

9.5 Single blade joint structure without tapering [5].

9.6 Double blade joint without tapering [5].

Cruciform structure

A cruciform structure is shown in Fig. 9.7 [5]. The 100% fill fiber of each flap goes through the intersection. Each flap consists of four layers and again they are connected by layer-to-layer interlock weave based on a plain weave. The typical construction of each layer is 16 × 16 ends and picks per inch in the case of 6K carbon fibers and 16 × 8 ends and picks per inch in the case of 6K carbon warp fibers and 12K carbon fill fibers. A double cruciform with all sides tapered was also woven using a modified narrow fabric shuttle loom.

I-beam structure

An I-beam structure is shown in Fig. 9.8 [5]. This I-beam was woven using carbon fibers and carbon rods. The I-beam can be used as a structural element in aircraft fuselage.

9.7 Cruciform structure [5].

9.8 I-beam structure [5].

Pi structure/intersecting Pi

A typical Pi structure [5] with all sides tapered is shown in Fig. 9.9 [5]. This Pi structure was woven from 6K and 12K IM7 carbon fibers. The base and clevis are eight layers and the upright legs are four layers. A layer-to-layer interlock weave based on a plain weave [2, 13, 14] was used. IM7-6K carbon fibers were used as warp and IM7-12K carbon fibers were used as fill at the base. IM7-12K carbon fibers were used as both warp and fill for the upright legs. The construction of each layer consisted of 16 × 9 ends and picks per inch. The gradual tapering was achieved by terminating one layer at a time.

The Pi structure had been woven from carbon, fiberglass, ceramic fiber, silicon carbide, and even from metallic wires at Bally Ribbon Mills. This Pi structure has been patented [7] and it is widely used in the aerospace industry for structural joints. The base, clevis and upright legs of the Pi structure can consist of any combination of layers based on the performance requirements.

The Pi and double cruciform structures designed at Bally Ribbon Mills were used in the NASA Composite Crew Module (CCM) [8]. One of the significant benefits of using 3D woven joints over traditional metallic structures is that they can be easily formed into complex shapes that may be structurally efficient.

An intersecting Pi structure is shown in Fig. 9.10 [5]. This intersecting Pi preform was woven as one piece on a modified narrow fabric shuttle loom. Each flap consists of four layers and they are connected using a layer to layer interlock weave based on the plain weave shown later in Fig. 9.37. in

9.9 Pi structure [5].

9.10 Intersecting Pi structure [5].

which IM7-12K carbon fibers were used. All the sides of this intersecting Pi are made up of woven edges.

Fillets (noodles)

Figure 9.11 shows a solid model of a fillet and a woven carbon fillet (noodle) is shown in Fig. 9.12. A bifurcate carbon fillet (noodle) is shown in Fig. 9.13. These fillets are used at joints to avoid resin-rich areas.

Panels

Figure 9.14[5] shows a 3D woven, 50 mm thick panel made from AS4-12K carbon fibers. An orthogonal weave architecture, as shown later in Fig. 9.43 [2, 13, 14], has been used for this panel. 3D orthogonally woven structures consist of non-interlaced reinforcing fibers that generally orient perpendicular or nearly perpendicular to each other in the X, Y, and Z directions. Orthogonally woven structures offer superior mechanical properties and corrosion resistance after infusing resin into a composite. The panel, as shown in Fig. 9.14, has 61 layers of warp, 62 layers of fill and Z fibers. The typical fiber volume of the panel is around 55% and X, Y, Z fiber volume fractions are around 40%, 40%, and 20% respectively. The typical Z fiber volume fraction can vary from 3% to 33% depending on the performance requirements of the end use.

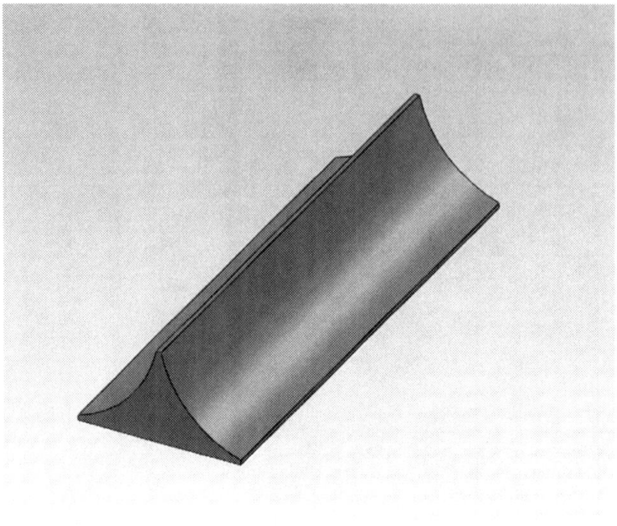

9.11 Solid model of a fillet (noodle).

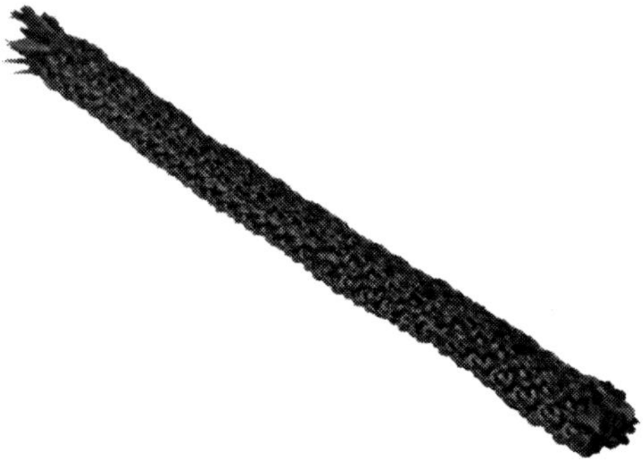

9.12 Carbon fillet (noodle) [5].

9.13 Carbon bifurcate fillet (noodle) [5].

9.14 50 mm thick 3D orthogonal carbon panel [5].

The panels can be woven using several different fibers to create a hybrid panel. The level of hybridization can vary, for example; panels can be woven with different fibers in each direction (X, Y, and Z), with different fibers in each layer of warp fiber, or the individual tows can vary within each layer. The Z fibers can also be made of different fibers. The level of hybridization is dependent on the intended application of the finished panel. The fiber concentrations in each direction can be varied to achieve the desired fiber volume fraction. A 50 mm thick 3D woven quartz preform is shown in Fig. 9.15 [5], where quartz fibers have been used in all three directions.

Figure 9.16 [5] shows a 50 mm thick 3D orthogonal pitch preform. It is difficult to weave a thick panel using only pitch carbon fibers due to their poor bending properties. Pitch fibers are therefore generally served with rayon or polyvinyl alcohol (PVA) fibers, which can withstand the friction and abrasion during weaving. The term 'serving' is used to describe the process of wrapping the core fibers with another fiber. In this case, the pitch fibers were wrapped with rayon. The selection of fibers, fiber volume and fiber volume fraction are determined by the performance requirements of the end product.

9.15 50 mm thick 3D orthogonal quartz preform [5].

9.16 50 mm thick 3D orthogonal pitch preform [5].

Truss cores

An interwoven triangular truss core with face sheets is shown in Fig. 9.17 [5]. A double interwoven truss core with face sheets is shown in Fig. 9.18 [5]. The integrally woven cells were made to reduce the delamination of the composite materials and to enable them to withstand multiple impacts.

Hollow structures

Figure 9.19 shows a hexagonal structure. Other hollow structures are shown in Figs 9.20 (hexagonal and diamond shapes) and 9.21 (diamond shape). Many types of hollow structures can be woven.

Figure 9.22 [5] shows a bifurcate woven vascular graft produced using 40 denier polyester yarns in both the warp and fill. Many woven 3D fabrics and structures using polyester, polyester/spandex, Teflon, polytetrafluoroethylene (PTFE), Spectra, Dyneema fibers, and Nitinol memory wires are currently used as implants in the human body [9].

9.17 Triangular truss core [5].

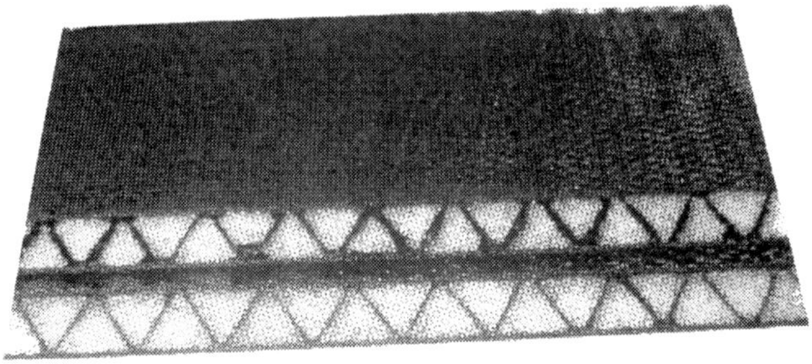

9.18 Double truss core [5].

9.19 Hexagonal structure [5].

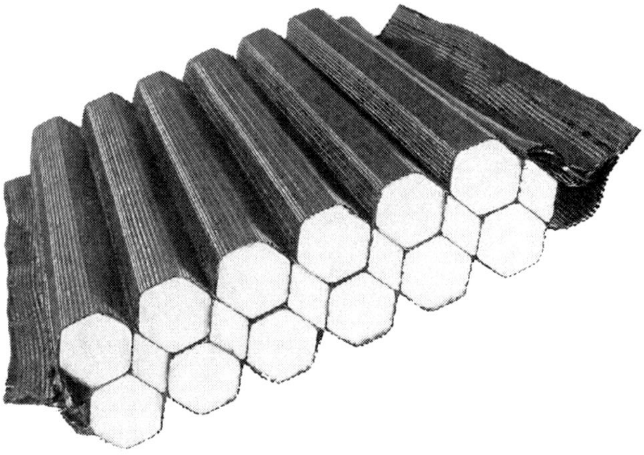

9.20 Multiple cell structure (hexagonal and diamond shapes) [5].

9.21 Multiple cell structure (diamond shape) [5].

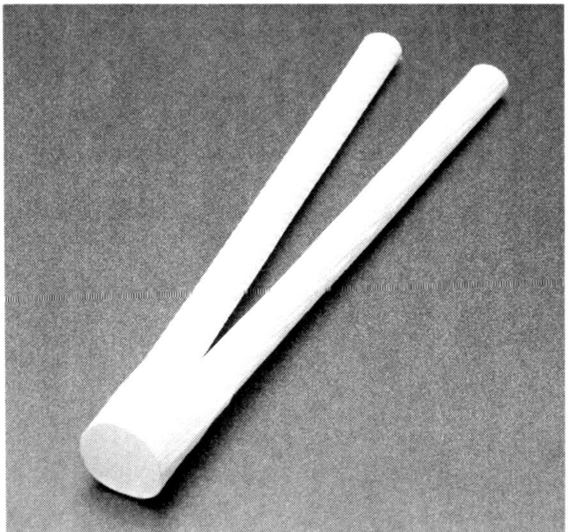

9.22 Bifurcate woven vascular graft [5].

Polar/spiral weaving

Polar weaving is a 3D weaving technique, which forms a two-dimensional (2D) flat fabric into a three-dimensional shape. A typical polar woven material is shown in Fig. 9.23 [5]. The take-up system is usually modified in order to weave a polar material. Modification of the typical orthogonal (0°, 90°) fiber orientation is required for the polar weaving process, which produces a circular woven material with the fibers orientated in the circumferential (hoop) and radial directions. The material produced on a loom, modified for polar weaving, produces a continuous woven cloth with a set inner and outer diameter. When producing a polar woven material, it will spiral on top of itself like a spring, and create a thick wall cylinder. The thickness of the cylinder will depend on the thickness of each polar woven layer and the number of layers making up the cylinder. Since the fill fiber is orientated in the radial direction, the material has a gradient from the inside diameter (ID) to the outside diameter (OD) and the fabric areal weight and material thickness will decrease from the inner diameter to the outer diameter. A solid model of a typical polar structure is shown in Fig. 9.24. Several manufacturers have developed techniques to reduce the above gradient by inserting partial picks across the width, creating several smaller zones and gradients. Multi-layer fabrics with Z fiber reinforcement can be polar woven.

3D multi-layer silicon carbide woven polar material is shown in Fig. 9.25 [5] with a very small ID (inside diameter) and a large OD (outside diameter). Several variations of polar weaving have been developed in order to create more complex shapes or preforms. Figure 9.26 [5] shows a curved T structure.

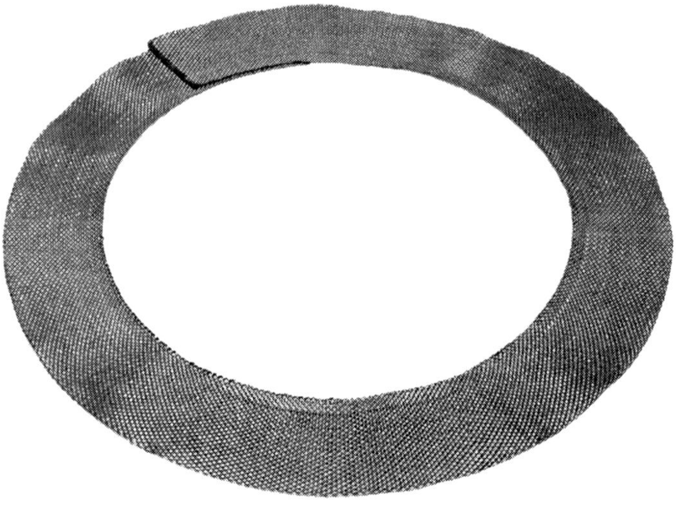

9.23 Typical polar structure [5].

9.24 Solid model of a typical polar structure.

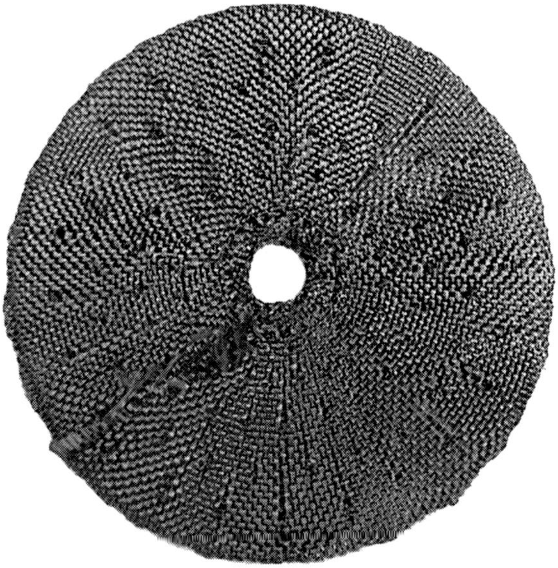

9.25 Silicon carbide polar structure [5].

One variation incorporates traditional orthogonal weaving with polar weaving (or ortho/polar weaves), which produces a single woven material with a section of orthogonal woven and another section with a polar weave creating a cylinder with a flanged end. This polar/spiral woven material changes typical 0°/90° woven materials into circular structures with a hoop and radial fiber alignment.

9.26 Curved T structure [5].

Polar/spiral structures

Figure 9.27 [5] shows a polar/orthogonal structure made from carbon fiber which can be used in a gear box. It has four layers and each layer is connected with the next layer using a layer-to-layer interlock weave based on a plain weave. Modified polar/spiral weave technology has been used to weave an aircraft window frame. A composite window frame is shown in Fig. 9.28 [5]. The window frame weaving technique was developed [5] to overcome the limitation of contour weaving. This technique allows the weaving of flanged 'L' preforms. The upright section of the L preform will always remain orthogonal, but the flanged section of the L preform changes from polar to orthogonal and back to polar. This technique also allows the weaving of rectangular or square flanged preforms with a corner radius.

Engine vane

An outlet guide vane (OGV) for a jet engine is shown in Fig. 9.29 [5]. A combination weave was used and the thickness varied in both width and length directions. Snecma (SN) and Albany Engineered Composites (AEC) have developed woven composite fan blades for jet engines [10]. 3D woven preforms produced using a layer-to-layer interlock pattern are processed into composites by resin transfer molding (RTM). The advantages of composite fan blades include improved damage tolerance and increased thrust-to-weight ratio due to the lower mass of the fan.

9.27 Polar/orthogonal structure [5].

9.28 Composite window frame [5].

Sine-wave structure

A woven sine-wave structure is shown in Fig. 9.30 [5] and a complex woven preform is shown in Fig. 9.31 [5]. The electronics can be placed into slots for various applications.

9.1.3 Applications

Joints entirely made from composites are being used in aerospace applications to alleviate the need for fasteners. Fillets are being used at the joints to avoid resin-rich areas. The aerospace industries are also looking into the possibility of composite window frames. Composite truss cores are used

9.29 Composite woven engine vane [5].

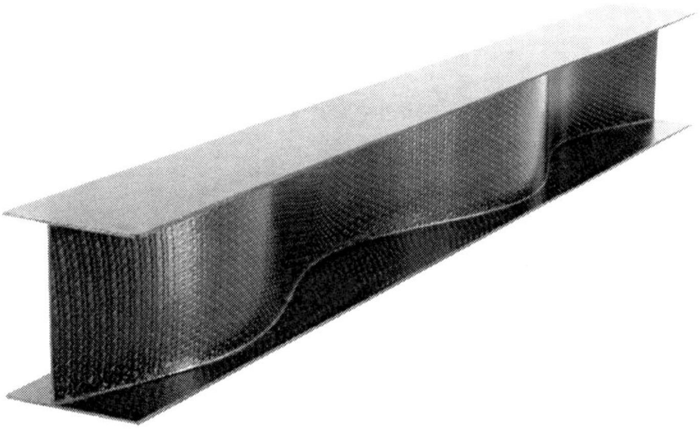

9.30 Sine-wave structure [5].

in fuselage areas and the composite vane and fan blades are used in the jet engine. Polar woven carbon material is used for carbon/carbon brakes. The automobile industries are using carbon panels and the construction and boat industries are using fiberglass panels. NASA, meanwhile, in partnership with Bally Ribbon Mills, is developing woven 3D structures to use as a thermal protection system (TPS) [11] for their future missions to Mars and beyond. In the medical industry hollow woven structures, such as, tubular

9.31 Complex woven preform [5].

and bifurcate tubular, tapered tubes and tapered bifurcates are being used as vascular grafts and stents and expandable tubes as penile grafts are being used as implants in the human body.

9.1.4 Advantages of woven 3D structures

3D woven structures are woven directly into net shapes. The advantages of woven 3D preforms are as follows:

- no plying, cutting, and stitching
- near net shape
- minimum machining to achieve required shape and size
- automated process
- consistent products
- minimized time for manufacturing
- increased production
- eliminates delamination, which is the primary damage mechanism in laminated composites
- reduces cost
- better performance.

The cost of manufacturing preforms is a major portion of the total cost. The benefits of using composites instead of metallic structures are that they are light weight, stronger and more corrosion free. The saving in weight is tremendous. Prichard [12] suggests that the use of 3D fabrics and structures could lead to a 30% reduction in aircraft weight and every pound of weight saved could save one million dollars over the life of the aircraft. The saving in weight also reduces the aircraft fuel cost. The greatest advantage of using 3D woven composites and structures is the elimination of delamination.

9.2 Weaves: basic and 3D

9.2.1 Basic weaves

The type of weave used is decided based on the stability, strength, flexibility, drapeability, permeability and shape, desired in the end product. Plain, twill and satin are the three basic weaves. All other weaves are modified, altered, combinations or derivatives of these. Various weaves have been discussed in Chapter 6. Three basic weaves are shown in the following figures for ready reference in order to understand 3D weaves.

A plain weave, in which each warp end passes over and under each fill fiber, is shown in Fig. 9.32. This types of weave is also called a 1/1 weave. Plain weave has the maximum number of binding points and is the most stable, but also has the potential for causing more fiber damage due to friction and abrasion during weaving. A 2/2 twill weave is shown in Fig. 9.33 where two ends are up and the next two ends are down. Twill weave has fewer numbers of interlacement than plain weave. A five harness satin and eight-harness satin weave are shown in Figs 9.34 and 9.35 and a four-harness satin (crowfoot) is shown in Fig. 9.36. There is only one binding

9.32 Plain weave.

9.33 2/2 twill weave.

9.34 Five-harness satin weave.

point in one repeat in one direction in a satin weave, which has typically the lowest number of binding points of three basic weaves.

Satin weave has the potential for minimum fiber damage during weaving, but it is the least stable weave. Twill is more stable and plain weave is the most stable. Three types of satin weaves (four-, five-, and eight-harness)

9.35 Eight-harness satin weave.

9.36 Four-harness crowfoot.

are common in the fabric ply used for textile composites. The maximum number of ends and picks can be inserted into satin weave due to it having the lowest number of binding points.

9.2.2 3D weaves

The weaves used to manufacture woven 3D fabrics and structures are considered to be 3D weaves. There are many types of 3D weaves and their selection is determined according to the end use and performance criteria of the 3D preforms. 3D weaves are generally based on the three basic weaves (plain, twill and satin) and are adapted using modifications, alterations, or combinations of these.

Layer-to-layer interlock weaves either by warp or filling

A typical layer-to-layer interlock weave based on a plain weave is shown in Fig. 9.37 [2, 13, 14] and another example based on a twill weave connection is shown in Fig. 9.38 [2, 13, 14]. Both figures show four layers joined together,

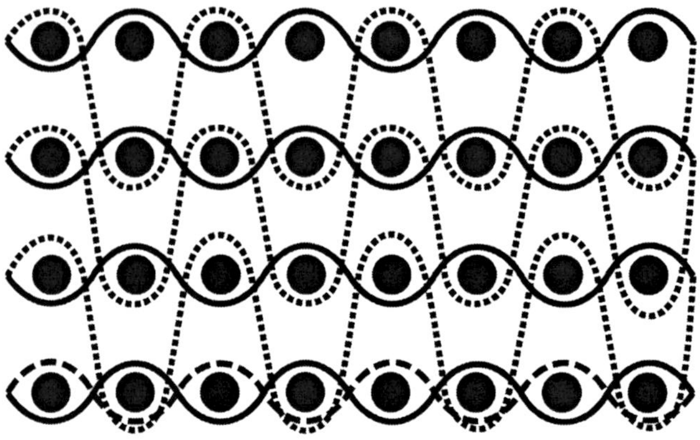

9.37 Typical layer-to-layer (ply-to-ply) interlock weave based on plain weave [2, 13, 14].

9.38 Typical layer-to-layer interlock weave based on twill weave connection [2, 13, 14].

but any number of layers can be joined together using the same technique. In this example the dark circles denote the warp ends and the curved lines indicate the filling path. The warp layers are joined by filling. The layers could however be joined by the warp ends, in which case the dark circles would denote the picks and the curved line would indicate the warp ends. The decision as to whether the layers should be joined either by the warp end or filling is based on the performance requirements of the 3D preforms.

Layer-to-layer interlock weave with Z fibers

A typical layer-to-layer interlock weave with Z fibers is shown in Fig. 9.39 [2, 13, 14]. Z fibers are added to the layer-to-layer interlock weave. The dark circles denote the warp ends and the curved lines show the filling path. Z fibers run through the fabric thickness. In the example figure, Z fibers are perpendicular to the warp, but Z fibers can also be inserted at various angles, such as 30°, 45°, and 60°.

Layer-to-layer interlock weave based on satin weave

Figure 9.40 [2, 13, 14] shows a typical layer-to-layer interlock weave based on satin weave. The curved line shows warp and the circles denote the fill. There are five warp layers and six fill layers, which are joined by the warp ends. The layers of 3D fabric as shown in Fig. 9.1 [5] are joined together using this method.

Layer-to-layer angle interlock weave either by warp or filling

A typical layer-to-layer angle interlock weave is shown in Fig. 9.41 [2, 13, 14]. In this type of weave, either the warp or the filling runs through the

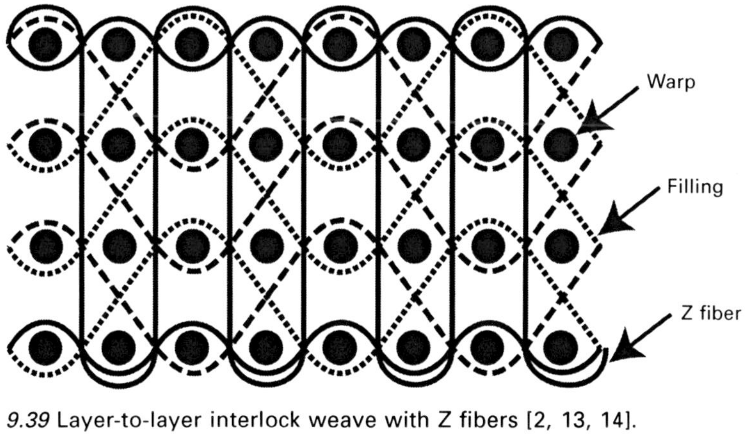

9.39 Layer-to-layer interlock weave with Z fibers [2, 13, 14].

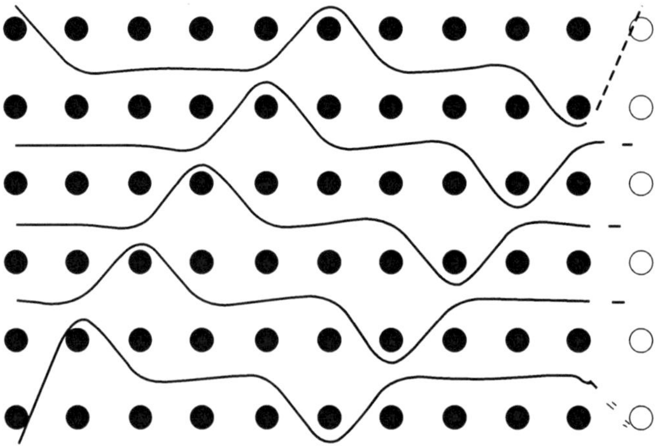

9.40 Layer-to-layer interlock weave based on satin weave [2, 13, 14].

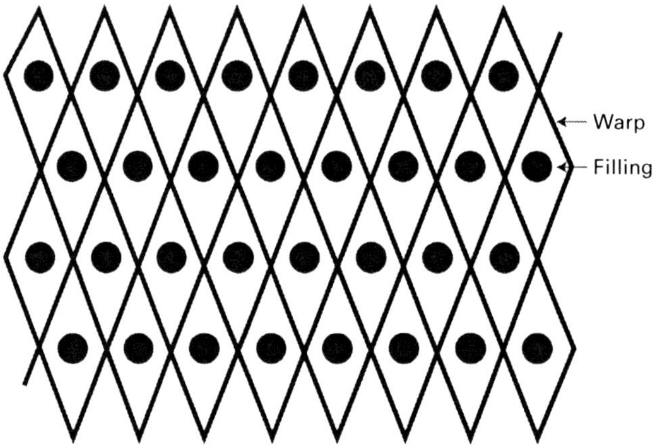

9.41 Layer-to-layer angle interlock weave [2, 13, 14].

fabric thickness at an angle. The joining of layers, either by the warp ends or the filling, is selected based on the performance requirements of the end product.

Layer-to-layer angle interlock with stuffer threads

Stuffer threads can be added to the layer-to-layer angle interlock structure to increase the thickness and tensile strength. A typical angle warp interlock with stuffer threads is shown in Fig. 9.42 [2, 13, 14].

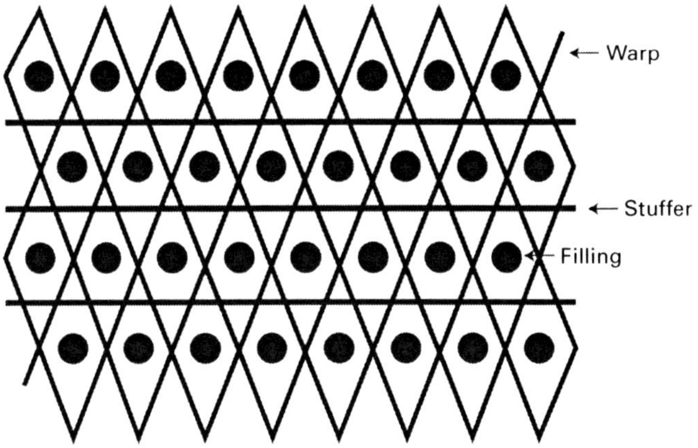

9.42 Layer-to-layer angle interlock with stuffer threads [2, 13, 14].

Orthogonal weave

In this type of structure, the warp and the filling run straight without crimping and the Z fiber binds all the layers. In a regular woven fabric there are crimps in the warp and the fill due to their interlacement. Such crimps are undesirable for maximum composite mechanical properties. The use of a Z fiber throughout the fabric thickness effectively eliminates delamination as a failure mode. An orthogonal weave with Z fibers is shown in Fig. 9.43 [2, 13, 14]. This type of structure is also called non-crimp fabric (NCF).

The composite industry prefers this type of orthogonal weave. The benefits of orthogonal weave architecture are that the fibers in all three directions run straight which makes the orthogonal structure rigid. Higher fiber volume can be achieved for this type of structure due to absence of fiber crimp in all three directions. However, the disadvantage of this type of weave is that, if the Z fibers break, there is nothing to hold all the layers together.

Different numbers of layers in various locations for complex shapes

Different numbers of layers are used in various locations to achieve various thicknesses in this type of structure. A typical four-layer weave architecture is shown in Fig. 9.44 [2, 13, 14], in which four-layers have been reduced to one. The circles show the warp fibers and the line indicates the filling path. The same technique is used to make a tapered structure. The gradual tapering in thickness is achieved by terminating one layer at a time. The base of Pi as shown in Fig. 9.9 [5] has eight layers and each layer was dropped from eight layers to a single layer.

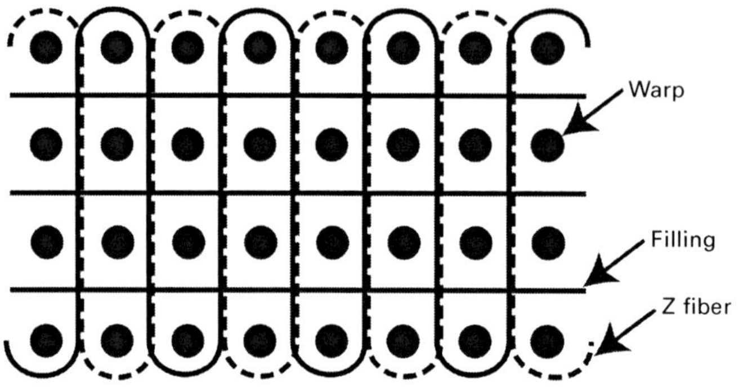

9.43 Orthogonal weave architecture [2, 13, 14].

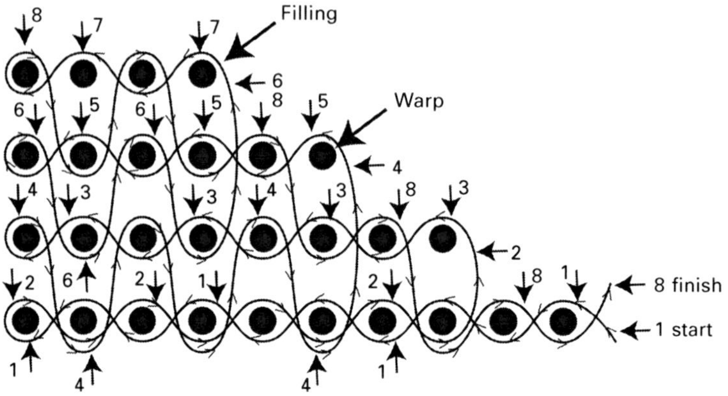

9.44 Weave architecture for tapered or complex shapes [2, 13, 14].

Combination weaves for complex shapes

A combination weave or modified weave can be used in order to create complex structures and to achieve specific properties. A combination weave made up of an orthogonal weave and a plain weave is shown in Fig. 9.45 [2, 13, 14]. The top and bottom layers are plain weave and the layers in-between are orthogonal weave. There are two straight warp layers between the top and bottom plain weaves and Z fibers bind all the layers. Layer-to-layer interlock weave is more flexible, but it has less tensile strength due to interlacement and more fiber damage during weaving. An orthogonal weave is more rigid, but has a higher tensile strength due to the lack of interlacement and lower fiber damage during weaving. Layer-to-layer interlock weave based on plain, twill or satin and orthogonal weaves can be combined in many ways. The sequence can be varied, for example, x number of layers joined by layer-to-layer, and then y number of layers orthogonal, or x layer, then y layer and so on. The optimum combination can be selected based on the performance requirements of the end structure.

9.45 Combination weave architecture [2, 13, 14].

9.2.3 Weave architecture

In this chapter, the weave architecture means the schematic view of a complete weave pattern. It shows the filling path in relation to warp and Z fibers. The structure means the shape of the woven material.

Weave architecture of T preform

The typical weave architecture of a four-layer T structure created using layer-to-layer interlock weave based on plain weave [2, 13, 14] is shown in Fig. 9.46 [7]. The warp fibers of both sides of T are shown spaced apart for clarity. The black circles show the warp fibers and the curved line denotes the filling. The four layers are joined together by a layer-to-layer interlock weave based on plain weave [2, 13, 14] and the upright flap is joined with the base flap by filling.

Weave architecture of Pi preform

The typical filling path for a four-layer base, four-layer clevis and two layer upright legs for a Pi structure without tapering is shown in Fig. 9.47. The warp fibers of the clevis are shown spaced apart for clarity. The filling path for an eight-layer base, eight-layer clevis and four-layer upright legs can be shown in the same way. The black circles show the warp fibers and the curved line denotes the filling. The four layers are joined together by a layer-to-layer interlock weave based on plain weave [2, 13, 14] and the upright legs are joined with the base by filling.

Figure 9.48 [7, 15] shows the weave architecture of a tapered Pi preform in which all the sides are tapered. The base and clevis are made up of eight layers and the upright legs of four layers. The gradual tapering was achieved

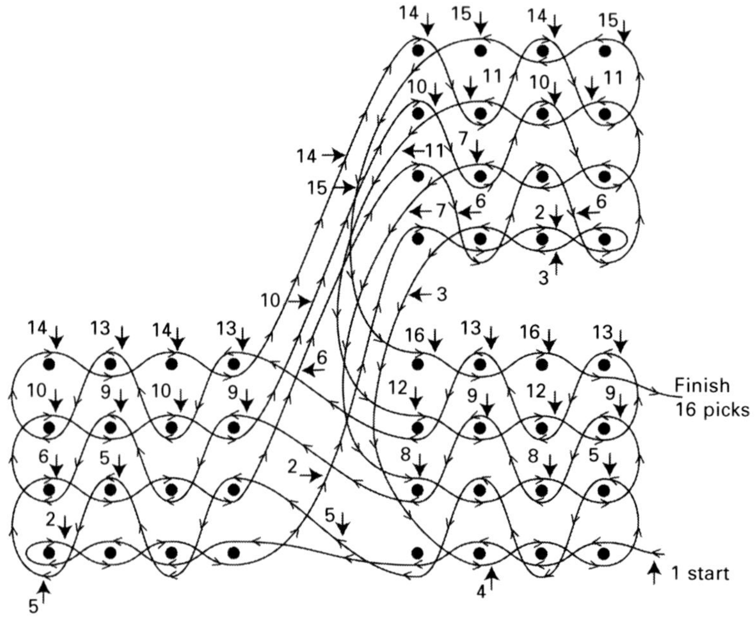

9.46 Typical filling path of T structure [7].

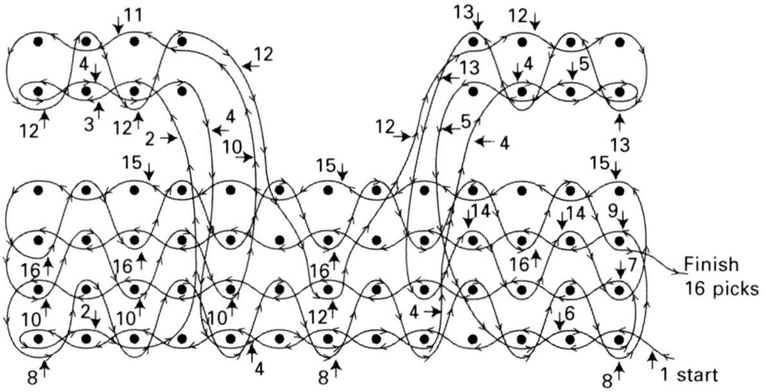

9.47 Typical filling path of Pi structure [7].

by dropping one layer at a time. All layers are joined together using a layer-to-layer interlock weave based on a plain weave [2, 13, 14]. The upright legs are joined to the base by the filling. The warp fibers at the clevis are shown spaced apart for clarity. Any combination of layers for a tapered Pi preform can be woven in the same way. The isometric view of the tapered Pi [7, 15] is shown in Fig. 9.49. The tracer fibers, comprising a colored strand and an x-ray opaque, shown by arrows, are used at the beginning of the tapers and

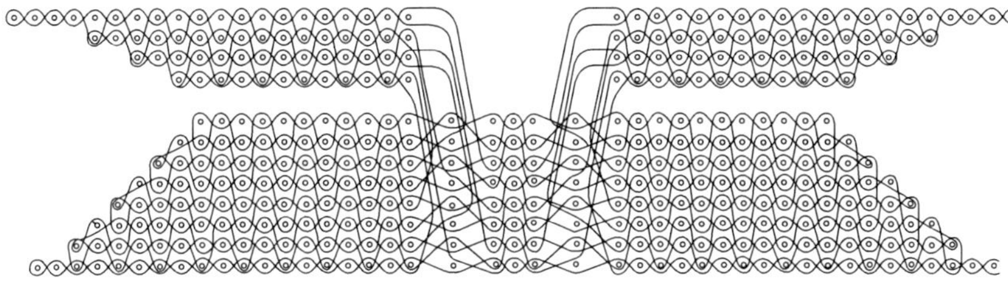

9.48 Weave architecture of a tapered Pi preform [7, 15].

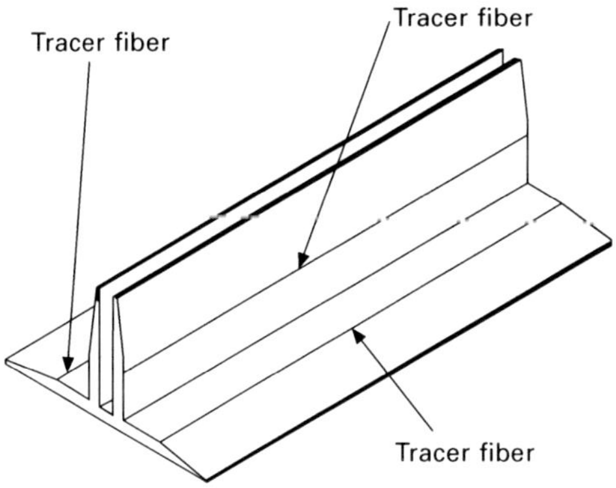

9.49 Isometric view of a tapered Pi preform [7, 15].

identify where the thickness of the legs or base begins to decrease for an accurate dimensional inspection of the preform.

9.2.4 Properties of 3D weaves

The selection of 3D weaves is very important. The selection of weave architecture depends on the end use of 3D fabrics and 3D structures, and their required performance criteria.

3D weave based on satin weave

If 3D fabrics have to conform to a three-dimensional shape, then the layers have to be joined using satin weave as shown in Fig. 9.40. Multi-layer satin weave fabrics are drapeable and easily conform to the shape. A multi-layer thick panel can be woven by joining layers using satin weave.

3D weaves based on plain weave

Multi-layer panels or 3D structures can be woven using a layer-to-layer interlock weave based on the plain weave as shown in Fig. 9.37. The warp end or fill is used to join the adjacent layers. If the joining threads are broken, then only a portion of one layer will fall apart, not the whole structure. This is one of the greatest benefits, but it means a lower tensile strength because the fibers are crimped by their interlacement. The likelihood of fiber damage during weaving is increased due to this high number of interlacements.

3D weaves based on orthogonal weave architecture

In orthogonal weave architecture the fibers in all three directions (XYZ) run straight, as shown in Fig. 9.43. Maximum fiber properties can therefore be achieved due to non-crimping. Orthogonal weave architecture is used to make thick panels or structures as shown in Figs 9.14, 9.15, and 9.16. The maximum number of warp fibers (X fibers), picks (Y fibers), and Z fibers can be inserted because of the lack of interlacements. That means, that a higher fiber volume percentage can be achieved from orthogonal weave architecture; however, the disadvantage is that the whole structure falls apart if the Z fibers break. Fiber damage during weaving is less likely because there are no interlacements.

3D weaves based on combination weave

Often, a combination weave is selected, based on the end product's performance criteria. An orthogonal weave is used to make thick panels and to achieve maximum performance properties so a combination weave made up of layer--to-layer interlock weave and orthogonal weave would be selected to achieve maximum performance properties without the layers falling apart.

9.3 Manufacturing technologies

9.3.1 Weaving preparation

Yarn preparation for regular fabrics has been discussed in Chapters 2, 3 and 4. The scope of yarn preparation for weaving 3D structures is not the same as that for a regular fabric. The following steps are used.

Yarns or fibers

The preference for a 3D woven composite structure is to use the fiber tows without any twist. Occasionally this is difficult to achieve, in which case a very small amount of twist is inserted. If the customer does not want any twist, fiber tows with a higher sizing percentage are used to facilitate weaving. As an example, carbon fiber tows usually come with a less than 1% sizing.

Some fibers, such as pitch carbon and silicon carbide, are difficult to weave. Pitch carbon fibers break easily due to their very poor bending rigidity. For this reason 'difficult to weave' fibers are usually served with polyvinyl alcohol (PVA) yarn or rayon to facilitate the weaving process. A machine called a 'serving machine' is used for this purpose. The serving machine wraps the 'difficult to weave' fibers with rayon or PVA. The rayon or PVA takes the abuse of weaving and the core fibers remain undamaged. Rayon is removed using heat and PVA is removed by dissolving it in warm water after weaving.

Winders

Fiber manufacturing companies usually supply fiber in large packages. Since these fibers are costly, it is necessary and economical to transfer fiber from larger to smaller packages on a suitable winding machine.

Creels

Unlike weaving regular fabrics, warp beams are not usually used for weaving 3D structures and instead, creels are the preferred method of feeding the warp. It is very important to maintain uniform tension on each package. A good creel is a prerequisite for weaving any 3D structures and complex shapes. A reciprocating creel, which takes out the slack during weaving, is used for more complex shapes.

Many thousands of ends are needed to weave a thick wide panel. Creels, which can accommodate such a large number of ends are very expensive and occupy a lot of space. In these cases, warp beams are used for the X or 0° fibers and a creel is used for the Z fibers. A combination of beams and creels can be used for tapered and complex structures.

9.3.2 Weaving technologies

The following machinery can be used in the production of woven 3D fabrics and structures:

- conventional 2D loom
- modified 2D loom
- 3D loom
- specially built loom or machine.

Conventional loom

Conventional 2D loom mechanisms have been discussed in Chapter 5. The shedding mechanism is one of the primary motions of a loom. A Jacquard shedding mechanism is used for 3D structures because the individual warp fibers are controlled independently by jacquard harness cords. A 3D fabric can be woven on a 2D conventional rapier loom, but there is a fabric thickness limitation when using any conventional 2D loom.

Loom set-up

Proper loom set-up plays a significant role in reducing warp breaks during weaving and is therefore key in increasing productivity. Loom set-up is even more important when weaving 3D fabrics and structures in order to minimize fiber damage during weaving to achieve maximum mechanical properties. The maximum translation of fiber or yarn properties to fabric for tensile strength [16] is dependent upon:

- the damage suffered by the fiber or yarn during weaving
- the fabric structure such as:
  - fiber or yarn size (linear density)
  - fiber or yarn absolute diameter depending on fiber density
  - ends and picks per inch
  - warp and filling yarn crimp
  - warp and filling yarn interlacement angle
  - cover factor
  - weave type and number of binding points between warp and fill.

Tensile strength realization in the fabric is primarily dependent on the weaving process, meaning that minimal fiber or yarn damage during weaving will improve the fabric tensile strength and weaveability. Research [16] shows that fiber damage during weaving can be minimized in the following ways:

- by setting up the harness on an elliptical path in order to keep the same warp tension
- by reducing friction and abrasion at every contact point of the fiber during weaving
- by reducing beat up force
- by using round eye heddles to reduce the coefficient of friction between the fiber and the heddle
- by reducing the warp tension variation
- by increasing the optimum open air space in the reed dent.

Fiber damage

In Fig 9.50, it appears that the damage to the warp increases gradually as the warp passes through the heddles and reed. There is a somewhat steeper increase in warp damage between the heddle and the cloth fell and a very

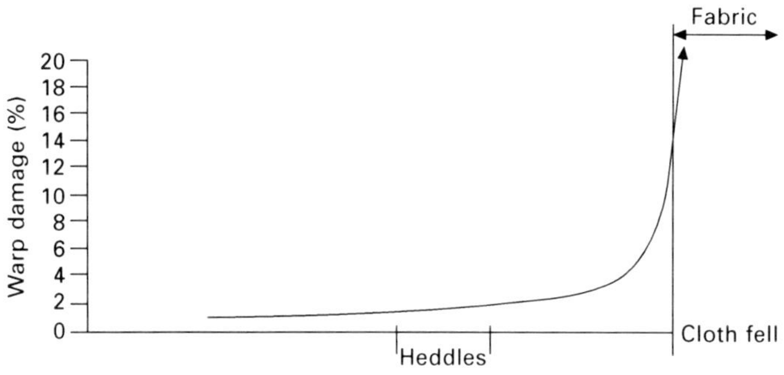

9.50 The warp damage [16].

large increase in warp damage between the region immediately in front of the cloth fell and the fabric itself.

The movement of the warp during the weaving process causes various stresses and strains [17, 18]. The let-off and shedding motions, together with the movements of the back rest roller and of the cloth fell, influence the total warp tension. Shedding is one of the main factors which contribute to the total warp tension during normal loom operation.

Taking into account the mathematical relationship between tension on the two sides of the heddle eye:

$$\frac{T_f}{T_b} = e^{\mu\theta} \qquad [9.1]$$

where: e = the base of Napierian logarithms, μ = the coefficient of friction between warp yarn and heddle eye, θ = the lap angle at the heddle eye in radians, T_b = the warp tension before passing through heddle eye, and T_f = the warp tension after passing through heddle eye.

Since tension = strain × elastic modulus, Eq. [9.1] can be written in terms of the strain ratio assuming that the strain is entirely due to shedding.

$$e^{\mu\theta} = \frac{W_f \text{ (Warp strain in front shed)}}{W_b \text{ (Warp strain in rear shed)}} \qquad [9.2]$$

Let P_f = AD/AB in Fig. 9.51 the normalized distance from the cloth fell to the heddle eye during shed formation, P_b = BD/AB in Fig. 9.51, the normalized distance from the back rest to the heddle eye during shed formation, H = CD/AB, the normalized shed height, F = AC/AB, the normalized front shed length, α = the front shed opening angle, and β = the rear shed opening angle. Then:

$$P_f = \sqrt{F^2 + H^2} \text{ in terms of F and H} \qquad [9.3]$$

$$P_f = \frac{\sin\beta}{\sin(\alpha + \beta)} \text{ in terms of } \alpha \text{ and } \beta \qquad [9.4]$$

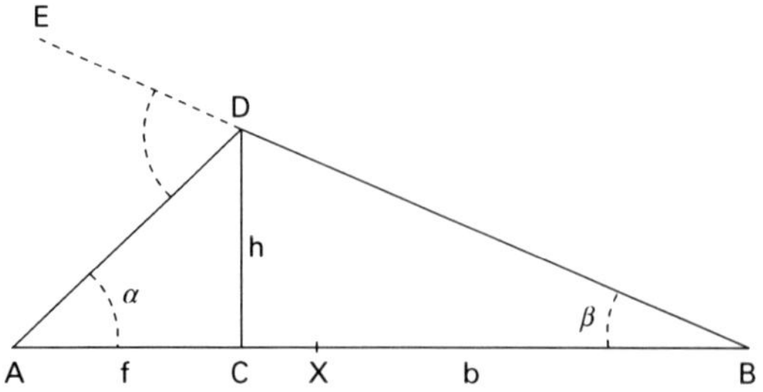

9.51 Geometry of the warp yarn path due to shedding [17].

$$P_f = \sqrt{F^2 + (F \tan \alpha)^2} \text{ in terms of F and } \alpha \qquad [9.5]$$

Similarly for P_b

$$P_b = \sqrt{(1 - F)^2 + H^2} \qquad [9.6]$$

$$P_b = \frac{\sin \alpha}{\sin(\alpha + \beta)} \qquad [9.7]$$

$$P_b = \sqrt{(1 + F)^2 + (F \tan \alpha)^2} \qquad [9.8]$$

Let X be the point on the closed shed line AB, which is in contact with the heddle eye at D. $F\mu$ = AX/AB, the normalized length of yarn which forms the front shed during shedding. The warp strain increments in the front shed $W_{f\mu}$ and rear shed $W_{b\mu}$ are given by:

$$W_{f\mu} = \frac{P_f - F_\mu}{F_\mu} \qquad [9.9]$$

$$W_{b\mu} = \frac{P_b - (1 - F_\mu)}{1 - F_\mu} \qquad [9.10]$$

From Eq. [9.2]:

$$\frac{P_f - F_\mu}{F_\mu} = e^{\mu\theta} \frac{(P_b - 1 + F_\mu)}{1 - F_\mu} \qquad [9.11]$$

Hence:

$$P_f - P_f F_\mu - F_\mu + F_\mu^2 = e^{\mu\theta} F_\mu (P_b - 1 + F_\mu)$$

Or,

$$F_\mu^2 (e^{\mu\theta} - 1) + F_\mu (e^{\mu\theta} P_b - e^{\mu\theta} + P_f + 1) - P_f = 0$$

$$F_\mu = \frac{-C + \sqrt{C^2 + 4(e^{\mu\theta} - 1)P_f}}{2(e^{\mu\theta} - 1)} \qquad [9.12]$$

where: $C = e^{\mu\theta}P_b - e^{\mu\theta} + P_f + 1$.

The warp strain increments using Eqs [9.9] and [9.12] are shown in Fig. 9.52.

If the front shed opening angle and the normalized distance from the cloth fell to the harness are measured or calculated, then the warp strain can be found from the equations and from Fig. 9.52. This figure is also a guide for how to set up the harness in order to maintain the same warp tension or to minimize the warp tension variation. It is very important to minimize the fiber damage for 3D woven composites.

The damage to warp can be minimized by reducing the strains caused by shedding motion, whether normal or distorted by traps. Damage can also be minimized by using round heddle eyes and by reducing strain due to friction and abrasion at all contact points. Minimum fiber damage during weaving will improve the fabric's tensile strength and weaveability.

Narrow fabric loom

The performance requirements, shape of the end product and cost dictate the selection between the shuttle and the needle loom in the narrow fabric industry.

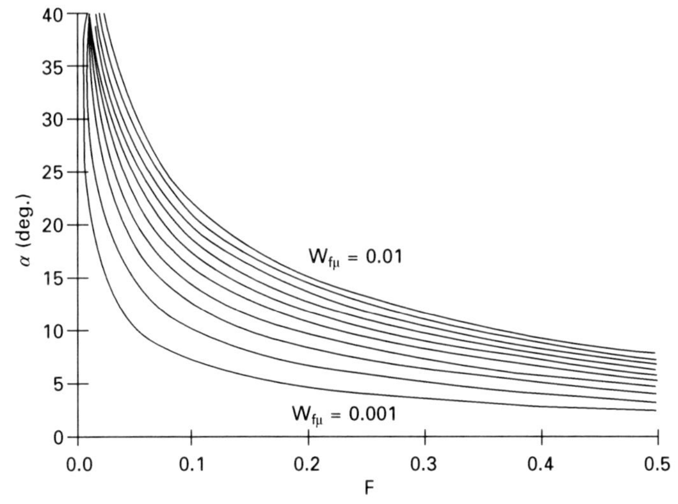

9.52 Dependence of warp strain increment ($W_{f\mu}$) in the front shed on α and F when μ = 0.30 at the heddle eye (Eqs [9.12] and [9.9]). Curves of constant strain increment at intervals of 0.001 [17].

A narrow fabric is defined by the American Narrow Fabrics Institute as any fabric, with a width not exceeding 12 inches (300 mm). The Narrow Fabrics Institute (NFI) is the industry organization for manufacturers of narrow fabrics and it is a division of Industrial Fabrics Association International (IFAI).

Narrow fabric shuttle looms are used to produce many 3D structures. There are different kinds of narrow fabric shuttle looms, such as single deck (horizontal and circular), double deck (upper and lower), two shuttles batten (both shuttles run together in the same station in two sheds), and multiple shuttles loom. Two shuttles batten, where two sheds are formed by dobby or three positions jacquard is used to weave tear tape, spacer fabrics, bifurcate and others. Two shuttles run simultaneously through the sheds to weave two separate fabric layers and Z fibers (binders) bind both layers. In a narrow fabric shuttle loom, the shuttle movements are positively controlled in a batten with rack and pinion unlike fly shuttle broad loom. The typical speed of a narrow fabric shuttle loom is 100 picks per minute; however, it is reduced to weave brittle and 'hard to weave' fibers. These looms are being replaced with high speed needle looms offering increased productivity. Table 9.1 presents the main differences between the characteristics of narrow fabric shuttle and needle looms.

There are multiple stations in a narrow fabric shuttle loom. That means the same structure in multiple stations or different structures can be woven on the same loom depending on the capacity of the jacquard hooks and the total picks per inch in the structure. It is difficult to produce 3D and complex shapes using carbon and brittle fibers on a needle loom. A narrow fabric needle loom [19] can also be used to produce some 3D structures. A typical narrow fabric needle loom is shown in Fig. 9.53 [5]. The selvedge on the side of the filling insertion needle is a true woven selvedge while the selvedge on the latch needle is a knitted selvedge. With the needle system, the fabric always has double picks. A narrow fabric shuttle loom is used for a perfect tube, uniform mesh size for filtration, medical textiles [9], 3D preforms, near net shape preforms, and complex woven structures. The narrow fabric shuttle loom with electronic jacquard head is used for weaving thick panel and complex 3D structures. With the advent of electronic jacquard head, it is much easier to change the weave design.

Table 9.1 Characteristics of shuttle and needle looms

	Shuttle loom	Needle loom
Speed	Low	High
Edge	Woven edges	Woven and knitted edge
No. of stations (typical)	Up to 84	Up to 20
Picks/shed	Single	Double

9.53 A typical narrow fabric needle loom [5].

The 3D structure on a conventional 2D loom is woven in a flat form. The first step in weaving of 3D structures on a conventional 2D loom is to figure out how to flatten [20] the structure. Once it is determined, it is followed by the subsequent set-up of the loom and the weave architecture. The filling path, based on the performance requirements of 3D structures, has to be carefully selected. Figure 9.54 [5] shows weaving of carbon tapered Pi structures in a narrow fabric shuttle loom.

Modified loom

The 2D conventional loom is modified to weave various 3D woven preforms. The kind of modification(s) that should be carried out depends on the nature of structures to be produced. The take-up and shedding motions are modified in many ways along with other tools that are also installed. Modified looms are used to produce thick panels, contour materials, polar materials, and complex 3D structures. An unlimited number of 3D structures and complex shapes can be woven on a modified loom with electronic jacquard. Figure 9.55 [5] shows a narrow fabric shuttle loom weaving 50 mm thick carbon panel. An orthogonal weave has been used with a dead pick motion so that X, Y, and Z fibers are perpendicular to each other.

3D looms

A few companies have developed 3D looms [21, 22, 24, 25, 26]. Mohamed *et al.* [21] and 3Tex [22] have developed 3D looms. Figure 9.56 [21] shows

9.54 Narrow fabric shuttle loom weaving Pi structures [5].

9.55 Modified 2D loom weaving 50 mm thick carbon panel [5].

the filling insertion from both sides. The warp fibers are fed either from bobbins positioned on a creel or from beams. They are separated into layers to allow for fill insertion. These warp layers are fixed open to form sheds without shedding mechanism.

According to Mohamed *et al.* [21], the warp ends are not required to be drawn through the heddles of the shedding mechanism, leading to less fiber damage during weaving. Multiple fill yarns or fibers can be inserted through the sheds (layers) by needles or rapiers from both sides simultaneously or alternately according to cross sectional shapes. The filling is inserted from one side in a regular needle loom [19].

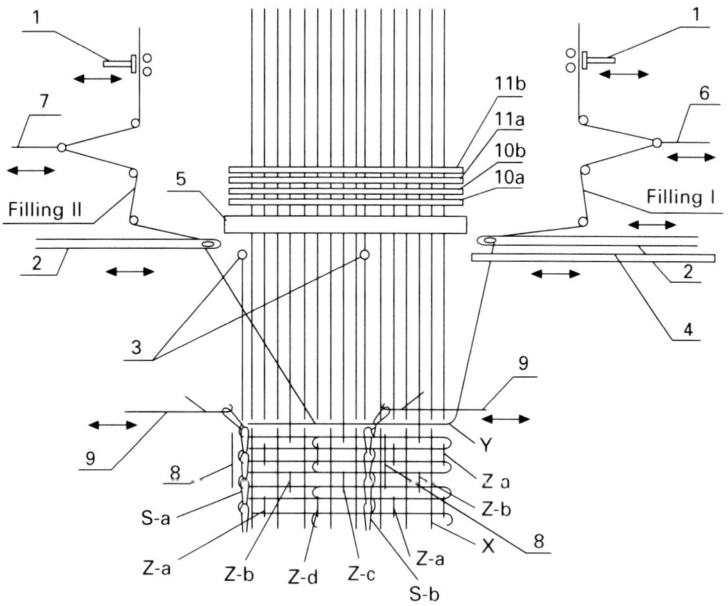

9.56 Filling insertion by needles from both sides [21].

Several needles (rapiers) containing double fill fibers are inserted horizontally through the layers in one motion as shown in Fig. 9.56. The fill fibers are temporarily held on the opposite side by a vertical selvedge needle. The rapiers are then withdrawn to their original position leaving a set of double fill fibers. The Z fibers are fed into the machine parallel with the warp and separated into two layers controlled by the harness movement. When the bottom Z fiber layer is moved to the top and the top Z fiber layer is moved to the bottom achieved by crossing the harnesses, a vertical fiber is added. A reed is brought forward for a beat up. The selvedge needles are lowered and the structure is moved corresponding to the pick spacing. The reed is moved back and the entire cycle is repeated.

Mohamed *et al.* [21] also developed a 3D loom where the warp (0°) fiber, the fill (90°) fiber, Z fiber and +/– bias fibers can be inserted. The bias fibers can only be positioned on the front and back faces of the 3D preform.

In Fig. 9.56, 1 denotes filling lock and selvedge lock, 2 shows filling insertion, 3 shows selvedge needles, 4 shows selvedge hold rod, 5 shows beat up, 6 & 7 show filling tension I & II, 8 shows loop forming rods, 9 shows selvedge latch needles, and 10 & 11 show harnesses. 3Tex [22] has built a commercial 3D loom. The number of layers is limited to 14 as reported in [23]. Bally Ribbon Mills [5] is producing panels with 61 warp layers. It can go up to 81 warp layers.

Biteam [4, 24] has developed a 3D loom where shedding can be formed in both horizontal and vertical positions. In a conventional weaving machine, the interlacement occurs between the two orthogonal sets of yarns, the warp and the fill. It does not produce the interlacement of three sets of orthogonal yarns. Biteam's loom provides a dual directional shedding method to form sheds in the column-wise and the row-wise directions of a multi-layer warp to enable interlacement of the multi-layer warp and two orthogonal sets of fill. Figure 9.57 [24] shows the preferred arrangement of the heddles in the harness frames for carrying out dual-directional shedding. In Figure 9.57 [24], 1 and 2 represent two mutually perpendicular harness frames, 3 shows heddle wire, 4ne and 4se denote heddle eyes, and 5 shows an opening created by the superimposition of harness frames 1 and 2, including the superimposed heddle eyes (4se); an end of multilayer warp yarn (6) is drawn through. All warp ends (6) will be disposed in columns 'A through I' and rows 'a through i'. The warp ends (6), which are labeled (6p), constitute the stationary or

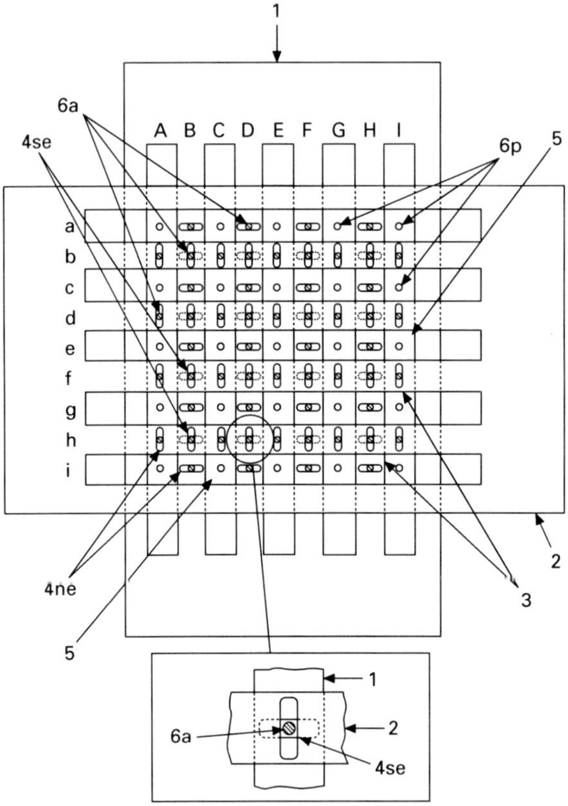

9.57 Preferred arrangement of heddles in the harness frames for dual-directional shedding [24].

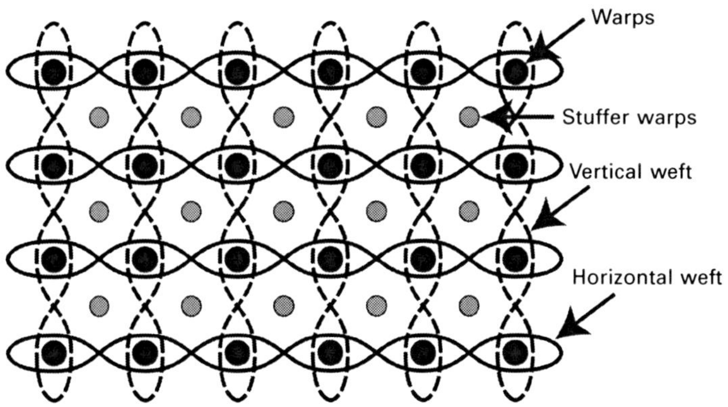

9.58 Weave architecture of 3D fabric [6].

passive warp ends, and the warp ends (6), which are labeled (6a), constitute the movable or the active warp ends. The Weave architecture is shown in Fig. 9.58 [6].

The vertical fills and horizontal fills are inserted into the created sheds of multi-layer warp. The warp fibers interlace with two sets of filling, thus creating an interlaced woven 3D fabric. The 3D fabric produced by this method may be cut into any desired shape without the risk of splitting, but the resulting structures have crimped fibers in all three directions. The fibers are likely to be damaged to a high extent due to dual shedding mechanisms. It is highly desirable for composite reinforcements to minimize the fiber damage.

Uchida *et al.* [25] developed a three-dimensional weaving machine including an improved bias yarn feeding apparatus as an integral part. Figure 9.59 [25] shows a side view of the 3D weaving machine. Figure 9.60 [25] shows a schematic perspective view showing a specific configuration of a 3D woven fabric produced by the 3D weaving machine. In Fig. 9.59 [25], the warp X, bias yarns B1 and B2, and vertical yarn Z are fed from each beam and guided to a frame (5) through a dancer roller and a split guide (6) in a cloth fell section (7). Y shows the filling (weft) between the warp layers. A filling insertion mechanism inserts the fill between the warp layers and outside the set of two biased yarn layers. The vertical yarns Z are inserted from top and bottom by (9A) and (9B). A press (10) is used on the downstream side of the cloth fell section (7). The 3D woven fabric (W) passes through a woven fabric shape retention section (11) and is wound around a fabric winding section (12). Short Brothers plc of Northern Ireland, part of Bombardier Aerospace, developed a narrow weaving machine [26] called five axis weaving to produce spatially reinforced composites (SpaRC). These spatially reinforced composites are designed to have fiber tows oriented in five axes, 0°, 90°, ±45°, and Z directions. A preform with one layer of

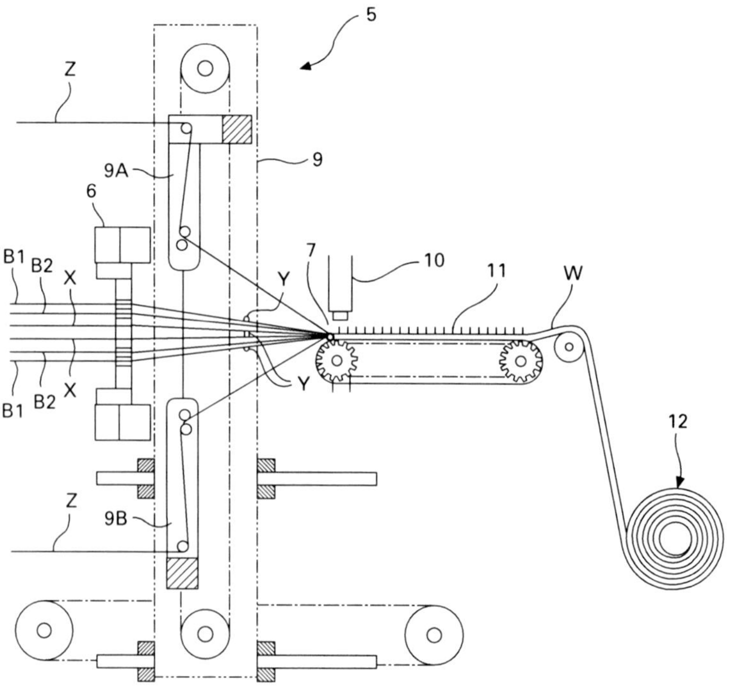

9.59 Side view schematically showing 3D weaving machine [25].

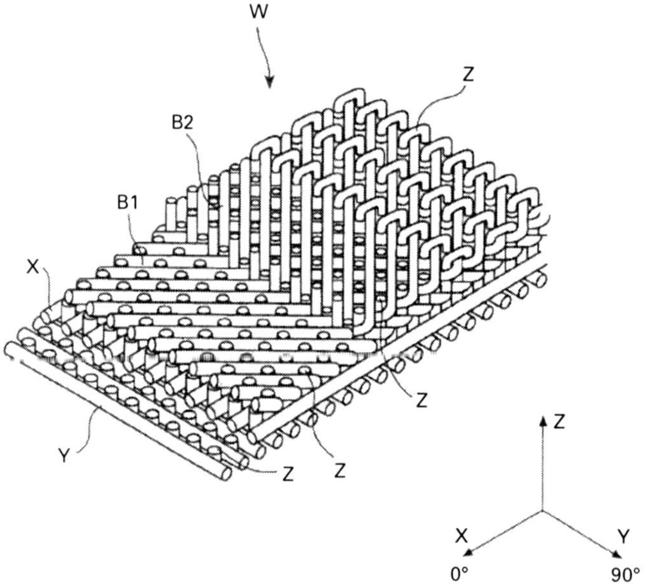

9.60 View of 3D fabric produced by 3D weaving machine [25].

±45° was woven on this loom. Quinn *et al.* [27] shows that, by increasing the amount of Z fibers in 3D orthogonal structures, the damage area after impact was reduced and low crimp, multi-axis, multi-layer woven structures offer improved damage tolerance.

Specially built machine

Bally Ribbon Mills (BRM) [5] has developed a patented weaving machine [28, 29, 30] where fibers can be inserted in X, Y, Z, +45°, and –45° directions. Various 3D structures, such as Pi, T, and cruciform, have been woven. The test shows improved results in Pi structures produced in this machine over woven Pi structure with +45°/–45° ply.

Figure 9.61 [5] shows the front section of the loom, whereas Fig. 9.62 [5] shows the rear section of the BRM's specially built machine. The X direction (O°) fibers were controlled by an electronic Jacquard harness which formed the shedding. Z fibers were also controlled by an electronic Jacquard harness. The filling insertion was carried out by a specially built shuttle operated by a servo motor. +45° and –45° fibers were also controlled by servo motors. The beat up was carried out by individual reed blade. The take up was controlled by a servo motor. There were multiple vertical columns for bias fibers. There were nine positions in each vertical column. Eight positions were filled with fiber bobbins and one position was left empty. The bobbins in each vertical column can move vertically (up and down) as well as move horizontally by servo motors. The bias fibers can be placed in any location up to eight locations between the layers and outside the structure.

9.61 Front section of the machine [5].

9.62 Rear section of the machine [5].

Many structures, such as Pi, T, and cruciform can be woven using fibers in five directions. BRM has built another machine with higher speed and it can produce various 3D structures.

9.4 3D weaving calculations

When designing 3D textile preforms for the composite industry, often the textile engineer is required to calculate the 3D preform construction from a cured composite specification, supplied by the composite manufacturer or the end user. The composite design engineer typically supplies the textile engineer with the following information about the cured composite:

- cured composite part thickness (CPT)
- fiber volume (V_f)
- fiber orientations or fiber percentages (%X, %Y, and %Z)
- fiber type/fiber density (ρ_f)/fiber linear density (yd/lb).

Often the end user will not have all items specified and it is up to the textile engineer to understand the composite part requirements in order to recommend a suitable preform construction. In particular, the fiber linear density is a variable which needs close review by the textile engineer. For high performance composite applications, the composite design engineer may specify fine denier fibers that exceed the capacity of the jacquards or, even worse, are physically impossible to weave. If there is a problem with the customer's supplied specification(s), the textile engineer must find a solution, often making compromises between performance and manufacturability.

From the cured composite specifications, the fabric areal weight (AW) is calculated, then multiplied by the specified fiber percentages to find the X, Y, and Z components of the total areal weight. By multiplying the areal weight components with the appropriate fiber yield, the 3D preform construction for each direction is calculated. Depending on the weave pattern and fiber crimp the 3D preform construction may need adjusting.

In imperial units, the 3D preform construction can be found with the following:

- CPT = cured part (or ply) thickness (in)
- V_f= fiber volume (%)
- %X = fiber percentages in the X direction (%)
- %Y = fiber percentages in the Y direction (%)
- %Z = fiber percentages in the Z direction (%)
- AW = areal weight (lb/in^2)
  - AW_X = X component of areal weight
  - AW_Y = Y component of areal weight
  - AW_Z = Z component of areal weight
- specified fiber (e.g., carbon, fiberglass, etc.)
  - ρ_f = density of specified fiber (lb/in^3)
  - specified fiber count or yield (e.g., 6K or 2220 yd/lbs)

$$CPT = \frac{AW}{\rho_f \times V_f}$$

$$AW = CPT \times \rho_f \times V_f$$
$$AW_X = AW \times \%X$$
$$AW_Y = AW \times \%Y$$
$$AW_Z = AW \times \%Z$$

$$\text{Total ends per inch (EPI)} = AW_X\left(\frac{\text{lb}}{\text{in}^2}\right)$$

$$\times \text{ Warp yarn yield}\left(\frac{\text{yd}}{\text{lb}}\right) \times 36\left(\frac{\text{in}}{\text{yd}}\right)$$

$$\text{Ends per inch per layer}\left(\frac{\text{EPI}}{\text{Layer}}\right) = \frac{\text{Total EPI}}{\text{\# of warp layers}}$$

$$\text{Total picks per inch (PPI)} = AW_Y\left(\frac{\text{lb}}{\text{in}^2}\right)$$

$$\times \text{ Fill yarn yield}\left(\frac{\text{yd}}{\text{lb}}\right) \times 36\left(\frac{\text{in}}{\text{yd}}\right)$$

$$\text{Picks per inch per layer}\left(\frac{\text{PPI}}{\text{Layer}}\right) = \frac{\text{Total PPI}}{\text{\# of fill layers}}$$

One of the easiest 3D weave patterns to design is the 3D orthogonal weave pattern, since all fibers are straight in the X, Y, or Z directions, fiber crimp is minimal and can be eliminated from the calculations. To calculate the total number of Z yarns required per square inch of 3D orthogonally woven preform, the preform thickness must be taken into account. The following equation assumes each Z fiber travels through the entire preform thickness.

$$\text{Z per in}^2 = \frac{AW_Z\left(\frac{\text{lb}}{\text{in}^2}\right) \times \text{Z yarn yield}\left(\frac{\text{yd}}{\text{lb}}\right) \times 36\left(\frac{\text{in}}{\text{yd}}\right)}{\text{CPT}}$$

For example, a 3D woven preform is needed to make a 1.5″ thick composite panel. The composite panel is required to have a fiber volume of 57% using a continuous multifilament standard modulus carbon fiber with an epoxy matrix. After a discussion between the composite and textile engineer regarding the part application and use, the textile engineer suggests a 3D orthogonal weave pattern with 45% of the fiber orientated in the warp (0°) direction, 40% of the fiber orientated in the fill (90°) direction and 15% of the fiber orientated through the thickness (Z) direction. A 12K standard modulus carbon yarn should be used in all three directions with a density (ρ_f) of 0.065 lb/in^3 and a yield of 567 yd/lb. The following calculations could be used to find the X, Y, and Z fabric count for the 3D orthogonally woven preform with 30 warp layers and 31 fill layers.

$$AW = 1.5\text{ in} \times 0.065\,\frac{\text{lb}}{\text{in}^2} \times 57\% = 0.0556\,\frac{\text{lb}}{\text{in}^2}$$

$$AW_X = 0.0556\,\frac{\text{lb}}{\text{in}^2} \times 45\% = 0.0250\,\frac{\text{lb}}{\text{in}^2}$$

$$AW_Y = 0.0556\,\frac{\text{lb}}{\text{in}^2} \times 40\% = 0.0222\,\frac{\text{lb}}{\text{in}^2}$$

$$AW_Z = 0.0556\,\frac{\text{lb}}{\text{in}^2} \times 15\% = 0.0083\,\frac{\text{lb}}{\text{in}^2}$$

$$\text{Total ends per inch (EPI)} = 0.0250\,\frac{\text{lb}}{\text{in}^2} \times 567\,\frac{\text{yd}}{\text{lb}} \times 36\,\frac{\text{in}}{\text{yd}} = 510\text{ total EPI}$$

$$\text{Ends per inch per layer}\left(\frac{\text{EPI}}{\text{Layer}}\right) = \frac{510\text{ Total EPI}}{30\text{ Warp layers}} = 17\left(\frac{\text{EPI}}{\text{Layer}}\right)$$

$$\text{Total picks per inch (PPI)} = 0.0222\,\frac{\text{lb}}{\text{in}^2} \times 567\,\frac{\text{yd}}{\text{lb}} \times 36\,\frac{\text{in}}{\text{yd}} = 453\text{ PPI}$$

$$\text{Picks per inch per layer}\left(\frac{\text{PPI}}{\text{Layer}}\right) = \frac{453\,\text{Total PPI}}{31\ \text{Warp layers}} = 14.6\left(\frac{\text{PPI}}{\text{Layer}}\right)$$

$$\text{Z per in}^2 = \frac{0.0083\,\frac{\text{lb}}{\text{in}^2} \times 567\,\frac{\text{yd}}{\text{lb}} \times 36\,\frac{\text{in}}{\text{yd}}}{1.5\ \text{in}} = 113\ \text{Z per in}^2$$

9.5 Applications and future trends

As 3D woven fabrics, structures and composites become more popular and inexpensive, they will be increasingly used in various industries for a range of new applications besides aerospace, such as, consumer products, sporting goods, boat and shipbuilding industries, automotive, construction and bridges, bicycle, alternative energy sector.

Although there are some simulations available for a few 3D structures, the weave design software currently available is limited. As a result, engineers have to spend a considerable amount of time figuring out how to weave the particular structure and coming up with a weave design. As software and the range of simulations improve, the time spent designing the weave structure will significantly decrease and the whole process of manufacturing 3D woven fabrics and structures will become much faster and cost effective.

The properties of 3D preforms are dependent upon fiber properties, weave architecture, and fiber damage during weaving. It is very important to minimize fiber damage in woven reinforcements for composites. Research is in progress to predict the properties of 3D structures based on the weave architecture. The loom set-up and fiber damage have to be considered in addition to the weave architecture for modeling the properties of the woven composites. Unfortunately, composite engineers lack textile knowledge and textile engineers lack composite knowledge. An improved understanding and knowledge of one another's subject areas would significantly improve the situation. Advanced research and developments in the areas of computerized weave design and simulation would also be beneficial.

9.6 Acknowledgements

The author would like to express his sincere thanks to Raymond Harries, President of Bally Ribbon Mills for allowing him to write this chapter. The author would also like to express thanks to Lorraine Hornig, Curt Wilkinson, and Mark Harries for their help with the line drawings and photographs.

9.7 References

1. Mohamed, M. H., 'Three-Dimensional Textiles', *American Scientist*, Vol. 78, Nov.–Dec. 1990
2. Islam, M. Amirul, '3-D woven near net shape structures', ASME International Congress & Exhibition, November 14–19, 1999, Nashville, Tennessee
3. Chen, X. and Hearle, J. W. S., 'Developments in design, manufacture and use of 3D woven fabrics', 9th International Conference on Textile Composites, Recent Advances in Textile Composites, University of Delaware, Newark, Delaware, October 13–15, 2008
4. Khokar, N., (Biteam) 'Second generation woven profiled 3D fabrics from 3D weaving', First World Conference on 3D Fabrics and their Applications, Manchester, April 10–11, 2008
5. Courtesy of Bally Ribbon Mills, Bally, Pennsylvania 19503, USA
6. Courtesy of Biteam AB, Gothenburg, Sweden
7. Schmidt, R., Bersuch, L., Benson, R., Islam, A., 'Three dimensional weave architecture', US Patent No. 6,712,099 B2, March 30, 2004
8. NASA Engineering and Safety Center (NESC) web site on Composite Crew Module
9. Islam, M. Amirul, 'Narrow fabrics for medical textile applications', Medical Textiles Conference, Clemson University, Clemson, South Carolina, September 9–10, 1997
10. Dambrine *et al.*, 'Development of 3D woven, resin transfer molded fan blades', 9th International Conference on Textile Composites, Recent Advances in Textile Composites, University of Delaware, Newark, Delaware, October 13–15, 2008
11. NASA Office of Chief Technologist Press Release on 'Development of new game changing technologies', November 2011
12. Prichard, A. K., Presentation at First World Conference on 3D Fabrics and their Applications, Manchester, April 10–11, 2008
13. Islam, M. Amirul, 'Designing of narrow woven fabrics: 3-D woven near net shape structures', Narrow Fabrics Conference of Clemson University, November 10–11, 1999, Charlotte, North Carolina
14. Islam, M. Amirul, '3-D woven near net shape and multi-axial structures', The Textile Institute 80th World Conference, April 16–19, 2000, Manchester
15. Schmidt *et al.*, 'Woven preform for structural joints', US Patent # 6, 874,543 B2, April 5, 2005
16. Islam, M. Amirul, 'The damage suffered by the warp yarn during weaving', Ph.D. thesis, University of Manchester, 1987
17. Islam, M. Amirul, 'Shed and shedding trap geometry', M.Sc. thesis, University of Manchester, 1985
18. Islam, M. Amirul, 'Designing of narrow woven high temperature fabrics', 25th International SAMPE Technical Conference, October 26–28, 1993, Philadelphia, Pennsylvania
19. Kipp, H. W., *Narrow Fabric Weaving*, Verlag Sauerlander, 1989
20. Greenwood, K., Islam, M. Amirul and Downes, L., 'Research into narrow fabrics opportunities', *Melliand Narrow Fabric and Braiding Industry Magazine*, Volume 29, 1992
21. Mohamed *et al.*, US Patent # 5,085,252, Feb. 4, 1992, and US Patent # 5,465,760, Nov. 14, 1995, North Carolina State University, North Carolina

22. 3Tex, Carry, North Carolina
23. Buckley, M., 'A Report on 3D Preform Technologies for Advanced Aerospace Structures', Airbus, UK
24. Khokar, N., (Biteam) 'Woven 3D material', US Patent # 6,338,367 B1, Jan. 15, 2002
25. Uchida *et al.*, 'Three dimensional weaving machine', US Patent # 6,003,563, Dec. 21, 1999
26. Wilson *et al.*, 'SPARC 5 Axis, 3D Woven, Low Crimp Preforms', Bombardier Aerospace, Short Brothers plc, Resin Transfer Molding SAMPE Monograph No. 3, 1999
27. Quinn *et al.*, '5 axis weaving technology for the next generations of aircraft-mechanical performance of multi-axis weave structures', 9th International Conference on Textile Composites, Recent Advances in Textile Composites, University of Delaware, Newark, Delaware, October 13–15, 2008
28. Bryn, L., Islam, M. Amirul, Lowery, W. L. and Harries, H. D., 'Three dimensional woven forms with integral bias fibers and bias weaving loom', US Patent # 6,742,547 B2, June 1, 2004
29. Bryn, L., Nayfeh, S. A., Islam, M. Amirul, Lowery, W. L. and Harries, H. D., 'Loom and method of weaving: three dimensional woven forms with integral bias fibers', US Patent # 6, 892,766 B2, May 17, 2005
30. Nayfeh, S. A. *et al.*, 'Bias weaving machine', US Patent # 7,077,167 B2, July 18, 2006

Part III

Applications of woven textiles

10

Woven textiles for automotive interiors and other transportation applications

J. M. HARDCASTLE, Consultant, UK

Abstract: Woven textiles are an important element within the car interior and also in other areas of transportation including bus, train, marine and aircraft. This chapter discusses these applications and gives a technical insight into the background of how woven textiles developed within the automotive industry. Also discussed are how and where within the vehicle they are used and the technical issues which the fabric producer has to overcome, and how fabrics have evolved to meet the varied but always very stringent requirements of the industry.

Key words: automotive, textiles, woven fabrics, yarns, test requirements, fabric constructions, transportation textiles.

10.1 Introduction

The transportation market covers a wide range of products each sector having its own special requirements in terms of design, colouration, fabric surface characteristics, technical requirements, price, weight, density, air permeability, abrasion, etc. Although using large quantities of decorative fabrics, the transportation market has been broadly classified as an industrial user of textiles. These products include woven fabrics to a very large degree, and the sector as a whole is the largest consumer, in terms of quantity, of all industrial uses of textile products, and has therefore developed worldwide into a very important part of the textile fabric market.

The transportation market can broadly be split into the following categories:

- Automotive – mainly cars, 4×4 vehicles, etc., but also including commercial goods vehicles such as trucks often made by a subdivision of the same companies. Automotive is by a large margin the biggest user of textile products within the entire transportation sector.
- Buses and coaches – predominantly use woven pile fabrics but also flat wovens. The fabrics are mainly decorative but have to be made to high levels of performance particularly in terms of fire retardancy.
- Aircraft – use decorative fabrics in the seating areas and industrial fabric types in other areas such as escape chutes, etc. Use is also made in aircraft of composite materials, some of which incorporate woven fabrics, to replace metal and other heavy weight items in the aircraft in order to reduce overall weight.

- Marine – use traditional furnishing fabrics where considerations of design and colour are very important in such areas as drapes and bedding, etc., in cruise liners to industrial fabrics in the more specialised craft.
- Rail – in terms of volume the use is largely confined to specialised decorative fabrics with special properties.

10.2 Automotive applications of woven fabrics

The automotive business is awash with statistics but by any measure it is a vast business and one which textile suppliers the world over eye with envy in terms of fabric consumption. However, it is a very difficult industry to supply successfully due to very high entry costs, particularly in terms of quality, very demanding delivery requirements, design, research and development plus very high quality requirements.

However, since it is by far the biggest fabric consumer within the transport sector, we shall devote most of the space in this chapter to examining the way in which woven fabrics have developed and have met the needs of the industry.

In 2008, over 52 million cars were produced worldwide (slightly down on 2007 which was a record peak year). However, this compares with just less than 40 million in 1999. Cars are broadly defined as passenger carrying vehicles with no more than eight seats and comprise around 85–87% of the total vehicles on the road. The remaining 13–15% or so comprise commercial vehicles (trucks, etc.) and buses and coaches.

The total amount of fabric used in the car is probably in the region of 20 kg plus, or between 45 and 60 m^2, of which approximately 18% by weight is seating fabrics, 22% is taken up by carpets, 30% by tyres, sound damping and other interior items and 30% by composite textile materials. Some of these replace items formerly made of metal or plastic. The textile items that are visible and appreciated as part of the interior are seating, carpets and headlining material, door panels and sometimes even dashboards. The seating areas and door panels are often produced from woven fabrics – either flat woven or pile woven – whereas for the headlining (i.e., interior roof space), a lightweight knitted structure is often used for its ability to be moulded easily to curved shapes. In all cases the fabrics are likely to be either *bilaminates* or *trilaminates*. This means that the face fabric has been laminated to a foam and sometimes to knitted scrim, a process we shall look at later in this chapter.

Wherever the fabric is used, it is subjected to severe testing procedures before approval and it is the form which these tests take that determines the types of fabric that can be used. These tests are reasonably uniform throughout the industry but the methods used do vary according to the standards laid down by the various OEMs (original equipment manufacturers, which in effect are the car manufacturing companies themselves).

The overall objective is to determine the suitability of the fabric for the intended end use. In the interior fabrics, which include most of the woven products used, the main objective would be to test the 'wearability' of the product. This term 'wearability' is a coverall for the ability of the fabric to survive the rigours of use during the lifetime of the vehicle. It has very many parts of which probably the most important ones are abrasion resistance, light and UV (ultraviolet) degradation in terms of colour fading and strength loss. It was this aspect which was the main factor in the near disappearance of nylon from the exposed areas within the interior of the vehicle (with the notable exception of seat belts and to some extent roof lining both of which are largely sheltered from UV exposure).

10.2.1 Test requirements for interior fabrics

This list of test requirements is not exhaustive, but covers the main areas in the standard test regimes of most OEMs.

Abrasion

The resistance of fabrics to surface abrasion is one of the most critical requirements, but the standards required are more severe than most of those required by traditional upholstery fabrics. Special tests have been devised which put the fabric under great stress. Different OEMs specify different test methods and these in turn tend to favour different fabric types. Three of the methods are described below:

- *Martindale test.* This is a fabric-to-fabric test where a standard cross-bred wool cloth is abraded, under a standard pressure, against the test specimen. The number of rubs required to create thread breakage is recorded. This test is particularly severe on pile surfaces but less so on flat woven cloths.
- *Taber test.* This uses an abrading wheel made of rubber with aluminium oxide abrasive particles which is rotated in pressurised contact with the fabric until wear occurs.
- *Schopper test.* This uses discs of specially graded emery paper rotating against the fabric. The test is less severe on pile surfaces which survive the test better than flat woven products.

In terms of woven fabrics there are many features which affect the results, such as the tightness of the cloth structure, the fibre denier, yarn construction and surface finishes, etc.

Pilling

This is the formation of small balls of fibre on the surface during wear. This is tested either by the use of the Martindale machine or a special ICI pill box where the fabric is wrapped around rubber tubes, face outermost, and placed inside rotating cork-lined boxes for a given time and the 'pills' compared to photographic standards on a scale of 1 to 5 (5 being the worst). Pilling can be a significant problem but can be reduced by back coating the fabric which is common practice with automotive fabrics.

Light fastness and degradation

This is an important test for interior cloths since they are subject to high amounts of light and heat (up to 120°C in hot climates), although different tests are used including 'weatherometres' which add moisture during the test procedure and also includes other features.

The fabric sample is prepared and exposed to very powerful arc lamps which simulate the different wavelengths of natural sunlight. Figure 10.1 shows the similarity of the xenon lamps, when fitted with suitable filters, to the strong sunlight in Miami. These lamps are fitted to machines in which the sample is exposed for specified amounts of time. In addition to light they create a lot of heat (and, in the case of weatherometres, moisture). The sample is then assessed against a grey scale 1 to 5 or a blue scale 1 to 8 (in the case of the blue scale a series of standard dyed wool samples are exposed

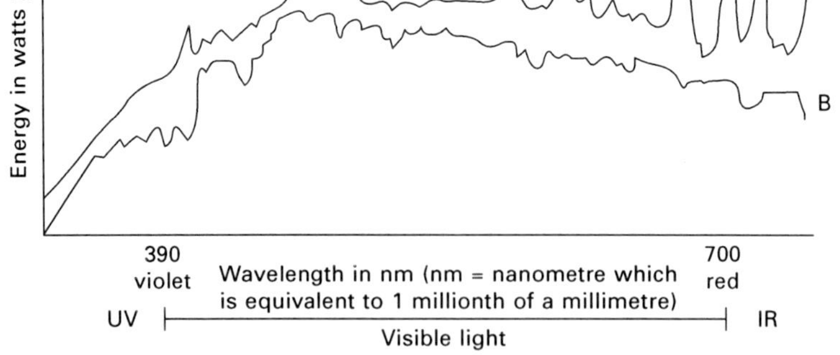

10.1 Comparison of daylight and xenon arc (Fung and Hardcastle, 2001). Used with permission from Woodhead Publishing Ltd.

at the same time and the test sample compared to these after exposure). In both cases, the higher the number the lower the amount of fading.

In some tests the samples are tested for abrasion and strength before and after the light test.

Flammability

The flammability test is less severe than the ones used for public transport, airliners and cruise ships. The reason for this is not very clear but could have to do with the fact that to pass the critical tests for public transport, it is often necessary to use inherently flameproof fibres, most of which are very expensive and they may not be able to pass the other rigorous tests required of automotive cloths. Another reason is that in most cases interior car fabrics can be non-flammable when coated with fire retardant or laminated to non-flammable foam and the resultant laminate can then meet the requirement. This could be because escape from a burning airliner or ship is more difficult, whereas private cars have two or four doors, and fire is seen as a lesser risk.

The test requires a horizontal strip of fabric to be ignited at one end and the length in centimetres burned in 1 min before it extinguishes is measured. This distance can vary according to the OEM specification and also according to where in the car the fabric is used.

Dimensional stability

This measures properties like stretch and recovery, shrinkage after immersion in water and drying out, and very often again varies according to the OEM and also to where the fabric is being used. These parameters are often specified for a particular car model and may vary significantly for a different model.

Fabric strength

There are four aspects: tensile strength, tear strength, bursting strength and seam strength. Each of these require their own test regime and the only one which is less relevant to woven fabrics is the bursting strength test which is applied mainly to knitted fabrics which are less robust than wovens in this respect and are often used in areas where extreme fabric stretch is required.

The tensile test reveals any problems with the fabric structure or yarns used. The seam test reveals problems due to loose weaves, which allow the yarns to slip over each other opening up gaps in the sewing, etc. It also reveals problems with yarns and the friction between them due to other factors, which will be discussed in Section 10.3. The tear strength can be affected by the coatings applied to the fabric and also the weaves used; the tighter the fabric structure the more susceptible it is to tearing.

There are many other tests applied such as handle and drape, breathability, soiling, snagging, environmental ageing, recovery from stretch, dimensional stability, anti-static, frosting and fogging, etc., but these are beyond the scope of this chapter. They can be referred to in detail in Fung and Hardcastle (2001). These tests define, to a large extent, the types of fabric, fibres and yarns which can be used in the car and in the following section we look at the use of woven fabrics in particular.

10.3 Woven fabrics in car interiors

The use of fabrics in cars goes back probably to the early years of the mass production of the motor vehicle. The first seat coverings were mainly leather or imitation leather, but also included wool and cotton before synthetic and other man-made fibres were available. The war years and afterwards saw the introduction of vinyl, both in the form of imitation leather and fabric made from fibres and yarns spun from a derivation of vinyl. This had the advantage of being able to be pigment dyed or spun dyed in the melt state before fibre extrusion to a very high standard of light fastness. Nylon made an appearance during and just after the war and also found its way into the car sometimes blended with other fibres. The technical requirements were nowhere near as stringent as they are today, but even then durability, the ability to be cleaned easily and fastness to light all were important.

During this time PVC (polyvinyl chloride) coated fabric was an important development when applied to woven and knitted base fabrics since it allowed 'leather look' fabrics to be produced, particularly when embossed to imitate the surface appearance of the real leather. The colours were light fast and the product had acceptable durability and a certain amount of stretch, particularly if the base cloth was a warp knit structure. These attributes meant that this product in various forms became very widely used in mass produced cars up until the early 1970s. However, PVC had some big drawbacks, among them being the 'stickiness' in hot weather and poor breathability, both of which led to it being somewhat uncomfortable when sat on for long periods, especially in hot weather. Nylon fabrics began to be used more widely again to try and add some much needed variation.

In addition to this, rising living standards were leading to an increasingly selective customer base demanding a greater amount of styling appeal both inside and outside the car, which was not being addressed by the limited textile products on offer. This was further emphasised when the Arab-Israeli war of 1973 created oil shortages and unprecedented increases in fuel costs which led to a concentration of effort on exterior styling to reduce drag, etc., and to increase fuel efficiencies. Greater use was made of glass and sleek, sloping windscreens and larger windows all of which increased the amount of light – and UV plus heat in many countries – inside the car, putting even greater strains on the fabric performance.

All these factors resulted in many cars being designed with very similar exterior lines which in turn caused the car companies to add more expertise and styling to the interior to differentiate their model line up, thereby creating a demand for more adventurous fabrics in terms of design and colour. At the same time the engineers were placing ever more severe performance requirements on the fabric suppliers.

A new approach was needed for the fabrics to be used within the car.

10.3.1 The new age of car fabrics

The driving forces mentioned above sent the fabric suppliers back to the drawing board to re-engineer the fibres, yarns and fabrics available to them to address this emerging requirement. There was no doubt that textile fabric had the edge over PVC-coated products in many respects, not only in terms of comfort and breathability but most of all in design and colouration potential. The use of nylon and polypropylene was becoming suspect after many failures in use of these products due to severe degradation in strength and abrasion caused by UV exposure. Many OEMs discouraged their use in areas which were exposed to sun such as upholstery, but not in other areas like seat belts, roof linings or carpets. The search for an 'all round' fibre at an economic price became urgent and was ultimately filled by polyester which, due to its suitability for almost all textile applications, has often been referred to as the universal fibre.

Table 10.1 shows the relative properties of the main fibres available to the manufacturer for transportation.

Yarn development

Polyester had many properties which made it suitable for many applications inside the car, including the major areas of seating and door panels. However, this was only part of the problem since every year the requirements of the OEMs were ramped up; whenever a requirement had been met, the process of continuous improvement meant that the requirement had to be increased. So ways of improving abrasion were looked at and dramatic improvements made by the development of polyester air textured continuous filament yarns using the Taslan process pioneered by Du Pont. This was a process in which continuous filament yarns were passed through a jet of high pressure air causing intimate entanglement of the filaments in the yarn cross section, thereby creating small continuous loops which were locked in position (see Fig. 10.2). The effect this had on the stability and surface abrasion resistance

Table 10.1 Basic properties of fibres (Fung and Hardcastle, 2001). Used with permission from Woodhead Publishing Ltd

	Density (g/cm^3)	Melting point (°C)	Tenacity (g/den)	Stiffness (flexural rigidity) (g/den)	Limiting oxygen index (% oxygen)	Abrasion resistance	Resistance to sunlight
Acrylic	1.12–1.19	150d	2.0–5.0 (HT)	5.0–8.0	18	Moderate	Excellent
Modacrylic	1.37	150d	2.0–3.5	3.8	27	Moderate	Excellent
Nylon 6	1.13	215	4.3–8.8 (HT)	17–48	20	Very good	Poor – good when stabilised
Nylon 66	1.14	260	4.3–8.8 (HT)	5.0–57	20	Very good	Poor – good when stabilised
Polyester	1.40	260	4.2–7.5 (HT)	10–30	21	Very good	Good – excellent when stabilised
Polypropylene	0.90	165	4.0–8.5 (HT)	20–30	18	Good	Poor – good when stabilised
Wool	1.15–1.30	132d	1.0–1.7	4.5	25	Moderate	Moderate
Cotton	1.51	150d	3.2	60–70	18	Moderate	Moderate

Table listing basic fibre properties which are important for fabrics intended for transportation. Note the low melting point for polypropylene which is not much above temperatures which can be reached in car interiors in tropical climates. Items marked 'd' indicate that this is the temperature at which the fibre starts to degrade since they do not melt in the traditional sense.
From this it can be seen that polyester, while not the best in every category, reaches a high and acceptable standard in all the important areas.

10.2 Air textured yarn showing the entanglement of the filaments which are locked in position to create a high abrasion resistant yarn but with limited stretch. See Fig. 10.3 for an illustration of different rates of overfeed of the component yarns to give varying surface texture effects.

of the woven fabric was dramatic and air textured polyester yarn quickly established itself firstly in Europe then gradually worldwide as *the* yarn to use for decorative applications within the car where abrasion was a critical issue.

Many methods and variations were developed – parallel textured, core and effect textured, multiple feedstock yarns with variable input rates of the feedstock yarns to create slub type effects, etc. (Fig. 10.3). The yarns were available in ecru for yarn or piece dyeing and also as colour spun using pre-dyed or melt-dyed filaments for the feedstock yarn.

Due to the complex process, and inevitable increase in yarn count caused by the bending of the fibres through the texturing process, the textured yarns were quite expensive and therefore alternatives were tried such as the 'false twist' process which had a similar effect in creating bulk by the twisting of the yarn > heat setting > untwisting at a lower temperature. The effect of this process was that the twist which had been heat set into the yarn is false since when untwisted the now parallel filaments try to assume the twisted or heat set position causing a bulking up of the yarn as they spiral around each other. In order to make the yarn weaveable it is passed through an air jet to create intermittent 'knots' or 'nips' along its length at intervals of around 1 cm, as shown in Fig 10.4.

This process created an attractive high bulk yarn via a cheaper route but the ballooning of the filaments between the 'nips' caused severe abrasion issues and could only be used if the fabric was very heavily back coated or the yarn was buried in the fabric or used on the back of the fabric and protected by a dominant air textured weft yarn. Despite this, however, it found a major use particularly in head liners where abrasion was less of a problem.

These yarns came in two main types FTF (false twist fixed, which underwent a setting process) and FT (missing the setting process). The FTF yarns were more stable but less stretchy and the FT yarns were unstable and had a tendency to shrink but had more stretch.

Staple spun yarns have always been used in automotive applications, usually in twofold form, particularly spun polyester, wool, polyester wool blends and even in rare cases blends with spun viscose and acrylic. Spun yarns almost always create a more friendly and comfortable handle than air

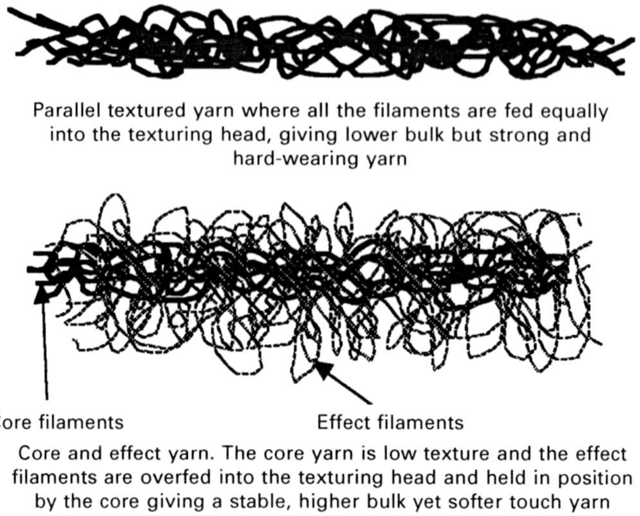

10.3 Comparison of core and effect yarns (Fung and Hardcastle, 2001). Used with permission from Woodhead Publishing Ltd.

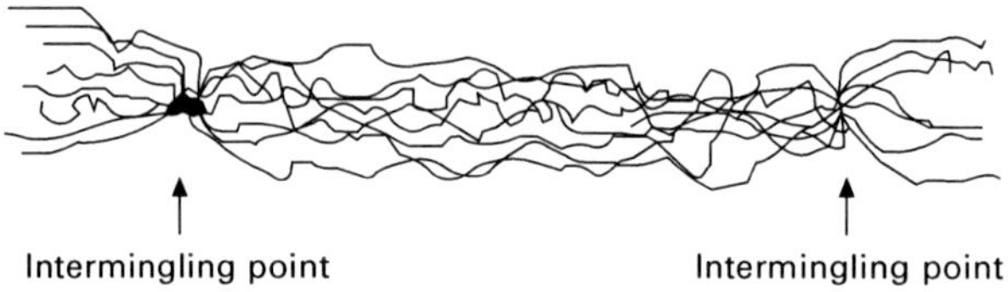

10.4 False twist yarn (FT or FTF) in relaxed condition showing how the crimp develops. The intermingling points are introduced to hold the yarn together and permit easier processing. Between these points the filaments are loose and can cause abrasion problems if the yarn axis is exposed on the cloth surface. Caution must be exercised when using these yarns in automotive structures. This method, however, is one of the cheapest ways of creating high bulk filament yarns. (Fung and Hardcastle, 2001). Used with permission from Woodhead Publishing Ltd.

textured yarns but apart from price issues they almost always have a lower abrasion result unless the fabric construction is increased. These factors mean that often they are used in combination with textured yarns or if used 100% they find their way to the top end of the market for higher trim levels in fabrics of high construction and short float lengths. Mercedes, for instance, used spun yarns in wool and wool blends for several trim levels.

Constructed or 'fancy' yarns are used but have to be specially developed for automotive fabrics to pass abrasion specifications. Boucle yarns have to be well bound to avoid loops being pulled out in the snagging test. Chenille yarns have to have their tufts well anchored into the core (sometimes by using low melt polyester in the core which melts during the stenter finishing process which binds in the tufts).

Flock yarns were used in large quantities by BMW for many years but again had to be almost totally reconstructed to pass the relevant tests. This proved to be one instance where polyester did not meet the criterion, since it has a much lower resistance to fracture (i.e. it is more brittle) than nylon which bends and recovers (one of the reasons it is used in blends with wool for carpets, etc.) so it proved to be unsuitable for use as the pile component or flock. Yarns had to be developed with nylon pile and viscose core where the nylon was treated to increase resistance to light, UV degradation and fading (see Fig. 10.7 for view of flock yarn compared with chenille). In almost all cases these 'fancy' yarns are used for decorative effect along with the ground yarns.

Introducing stretch

Another requirement becoming very apparent was the need for fabrics used in interiors to have flexibility and stretch. This was to make seat fabrication easier and also to address the problem covering the deep draw door panels which were becoming increasingly three dimensional and moulded into quite remarkable shapes. An example of this is illustrated in Fig. 10.5, but almost all cars have this feature to a greater or lesser extent. This presented a serious problem for woven cloths, whether flat woven or pile woven velvet types, and if they were not to lose market share to knitted structures it was important that it was solved.

10.5 Moulded door panel and armrest on VW Golf showing the stretch required in the fabric to mould around the deep contours. A flat woven fabric is used in the illustration.

Woven cloths comprising warp and weft yarns laid straight in the cloth at right angles were in fact the most stable structures known, so either they had to be modified through the finishing process or new yarns had to be researched and developed which were more suitable for this end use. However, it was necessary to define what kind of stretch was required. Did it need to be two way (i.e. stretch *and* recover)? Or was one way sufficient (i.e., the 'chewing gum' effect) where the yarn or fabric stretched but did not recover?

Two-way stretch solutions

The use of false twist yarns was one partial solution since these had a lot more stretch than air textured yarns, but had the abrasion problem. Elastomeric fibre content (e.g. Lycra®) was the obvious solution but again the elastomeric component had to be small and well hidden due to high cost and poor abrasion and light degradation. However, despite the disadvantage it did give high amounts of recoverable stretch and has been widely used, usually incorporated into the yarn before weaving.

Fabric finishing was another way in which some low amounts of extensibility could be induced either by totally relaxing the cloth to induce high yarn crimp or designing special properties by weaving tensions and finishing technique. Figures 10.6 and Table 10.2 show the theory behind this. The technique is based on altering the fabric geometry during the finishing

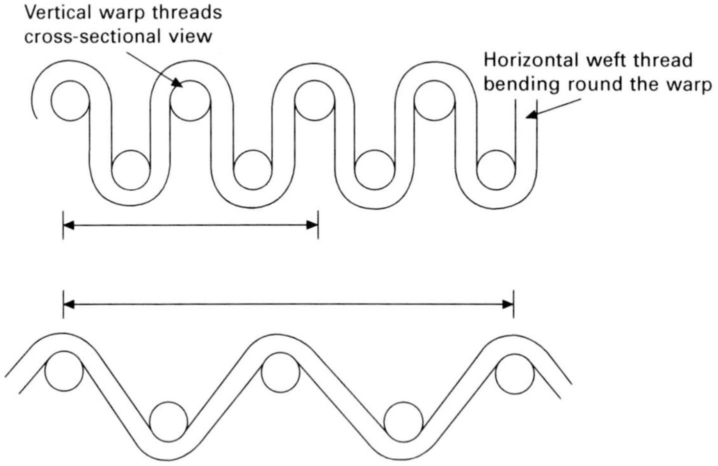

10.6 Illustration of how the introduction of high yarn crimp in a fabric by weaving and finishing techniques can help to increase extensibility into the fabric to make it more manageable in the engineering of car interior components. (Fung and Hardcastle, 2001). Used with permission from Woodhead Publishing Ltd.

Table 10.2 Weaving and finishing tensions (Fung and Hardcastle, 2001). Used with permission from Woodhead Publishing Ltd

		Weaving tension	Greige stretch	Finishing tension	Finished stretch	Effect
Warp	A	High	Low	High	Lowest	Lowest warp stretch (Length gain)
	B	High	Low	Low (or overfeed)	High	Crimp interchange (Length loss)
	C	Low	High	High	Low	Crimp interchange (Length gain)
	D	Low	High	Low (or overfeed)	Highest	Highest warp stretch (Length loss)
Weft	E	High	Low	High	Lowest	Lowest weft stretch (Width gain)
	F	High	Low	Low	High	Crimp interchange (Width loss)
	G	Low	High	High	Low	Crimp interchange (Width gain)
	H	Low	High	Low	Highest	Highest weft stretch (Width loss)

process to interchange crimp and therefore potential stretch. This technique was not really a total solution but was useful for making cloths easier to fabricate.

Another method similar to the introduction of elastomerics was to use stretch polyester made from special polyester polymer known as PBT (polybutylterepthalate). This had stretch and recover properties and was capable of being dyed, albeit at lower temperatures. The stretch potential and power of recovery of PBT were very much less than elastomerics but capable of filling certain requirements.

One-way stretch solutions

This is an important property of woven fabrics when applied to door panel material and also to head linings (see Fig. 10.5) due to the requirement to apply the fabric to the casing in a three-dimensional mould often under pressure. The fabric had to be capable of being easily stretched around the curves but once in place there was no requirement to recover from the deformation. This problem was addressed by the use of yarns composed of partially orientated filaments; the yarn is known as POY (partially orientated yarn).

When any synthetic filament, including polyester, is extruded and the filaments are drawn together to create the yarn (before further processing such as texturing) they are 'drawn' or stretched out. This drawing process reduces the thickness and denier of the filaments and at the same time increases the strength and reduces the potential to stretch further afterwards. In other words, it stabilises the yarn to a large extent. The extent of the drawing process is carefully controlled and when completed the yarn is known as FOY (fully orientated yarn). If this drawing process is only partially completed it creates the POY version, leaving a potential for the yarn to be stretched or drawn out further. This potential can be realised either in yarn form or in fabric form after weaving with no danger of the fabric trying to recover afterwards.

This, therefore, is an ideal fabric to use in moulding applications. Stretch fabrics referred to earlier can also be used but they tend to be more expensive and always have the danger of creep, creating faults if the moulding process has not entirely removed the potential recovery property.

10.4 Fabric constructions and finishing processes

10.4.1 Cloth and yarn structures

We have taken some time to study the yarns used for automotive applications, as they are really the most important element when considering the development of these fabrics. The actual fabric constructions are often specified based on past experience of successful cloths which are reliable – OEMs don't like dramatic changes, better safe than sorry! So the introduction of a brand new yarn is quite a significant change in this application. Design and colour, however, are a different matter.

These days colours and dyestuffs can be changed without too much risk due to years of development by dyestuff manufacturers such as BASF and Ciba-Geigy and Sandoz, both now merged as Novartis, all of whom did early pioneering work on creating dyestuffs which could meet the increasingly strict lightfastness requirements.

The air texturing process itself can create interesting colour effects by using coloured feedstock yarns where the filaments are dyed in the melt before fibre extrusion. By this method multi-colour marl and mixture effects are possible, without changing the basic performance of the yarn.

By using colour and weave effects, subtle and interesting fabrics can be created. Jacquard woven fabrics are of course widely available as piece dyed or colour woven. In the early days of the development of air textured yarns and fabric, the jacquard option was not thought to be possible due to the fibrous nature of the yarn, making weaving extremely difficult through the traditional jacquard harnesses, but improvements in both harness design and careful yarn development resulted in successful fabrics being developed. This aspect is described in more detail in the section on jacquard flat woven cloths below.

The actual constructions used often vary from one manufacturer to another but a few well tried and tested examples are shown in Table 10.3.

Table 10.3 A selection of fabric structures used for automotive interiors. Note that most of the yarns would be polyester and frequently air textured (Fung and Hardcastle, 2001). Used with permission from Woodhead Publishing Ltd

Warp	Ends cm (inch)	Weft	Picks cm (inch)	Singles g/m^2 (oz/yd^2)
420 dtx (14cc or 2/28cc)	20 (51)	830 dtx (7cc or 2/14cc)	17 (43)	236 (7.0)
540 dtx (11cc or 2/22cc)	20 (51)	830 dtx (7cc or 2114cc)	16 (41)	254 (7.5)
420 dtx (14cc or 2128cc)	40 (101)	540 dtx (11cc or 2/22cc)	20 (51)	289 (8.5)
540 dtx (11cc or 2/22cc)	25 (64)	830 dtx (7cc or 2/14cc)	18 (46)	298 (8.8)
420 dtx (14cc or 2128cc)	30 (76)	830 dtx (7cc or 2114cc)	20 (51)	307 (9.0)
830 dtx (7cc or 2/14cc)	20 (51)	830 dtx (7cc or 2/14cc)	16 (41)	315 (9.3)
1300 dtx (4.5cc or 2/9cc)	15 (38)	1300 dtx (4.5cc or 2/9cc)	12 (30.5)	368 (10.8)
830 dtx (7cc or 2/14cc)	25 (64)	830 dtx (7cc or 2/14cc)	18 (46)	375 (11)
600 dtx (10cc or 2120cc)	40 (101)	600 dtx (10cc or 2120cc)	30 (76)	441 (13)

Preferred yarn and fibre types

Due to ever increasing standards of (mainly) abrasion, strength and resistance to sunlight and the ever present need to keep costs down, the polyester fibre rose to prominence along with the air-jet texturing method of yarn production.

Table 10.1 shows the main basic properties of the various fibre types and it can be seen that, while polyester is not the best at everything (e.g., nylon is stronger, hence used in seat belts, and acrylic has better lightfastness), when considered overall it possesses all the main attributes required of automotive trim cloths to a standard which is 'good enough' for the job while at the same time retaining a low cost base. It is for this reason that polyester has risen to be *the* preferred fibre for use in these applications. However, this is mainly for the seating areas and other areas exposed to light and requiring high strength. Other areas such as headlinings, carpets, seat belting, sound damping, etc., do not require these properties to the same extent and therefore other fibres are often used, such as polypropylene for carpets, nylon for seat belts and often headlinings.

Wherever a spun yarn is used it is almost always in the two-fold form to maximise abrasion resistance. In yarns the air textured route probably offers the ultimate in abrasion strength and resistance to thread slippage during seat manufacture. However, this does not mean that other yarn structures are not used to create special effects and reduce prices. For cost reduction the false twist method is frequently used, often as a warp yarn but also occasionally as weft, but in all cases it is kept well hidden and away from the fabric surface in woven structures.

In the past the use of viscose and cotton was rare, although in lower cost models acrylic was sometimes used. Environmental concerns, however, have aroused interest in trying to incorporate natural fibres and those not using hydrocarbons during manufacture.

In higher trim levels, and more expensive car models, wool is often used to add greater comfort, mainly due to better moisture absorption, but even then it is often used in conjunction with, or blended with, polyester or occasionally nylon.

Pile yarns are often used in woven structures since they produce a pile surface which naturally has high abrasion and are cheaper than true woven velvet and compete mainly with weft and warp knitted pile fabrics in automotive trim uses. There are two main types of pile yarns: chenille and flock.

Chenille yarns are constructed by twisting two or more fine core yarns together while at the same time introducing at right angles the pile yarn and securing them in position by the twist of the core yarns. The pile yarn is cut to between 1 and 2 mm and thereby creates a pile surface to the yarn. This can be seen in Fig. 10.7. There are various techniques for making these

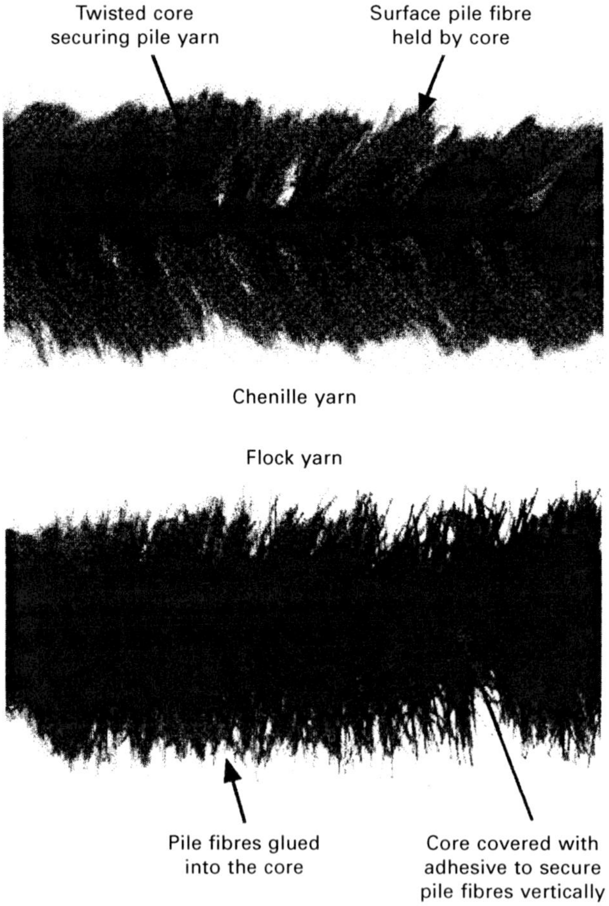

10.7 Comparison of chenille and flock yarns (Fung and Hardcastle, 2001). Used with permission from Woodhead Publishing Ltd.

yarns, which is an art in itself, and because of this and the heavy nature of the yarn, they tend to be expensive. The main problem is the tendency for the short pile tufts to pull out of the yarn during use. However, ways have been devised to introduce low melt yarns into the core which melt and fuse the pile to the core during subsequent heat processing.

No such problems exist in the other main pile yarn which is the flock yarn. In this the pile is chopped up into short lengths of 1–2 mm and given an electrostatic charge to create a positive and negative charge before depositing onto the core which is also charged (and incidentally is often viscose coated with an adhesive). Due to the charges, the short pile fibres all assume a position at right angles to the axis of the core yarn and the result can be seen in Fig. 10.7. This creates a more rigid yarn than chenille but with higher abrasion. As described above, when discussing BMW, the flexural rigidity of polyester as compared with nylon means that nylon is often used in these yarns as the pile but with extra UV stabilisation.

The yarn counts which are used have over the years become somewhat standardised, although technically there is really no limit to the range which can be produced. In filament yarns generally, and also to some extent in man-made spun yarns, the denier of the individual filaments (known as DPF) can affect both handle and abrasion and it is usually a trade off.

- Finer DPF (2 and below) = better handle = poorer abrasion
- Coarser DPF (3 and above) = coarser handle = better abrasion.

830 Decitex with 3 den per filament feedstock yarns air textured either parallel or core and effect is often used as is 420, 600, 1340 and 3000. False twist yarns are usually available in 100 or 167 decitex but special yarns of alternative counts are also available.

Woven fabric structures

Table 10.3 indicates various flat woven fabric structures which have been used for car seats. The ultimate suitability of these structures is determined by a combination of:

- yarn properties: count, DPF, strength, stretch, fibre types, thickness, surface characteristics including texture
- weave of the fabric which is the way in which the warp and weft threads interlace
- float length (i.e., the number of threads warp/weft yarns float over in the weave)
- fabric density or setting (i.e., the number of warp and weft threads per unit length)
- subsequent finishing techniques.

Although originally most automotive fabrics were plain or dobby woven, they are now available in many jacquard woven patterns and designs, so much so that the car companies or OEMs employ qualified textile designers to create the designs and colours.

Frequently the interior fabrics are designed to offend the fewest and this means that they are often understated in terms of design and colours. In jacquard fabrics smaller scale designs are preferred since these do not clash with other interior features and are also less troublesome to make up into the seats and door panels. This type of jacquard also adds a feeling of slight opulence to the interior and is often used in higher trim levels. This is illustrated in Fig. 10.8 which shows a design which has been used in the Mercedes E class. More brash designs in brighter colours and larger repeats are sometimes used in cheaper models to brighten up what could be a drab interior or to target a particular segment of the market.

Jacquard flat woven cloths

When the 'new age' of colour woven car fabric using pre-dyed air textured polyester yarns began to emerge in the 1970s, the design aspect took most of its input and inspiration, and to some extent colouration, from the apparel trade – men's, ladies suitings, etc. – and fabric suppliers began researching the latest trends and colours. By adapting these to approved automotive structures, special automotive ranges were incorporated into travelling presentations to the individual car companies worldwide. However, most of the designs were of a simple or geometric type woven on dobby looms, generally typical of suiting type designs, making good use of colour and weave effects.

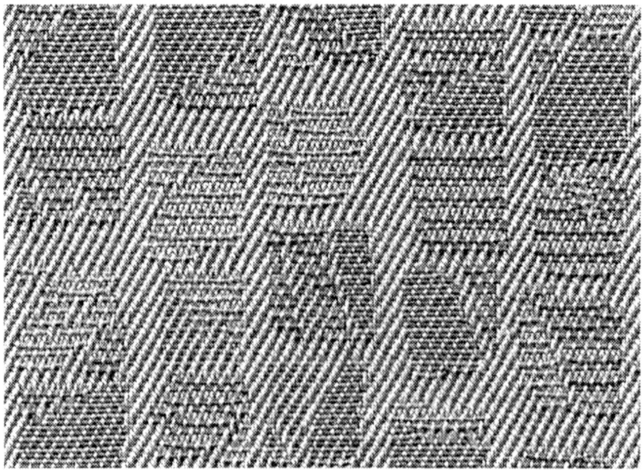

10.8 Typical automotive jacquard seating fabric. Main features are understated design in muted self colour and small scale to avoid dominating other interior features of the car.

Courtaulds Automotive Products (CAP) was an English company situated in Manchester, which was an early entrant into this business and particularly the use of air textured yarns. As the decade gave way to the 1980s new directions were investigated and one of these was the possibility of developing jacquard fabrics. Jacquards had been extensively used before in automotive and also transport textiles in general, but as described earlier the move to air textured yarns for performance reasons caused quality problems during manufacture due to the fine jacquard harnesses. An intensive development programme which the author was very much involved in, saw the development of new parallel textured yarns which performed well and one of the first jacquard ranges in these newly developed structures was put into production by the Rover Group.

Following this, the use of jacquards in the new fabrics for automotive applications using air textured yarns, including warp yarns, exploded all over Europe and then the world. Table 10.3 illustrates some fabric structures which have been approved in the past for jacquard fabrics. The more technical aspects of developing these fabrics is beyond the scope of this chapter but is a very interesting part of the story.

Yarn and fabric colouration

The colouration of automotive and transport fabrics in general falls into one of two categories (and very occasionally a combination of both):

- Colour weaving from pre-coloured yarns. This in turn can be subdivided into package dyed yarns and colour spun or spun dyed yarns.
- Fabric (i.e., piece) dyeing where the fabric starts life in the ecru form.

As has been described earlier, car interior fabrics have to withstand a lot of wear and misuse and the yarn air-jet texturing technique has proved over the years to be the most able to withstand this. It also has another advantage in terms of ease of colouration mainly by the colour weaving route as detailed above.

It is always necessary when designing trim fabrics to have the ability to 'fine tune' the colours so that either all, or at least a major part, of the fabric matches or coordinates well with the other interior items such as plastic trim, dashboard, leather, which may also be part of the interior or even the same seat. This matching is a big issue when different materials are involved and the OEMs place stringent requirements on the suppliers to achieve a non-metameric match.

Metamerism is the ability for a single colour to appear different when viewed under different lighting conditions, e.g., in daylight and in tungsten light, so it is possible for a metameric seating material to match the dashboard in daylight but be a total mismatch when viewed by the car interior lights. Dye technology has advanced to a point where this can to a large extent be predicted and avoided, but it all adds time and cost to the development process.

The use of air-jet textured yarns makes this problem easier to deal with by the employment of yarn package dyeing, in which packages of the yarn are impregnated with dye in a pressure vessel. This enables small samples to be prepared to achieve a good match on a short timescale. The package dyeing technique also offers great potential in the creation of a virtually unlimited colour range which is a massive advantage to the designer.

Another yarn colouration technique is the use of colour spinning or the colouring of the melted dope before it is extruded into filaments. This is often referred to as 'spun dyed' or 'melt dyed' in that the filaments are dyed before being brought together to form the yarn. This is the cheapest method of colouring yarn and also gives probably the highest lightfastness, but it requires large volumes to be economic and is therefore used in carefully controlled ways. One of these is to tap into a standard colour range produced by a fibre manufacturer or to develop the fabric via the package dye route and, when the fabric is running in volume production, have the yarn colours matched through the spun dyed route.

A third way is to combine several spun dyed feedstock yarns (usually 167 decitex) within one air textured yarn. For example an 830 dtx air textured yarn could comprise four or five individual feedstock yarns each of which could be a different colour. This would create a tweed or heather mixture yarn and these are very popular with designers. Since the colours of the feedstock yarns are fixed for reasons mentioned earlier, subtle colour changes can only be achieved by changing texture levels or changing one of the feedstock yarns.

Although colour weaving is a very convenient way of creating fabrics for interior uses, the piece dyeing technique is also used particularly where the fabric is of a plainer character – seat backs for instance – and can work out cheaper in certain circumstances, but lacks the versatility to create the very subtle design and colour effects often required. Sometimes the two methods are combined in which partly coloured fabric is overdyed but this is only used in special cases.

The piece dyeing of polyester has to be carried out at around 130°C or higher and therefore requires a pressurised dyeing vessel.

10.4.2 Finishing processes

All fabrics go through some form of finishing and exactly what is involved in this process depends to a large extent on which part of the vehicle the fabric is to be used. If the fabric is to be coloured by piece dyeing, then this will be the first process to be followed by the others, which also are used on colour woven fabrics. All fabrics are either scoured to relax the structure and/or stentered at or around 130°C to heat set the fabric to remove creases and prevent subsequent shrinkage. The next process is to apply some form of acrylic back coating usually by the foam method often with an additive to impart some flame retardancy (FR). The foam method is simply a way of mixing the compound with air to create a bubble-filled foam and is often used since it is a very economical way of applying compounds to fabrics and it allows great control of thickness and consistency. The process is illustrated in Fig. 10.9.

The coating process has several uses: it allows just enough FR compound to be applied to ensure that the fabric passes the fire retardancy requirements, and makes a substantial contribution to improving face abrasion and snagging resistance by anchoring the filaments of the yarns into the structure.

After, and sometimes instead of, coating, the fabric is often laminated to a polyurethane or polyester foam and nylon scrim to create what is known as a trilaminate structure. If the foam is a special fire retardant variety, the FR coating can be dispensed with since the FR foam will be sufficient to apply the necessary FR property to the fabric. This process is illustrated in Fig. 10.10. The reason for creating these trilaminate (or sometimes bilaminate) structures with scrim on the back is to allow the fabric to be tightly stretched over the seating 'bun' (also made from polyurethane foam) and manipulated into place since the scrim is able to slide over the foam bun.

A bilaminate without the nylon scrim and with thinner foam may be used as door panelling to give a soft and resilient touch.

Other finishing processes include spray on anti-static agents or sometimes anti-slip compounds to ensure that the warp and weft yarns do not slip at the seams.

10.5 Other transport applications

Although the automotive application is by far the largest in terms of volume of fabric used, almost all other forms of transport use textile products

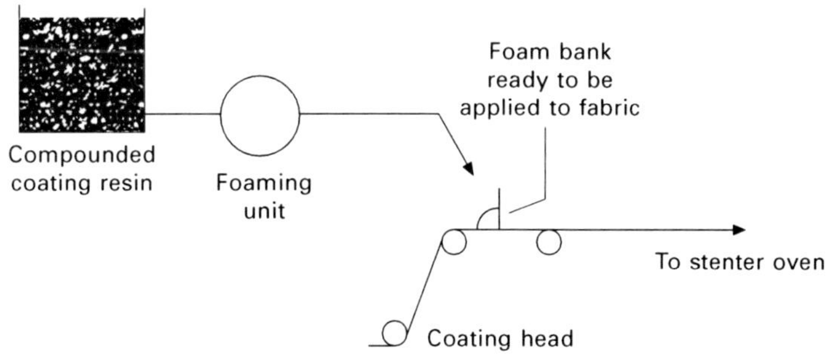

10.9 Schematic of knife over air foam coating (Fung and Hardcastle, 2001). Used with permission from Woodhead Publishing Ltd.

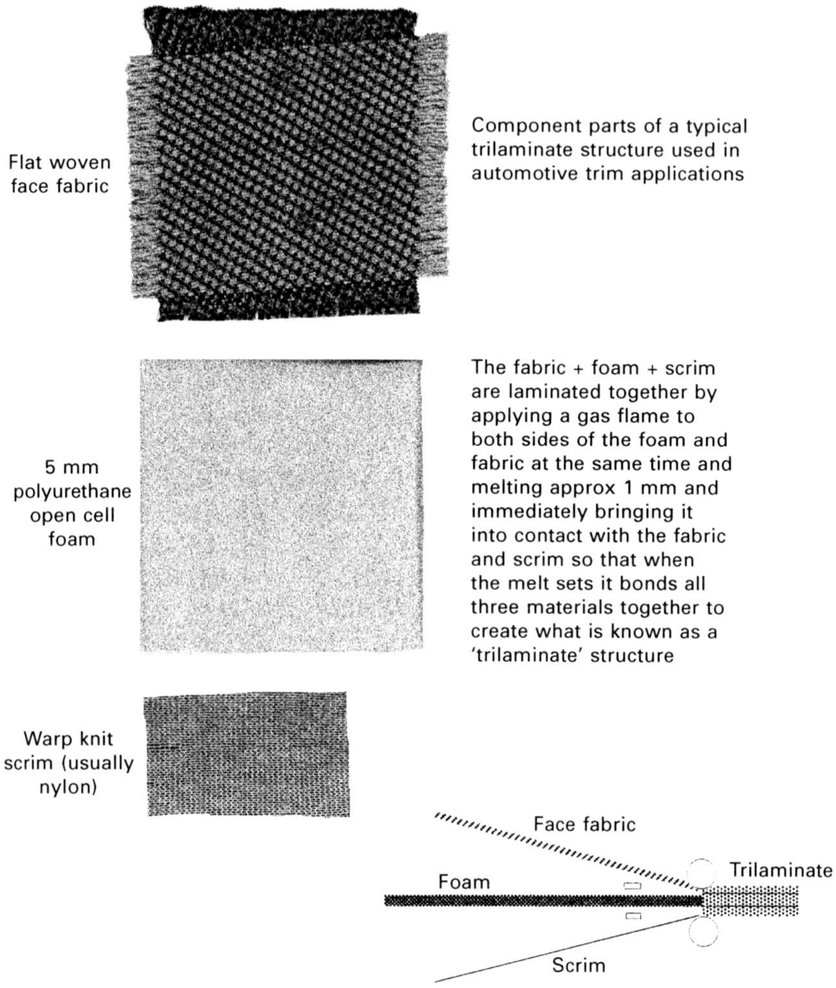

10.10 Automotive trilaminate fabric.

including woven fabrics in some degree. The fabrics all have to perform similarly in term of abrasion strength, etc. The one area which is far more critical in public transport is fire retardancy where very different and very much more stringent requirements are placed on the suppliers to the point where often inherently flameproof or retardant yarns and fibres have to be incorporated.

10.5.1 Commercial heavy goods vehicles (HGVs)

Often these are made by the same OEMs who manufacture cars and frequently use the same interior trim fabrics, although special fabrics made from 3000 Dtx yarns to create heavier materials with enhanced wearability are sometimes specified, although they are subject to the same performance test requirements.

However, HGVs also use very large quantities of tarpaulin fabrics in their side curtains. These fabrics are often flat woven structures made from either high tenacity polyester or nylon in base fabric weights ranging from 100 gsm to 300 gsm. However, these are then subject to substantial post-treatment of coating and lamination. PVC multi-layer coating is often applied to both sides of the fabric ending up with fabric total weight of 1,000 gsm or even more. At the better quality end, an additional polyurethane or acrylic lacquer with UV additives is often applied as a final coat to improve lightfastness, abrasion and resistance to soiling.

10.5.2 Buses and coaches

The interior seating and trim fabrics used in this market are in fact different from the automotive product both in terms of fabrics and also design and colour. Also, due to the heavy use over the lifetime of the vehicle, a different emphasis is applied to the physical requirements. Shampooing is often carried out to keep the seats clean and therefore colour fastness to this procedure is an important test requirement – this hardly features in automotive cloths.

Abrasion testing is usually carried out by Martindale where 80,000 rubs or more is needed. Fire retardancy is also stricter and sometimes requires the use of fire blocker interlining fabrics. Toxic fumes from the fire source is an important issue and the materials have to be selected carefully. Comfort of the passenger is of greater importance in buses and coaches and this requires the material to absorb moisture without feeling damp, particularly in hot climates.

When considering all these issues together, fabrics which tend to meet the overall requirements best are those containing wool, particularly when blended with nylon with the comfort aspect best served by the use of pile fabrics. Mainly for these reasons, a pile comprising 85/15 blends of wool/nylon in moquette or woven velvet fabrics were first developed to meet these requirements and are still an important segment of the market. The fabrics are invariably back coated and total weights are in the 800 to 900 gsm range. However, fibre developments, cost pressures and the fact that polyester is inherently a cheaper fibre than nylon, have seen other blends appear such as wool/polyester in 80/20 and even 100% polyester and also polypropylene have entered the market particularly from overseas manufacturers.

Frequently quite bold design and colour work is used to make specific statements about the coach or bus company and this is often met by the use of bright multi-colours and the velvet jacquard machine. The Belgian company Michel Van de Wiele is in the forefront of supplying weaving machines for this market.

10.5.3 Railways

For the same reason as for coaches, the 85/15 wool/nylon velvet at around 800 gsm is a large part (but by no means all) of the market. Again FR requirements are very stringent and sometimes these are more easily met by the use of inherently flameproof fibres such as Trevira CS (this is a flameproof polyester fibre produced by Trevira GmbH, previously a subsidiary of Farbwerk Hoechst AG) either alone or as blends with wool. The fabric types are generally woven and either cut pile velvet or sometimes moquette, which is a form of velvet traditionally woven as a single fabric with the pile yarns being lifted over wires during weaving to create a raised loop pile surface as illustrated in Fig. 10.11. Using this technique it is possible to create both cut and uncut pile in the same cloth and to use the difference in appearance to create the design effect. However, since it is a single cloth construction, it is only used either for loop (uncut) or a combination of both since it is too expensive to be used as a way of creating only cut pile. This can be produced much more cheaply on the face-to-face principle where two cloths are woven simultaneously one above the other and separated by a traversing knife.

Apart from high non-flammability requirements, fabrics used for trains are often subject to vandalism and ways of combating this, have been developed, albeit at high cost, by using cut resistant fabrics made from high modulus polyethylene fibre such as the Spectra® produced by Honeywell. Although mainly developed for military uses and body armour, they are one possible solution to combat wilful damage. However, Spectra® fibres have a disadvantage in terms of design and colour potential, and of course cost.

It is interesting to note that, unlike the automotive and HGV applications, both bus and coach and also railway segments have been noted for the creation of iconic designs for seat materials in both velvet and moquette. Many of these

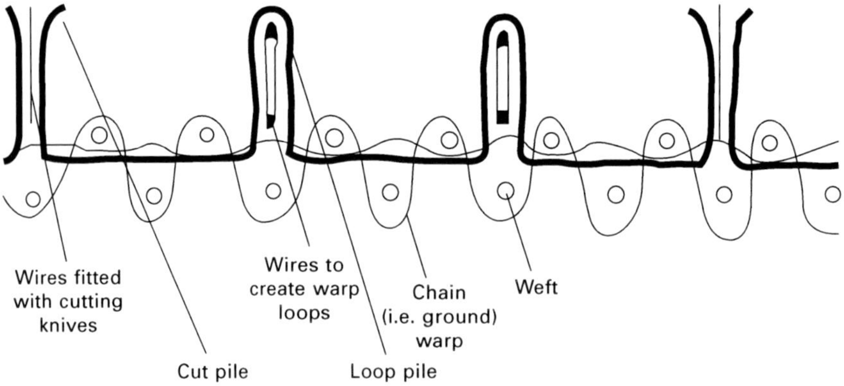

10.11 Illustration of moquette cut and uncut pile.

were produced over the years by both Firths and also John Holdsworth, West Yorkshire, who have been producing transportation fabrics for over 200 years. The bus, coach and rail segments seem more prepared to push the boundaries in terms of design and colour than the more conservative automotive sector. They also use colour for practical as well as aesthetic purposes. An example of this is London Transport, who in the 1990s refurbished their trains with moquette upholstery allocating a different design to each Underground line and where the predominant colour in the designs reflected the colour of the lines on the original Harry Becks underground map (Yellow for circle, turquoise for Victoria, etc.). Other innovations have been tried by London Transport using the fabric designs to convey a message or advert such as a trial in 1998 where a train was refurbished in a moquette design specially commissioned to advertise the London Yellow Pages telephone directory. Further information and illustrations of the moquette designs and much more can be seen on the London Transport Museum website at <ltmuseum.co.uk/> and typing 'moquette design' into the search box.

10.5.4 Aircraft

The use of colour woven fabrics with high design and colour content is increasing in aircraft, although in volume terms it hardly registers when compared with automotive. Increasing competition drives manufacturers and airlines to try to differentiate their interiors and make them more passenger friendly, but it is often the safety aspect, particularly the non-flammability and low toxicity properties of the fabrics, which occupies a lot of development time bearing in mind that fire and smoke inhalation causes over 25% of possibly survivable deaths in air accidents.

Weight reduction is also an important issue but the production of extremely hard-wearing fabrics required of the rail, coach and car businesses is not so important, hence softer and more aesthetically pleasing flat woven fabrics are possible. It is probable that the first textile fabrics made specifically for aircraft were developed in the early 1940s by Collins and Aikman in the USA for the Douglas Aircraft Company.

An important property of textile fabrics is their ability to carry design and colour, enabling corporate logos and colours to be incorporated. Much effort is being put into making the cabins appear more spacious with softer surfaces and woven textile fabrics, including jacquard woven fabrics.

All fabrics have to pass the vertical burn test such as BS3119 or DIN 53906 or the International FAR25.835b and c in which the product has to be self-extinguishing in 15 seconds after removing the flame. Fabric foam seat composites have to resist a burner applied to the seat for 2 minutes. In order to pass this very rigorous test, fire blocker fabrics have sometimes to be included as part of the assembly along with fire resistant foam. The role of the fire blocker fabric is to enclose the foam and considerably enhance the overall fire retardancy of the seat and for this reason the fire blocker fabrics are made from non-flammable or extremely low flammable, and frequently woven, fabrics.

The main fibre used in fabrics for aircraft is wool which has been ZIRPRO treated. This is a treatment developed by the IWS (International Wool Secretariat) to improve flame retardancy of wool fabrics. A special 'low smoke' version has been developed specifically for aircraft manufacturers.

Other uses for woven fabrics in aircraft are for escape chutes which are mainly woven polyester or nylon which has been polyurethane coated.

10.5.5 Marine

Among the traditional marine applications for woven products are: specialised polyester for the sails, coated heavy fabrics for inflatable life rafts, boat coverings, awnings, composite materials using woven textile fabrics as part of the composite and used in the actual structure of the vessels. Again strength and reduced flammability are some of the more important properties in these applications. Most of these are quite specialised and relatively low volume.

Another rapidly growing sector, which has strong connections with the past, is the cruise liner business. Nowadays these are often massive constructions possibly the largest to date is the 'Allure of the Seas' owned by Royal Caribbean at over 220,000 gross tons around 1/4 mile long and holding over 6,000 people on 18 passenger decks.

There are many more, which are not a lot smaller, cruising the world's oceans and all need furnishing with carpets, curtains and bedding, most of which will be woven fabrics. This is a very large fabric requirement which needs replacing on a regular basis and at the moment it is growing. The fabrics are similar to and as diverse as the domestic and contract home furnishing market but with the added requirement of stringent flame retardancy and low toxicity of the emissions. The fibres used are: cotton with fire retardant 'Pyrovetex' treatment, also Trevira CS with permanent FR properties and ZIRPRO-treated wool, particularly for carpets.

10.6 Future trends

Environmental concerns are a big issue which the automotive industry has been trying to deal with for many years. It is becoming even more important with the rising price of oil and the need to reduce fuel consumption and address the recycling problem. Weight reduction has been one way to address this with the increasing use of lighter materials such as aluminium, magnesium and also titanium and carbon fibre composite. Some are more successful than others and all have problems not least of which is the cost.

In addition it is also possible to redesign the car body to use less material, but it always has to meet strict safety standards and minimum strength, and also be capable of economic manufacture. The use of woven textile fabric in car bodywork is not new, soft top convertibles have been used for decades. However, the use of a textile fabric to totally replace the metal panels in the whole of the body is, or was, an innovative approach and was initiated by a well-known German-based automotive company and launched in 2008. It was an experimental project and has not progressed through to production but it did illustrate the uses to which a textile material could be put with careful design of both the material and car. The material was a specially developed coated polyurethane-polyurea copolymer stretch material and was stretched over metal hoops to form the body shape.

Other ways of reducing the carbon footprint and also the reliance on oil are being tried. For example, some automotive manufacturers have developed so-called bio-fabrics which are modifications of the normal oil-based synthetic products like polyester by incorporating as part of the process plant-based raw materials (e.g. corn and terephthalic acid) which can use up to 15% less energy through the production process.

The Green agenda, in addition to driving the search for new synthetic fibres, is also seeing the re-introduction of natural fibres such as linen, flax and hemp. This can be either as a replacement for some of the synthetic fabrics with possible performance reductions or as a composite with other materials to replace some of the heavier items within the car.

New fabrics have also been seen in the development of inflatable seat belts which, like air bags, are blown up on impact by explosive gas to protect the occupants.

Whenever a new vehicle is developed with new propulsion systems, such as electricity or hydrogen or even hybrid variations, it is becoming axiomatic that the OEM will want to use textile products which follow the same thought process of carbon reduction, and it is up to the fibre and fabric manufacturers to be developing answers to these requirements.

'Smart textiles' also need to be borne in mind. These are textile-based structures with the ability to sense and react to stimuli. This could be a traditional woven cloth incorporating a conductive filament which can sense changes and transmit data. How could they be used inside a vehicle? This is up to the designer and engineer to determine in the never-ending quest to create a competitive edge for their company.

The automobile industry is well known for its desire to continually push the boundaries and address concerns of safety, weight reduction, carbon footprint, recycling, and explore all ways of reducing the environmental impact and dependence on oil throughout the life of the vehicle. In this scenario the future of textile products within the automotive and transportation industries can only be good provided manufacturers keep themselves well informed and up to date about the latest thinking and developments of both the OEMs and the major fibre manufacturers.

10.7 Acknowledgements

The author thanks the following people and organisations in kindly providing information, permissions and also in helping to check the manuscript. Textile Institute, Woodhead Publishing, London Transport Museum, Tiffany Daneff, Saga Magazine, Karlheinz Merkens and Walter Fung.

10.8 Sources of further information and advice

Behera, B.K and Hari, P.K. (2010) *Woven Textile Structure*, Woodhead Publishing, Cambridge.

Fung, W. and Hardcastle, M. (2001) *Textiles in Automotive Engineering*, Woodhead Publishing, Cambridge.

Horrocks, A.R. and Anand, S.C. (2000) *Handbook of Technical Textiles*, Woodhead Publishing, Cambridge.

Schindler, W.D. and Hauser, P.J. (2004) *Chemical Finishing of Textiles*, The Textile Institute and Woodhead Publishing, Cambridge.

The Textile Institute (1993) *Textile Terms and Definitions*. The Textile Institute, Manchester.

Tortoro, B.J. and Collier, P.G. (1997) *Understanding Textiles*, Prentice-Hall, Englewood Cliffs, NJ.

http://www.motortrend.com/future/concept_cars/112_0806_bmw_gina_light_visionary_model/viewall.html

http://www.teijinfiber.com/english/products/specifics/biofront.html

http://autos.yahoo.com/articles/autos_content_landing_pages/1158/ford-to-put-air-bags-into-back-seat-belts/

http://www.ltmuseumshop.co.uk/for-home/furniture/our-moquette-range.html

10.9 Reference

Fung, W. and Hardcastle, M. (2001) *Textiles in Automotive Engineering*, Woodhead Publishing, Cambridge.

11
Woven apparel fabrics

N. A. REDMORE, University of Huddersfield, UK

Abstract: This chapter considers the different woven manufacturing processes used in the production of apparel fabrics. It details the main apparel fabric types and looks at the key performance requirements of those fabrics, in relation to both the weave structure and the fibre type. The chapter then goes on to briefly describe important considerations in the design process and the various end uses for woven fabric. Application examples detailed towards the end of the chapter include fabrics that are timeless classics and fabrics that are established fashion favourites.

Key words: apparel, design, applications, performance, fabric aesthetics.

11.1 Introduction

Woven fabrics for apparel are extensive in range and diverse in nature, and have developed over the centuries to meet a wide range of requirements and end uses. Functional or decorative, aesthetically pleasing or high performance, the combinations of fibre, yarn, texture, weave, colour and finish provide a vast range of possible options. This chapter will focus on the more widely used woven apparel fabrics, their characteristics, performance and applications.

11.2 Performance requirements of apparel fabric

11.2.1 Overview of the manufacturing process

The majority of woven apparel fabrics are produced on one of three loom types: a tappet loom, dobby loom or a jacquard loom. Tappet looms are used to produce designs that require 12 shafts or less and are more economical than dobby looms. Dobby looms are used for small-scale patterns beyond the capability of the tappet loom and are generally used for designs that require over 12 shafts and up to 32 shafts. Fabric types produced on these two loom types include fabrics of all weights and in any type of fibre from natural to man-made fine voiles, silk satins and cotton shirting, through to heavier weight denim and woollen tweed jacketing. Jacquard looms are used to create larger figured designs, which cannot be produced on a dobby loom and where the warp threads are controlled individually through a patterning mechanism which can be a mechanical one or linked to a computer aided design (CAD) system. Fabric types produced on this type of loom include the ornate and decorative, brocade, damask and cloqué. The fabrics produced on both of these loom types are termed flat-woven but there is another category of apparel fabrics that require specialist looms in order to form a pile on the surface of the cloth.

Pile woven fabrics, including velvet and corduroy, consist of a base ground fabric with an extra layer of loops formed in either the warp or the weft direction by an extra set of yarns. The loops formed by these extra yarns may be left intact, i.e. uncut, as in the case of terry towelling or cut (either on or off the machine in the finishing process) creating a raised surface of yarn tufts or pile for fabrics like velvet and velveteen.

Apparel fabrics may be woven in ecru (un-dyed, natural colour) ready to be piece-dyed on demand, or colour-woven (fibre or yarn-dyed), as is the case with tweed and tartan fabrics. Woven fabrics also make up a large percentage of the base cloths used in the printing industry, where their inherent stability makes them an ideal substrate for achieving good print registration and clarity of design.

11.2.2 Trends and design

The popularity of woven fabrics for fashion is subject to seasonal trends influenced by catwalk collections and by the trend prediction companies that specialise in creating key directions for the colour, texture and patterning. The demand for the cloth types that make up the clothing staples of shirts, suits, jeans and coats remains fairly constant, but are subject to minor seasonal changes in terms of colour, fibre composition, and finish. This continuous development keeps them up to date and in line with current taste.

The design content of woven fabrics can vary, dependent on the fabric type and the desired end use. Plain or semi-plain fabrics that have been in use for hundreds of years require little design beyond the minor changes in construction or finish, but change may be driven by a change to the proposed fibre or yarn type. The development of a fabric woven from a new sustainable fibre, for example, may require some special loom settings, fabric sett, weave structure and finishing route before the optimum result is achieved. At the other end of the scale, jacquard woven designs and colour-and-weave fabrics (a Prince of Wales check, for example) may necessitate multiple trials of possible weave design and/or colour options before a new design is signed off. Fabric options may be offered in multiple colourways in a sample book, or in a design blanket (as is often the case in the worsted suiting industry).

The majority of the manufacturing for apparel fabric sold to the British high street is now situated overseas in order to be able to produce material at the price point and quantity needed in order to remain competitive. The companies that do retain a manufacturing and weaving presence in the United Kingdom, are usually specialists in the production of high-end, bespoke or high-performance fabrics for end uses including suiting, coating and ties. These specialists retain a presence in areas of the country that traditionally concentrated on the production of worsted, woollen, tweed, cotton, or silk manufacture, drawing on hundreds of years of craftsmanship in those regions.

11.2.3 Fabric specification

When specifying or selecting a fabric for apparel end use, a number of considerations need to be taken into account. The fashion designer, having designed or selected a specific garment shape, will be looking to maximise certain characteristics through the careful selection of a fabric that will compliment and fulfil the vision for their collection. A sensitivity to and knowledge of the characteristics of fabrics, along with a creative approach to their use are essential and central to the sourcing of the appropriate cloth for a given garment. Consideration of how the fabric will drape on the body and how it will be affected by the cut of the garment should be taken into account, along with inherent properties such as how open or dense, stiff or floppy, fluffy or smooth it is. Fabric comfort can be described as the extent of stretch and its recovery from extension. Woven fabrics are not naturally as stretchy as knitted constructions, but 'stretch woven fabrics' can be created through the use of elastomeric yarns.

The target customer and garment style needs to be considered when selecting a fabric, in order to match the performance of the cloth to its intended end use. How much will the garment be worn, in what situations will it need to perform, how long is it expected to last and how will it be laundered? These are all key questions to be answered along with the cost constraints of the range.

Fabric selection

When choosing the type of fabric, it is important to match the performance of desired fabric properties to the end purposes. Fabrics for work-wear need to have good durability, abrasion resistance and cleanability. Outerwear needs to protect, insulate, and be hardwearing. Conversely, finer and lightweight fabrics for eveningwear or lingerie are selected for their aesthetic qualities, drape and handle. The combination of yarn and fibre type along with the weave, gives the designer further options to engineer a fabric to meet the garment requirements. A fibre such as wool has better extension than say flax (linen cloth), which is inherently stiff and has poor crease recovery. More densely packed threads in the warp can help to improve tensile strength than a higher pick rate in the weft, and a higher twist in the yarn will help give further strength to cope with the stresses and strains during weaving.

Weave properties and characteristics

Fabric properties are diverse, with weave structure, fibre type, fabric sett, yarn type and finishing route all affecting the characteristics of a given cloth. Some properties are inherent to a fibre type such as its stiffness, thermal properties, handle or drape. The weave type, including the length of floats in the structure, will also influence these properties. The amount and direction of the twist in a yarn affect the handle, drape and other surface and physical properties.

One of the simplest weave structures, plain weave, produces a plain, non-directional surface that may be fine or coarse depending on the linear density of the warp and weft yarns. Taffeta is a good example of a fabric woven in plain weave, a weave that is used for many lightweight and sheer fabrics. The high number of binding points (also known as interlacements) in this structure gives firmer and stronger fabrics than that in a twill weave structure using the same fabric sett and yarn linear density. Plain weave is less susceptible to distortion of the threads in the fabric.

Twill weaves in which the yarns are usually more loosely bound than plain weave provide fabrics that are more supple and have a degree of natural stretch, especially when cut on the bias (45° to the selvedge direction). Twills have a distinctive appearance as in the example of cavalry twills, and the double-twill in this cloth not only defines the look of the cloth, but also the way in which it handles. In the case of denim, the longer floats in the warp direction of the twill (3/1 twill is most common) can be used to hide the un-dyed weft.

Hopsack has a greater resistance to tearing compared to a fabric constructed from plain weave, as the pairs of ends work together in the structure. Hopsack fabrics can, however, create problems with seam fatigue, where it is more difficult to secure the more loosely bound ends and picks with the sewing thread. Loose weaves with long floats do have better tear strength as the threads slip and move past each other rather than ripping straightaway.

The smooth face of satin woven fabrics creates an ideal surface for printing, embossing or embroidery applications. The durability and appearance retention of cord and velvet fabrics should be considered when designing as they have a tendency to wear out through pile loss or flattening of the pile. In the case of velvet, a 'w' weave, where the tufts are locked in place by two picks rather than one, will help the pile retention.

11.3 Types of woven apparel fabrics

Apparel woven fabrics can be categorised according to their end use whether it be for intimate apparel, dresses, shirting, casual wear, outerwear such as suiting, coating or for a performance or sports wear (see Table 11.1). An

Table 11.1 Key characteristics of woven fabrics

Fabric weight	Fabric type	Applications	Properties
Lightweight	Foulard Gauze Habutai Mousseline Muslin Organdie Organza	Dress fabrics Evening wear Wedding dresses and veils Trimmings Scarves Blouses	Fluid Transparent Soft Sheer Stiff
Medium lightweight	Lining Madras Oxford Sateen Satin Taffeta	Shirting Dress fabric Lining Intimate apparel	Smooth Silky Supple Airy
Medium weight plain	Calico Crepe Flannelette Grosgrain Percale Poplin Seersucker Terry	Toiles Night wear Shirting Dress fabric	Ribbed Crinkled Puckered Textured Smooth
Medium weight patterned	Brocade Cloqué Damask	Evening wear Ladies jackets	Figured Textured
Medium weight suiting	Barrathea Serge Worsted suiting	Suits Trousers	Uniform Smooth
Jacketing	Flannel Gabardine	Coats and jackets	Felted Smooth
Workwear	Calvary twill Chino Denim Drill Moleskin	Trousers Jeans Jackets	Dry Stiff Uniform
Pile and texture	Bedford cord Corduroy Piqué Velvet Velveteen	Evening wear Trousers Jackets	Raised Pile Soft Textured
Heavy weight	Loden Tweed Harris Tweed	Coats and Jackets	Soft Textured Hairy
Performance	Goretex® Ripstop	Coats and trousers	Crisp Smooth

easier approach to classifying the broad range of fabrics for fashion is to group them according to their weight as followed in this chapter.

11.3.1 Lightweight fabrics

Supple, lightweight, silky, smooth, transparent, open, cool and crisp are all words that describe fabrics that go into garments ranging from lingerie, to shirting and into evening wear. Fabrics in this category include: foulard, gauze, habutai, muslin, organza. These are discussed below.

- **Foulard** is a soft, lustrous, lightweight fabric, traditionally woven in silk yarns and using a 2/2 twill weave, making it suitable for ties, scarves and linings. More often than not, this apparel fabric is printed and is now also widely produced using synthetic yarns such as viscose, polyester or acetate.
- **Gauze** is used for veils and trimmings; this is another fabric with an open semi-translucent construction that imparts a fluid, floppy handle. Constructed in plain weave or a leno construction that gives a more stable fabric (the warp ends using special heald wires cross over one another to trap the weft).
- **Habutai** the name of this soft, silk cloth, originates from the Japanese word, meaning lightweight. The fabric is woven with silk yarns that still contain the natural gum from the silk cocoon, and using plain weave to create cloth suitable for dresses, blouses and scarves.
- **Muslin** is a lightweight cloth suitable for dresses, scarves and blouses. The fabric is woven in an open construction to form a soft, fine fabric, and is usually made from cotton. Muslin fabrics are usually woven in grey form (un-dyed ready for bleaching, dyeing or printing) and the construction used is either plain weave or a leno weave.
- **Organza** is used primarily for evening and wedding dress fabrics, where a stiff, sheer fabric is required. Plain weave in construction with an open sett, organza can be made from silk or man-made fibres that have a high-twist. Additional stiffness is imparted into the fabric through the application of gum or resin in the finishing process.

11.3.2 Medium lightweight

Functional, textural, plain, semi-plain, checked or finished to a high shine, these fabrics are used extensively for shirting, dress fabrics and nightwear. Mechanical finishing processes such as mercerisation or a schreiner finish give an even sheen and smooth appearance to satin and taffeta type fabrics. Fabrics in this category include: acetate lining, madras, oxford, satin, sateen and taffeta. These are discussed below.

- **Acetate lining** – acetate fabric is as the name suggests, made from acetate fibres or yarns and is primarily used for lining fabrics, dress fabrics and some lingerie.
- **Madras** was originally produced in the Indian city of the same name (now called Chennai), and is now synonymous with shirting fabric that is plain woven, striped or checked. Madras has a fine, lightweight and airy handle and often colour woven, an effect that was traditionally created with vegetable dyes. The tendency for these dyes to bleed has led to the use of colourfast yarns being employed in the production of today's woven fabrics.
- **Oxford** is an inexpensive cotton or cotton/polyester shirting fabric named after a shirt worn by the undergraduates at the University of Oxford; this cloth is soft with a slight lustre. Two warp yarns are woven as one, in hopsack weave construction and typically with a white ground and coloured stripes or checks. The easy care nature of this fabric makes it suitable for casual shirts and blouses.
- **Sateen** is a completely weft-faced structure that is named after another of the four elementary weave structures, and is commonly used as a lining fabric for suits. The weave structure resembles the back of a satin cloth, but differs in a lower density of warp threads and a high number of picks per centimetre. Five-end sateen is the usual weave structure used in conjunction with a weft yarn that is usually coarser than the warp and may be mercerized (in the case of cotton) or given a 'schreiner' finish to give a smooth lustrous look.
- **Satin** derives its name from the weave structure used in this fabric, where the longer warp floats produce a smooth even surface with no obvious patterning. The longer the warp threads, as in the case of an 8-end satin (each warp thread floats over seven weft picks), the more luxurious the cloth. Duchesse satin is an example of this. Traditionally made from silk yarns, but now also produced in a range of man-made fibres and other natural fibres such as cotton, there are a wide range of qualities and weights that are produced to meet many different end uses. Satin cloth for intimate apparel is typically lustrous in appearance as a result of using high lustre yarns in either silk or a man-made fibre such as polyester.
- **Taffeta** is a medium weight fabric, constructed in plain weave that forms a slight ribbed appearance in the weft direction, resulting in a crisp hand and lustrous appearance. Taffeta is most often used for formal wear, and is usually produced in a solid colour but is also woven with contrasting colours in the warp and weft to give a changeant effect.

11.3.3 Medium weight plain

Calico is one of the staples of the fashion industry through its extensive use for toiles and mock-ups of garments. Texture features heavily in this category in fabrics like seersucker that use chemical finishes or differential shrinkage to achieve surface interest and crepe which uses the weave structure to create a pebbly feel to the cloth. Fabrics in this area include: calico, crepe, flannelette, grosgrain, percale, poplin, seersucker and terry. These are discussed below.

- **Calico** is often woven in unbleached cotton, characterised by the presence of flecks of seed matter from the plant. Plain weave is used for this utility fabric, a fabric that creases easily and is frequently sized to give it an added stiffness. Used in its plain state as toiles in dress making, it is also seen printed with small bright floral patterns in the United States.
- **Crepe** is a lightweight fabric of silk, rayon, cotton, wool, man-made or blended fibres, with a distinctive crinkled surface. This fabric type has a number of variations including crepe de chine, crepe marocain, crepe chiffon and moss crepe. The crinkled surface is obtained through the use of either crepe yarns (yarns that have such a high twist that the yarn kinks), a combination of yarns with alternating 's' and 'z' twist, by a chemical treatment causing areas to shrink and pucker, embossing, or through the weave itself (usually with thicker warp yarns and thinner filling yarns).
- **Flannelette** is frequently used in nightwear for both men's and women's pyjamas. This fabric is usually constructed in a twill or plain weave, using cotton yarns where the surface fibres have been raised in finishing to give a soft handle. Flannelette was traditionally used for sheets, but is less commonly used since the introduction of central heating.
- **Grosgrain** fabrics are traditionally used for ties and the appearance of this fabric is described by the translation of grosgrain, which literally means 'large cord'. A strong cloth usually made from silk or man-made continuous filament, it can be quite expensive due to the weight of the heavier weft yarns and highly sett warp yarns (Fig. 11.1). Woven in a plain weave construction, the combination of sett and yarns makes this an ideal fabric on which to create moiré effects popular for dresses and eveningwear. A moiré fabric has a watermark effect on the surface created by one of two methods. The rib or corded nature of this type of fabric is essential in creating a true moiré fabric, which is a mix of flattened and still rounded areas of the cloth, that result after two fabrics are stitched face-to-face and pressed through two heated, smooth and heavy rollers. The other method for creating a moiré effect uses an embossed roller

11.1 Grosgrain fabric.

with the characteristic watermark pattern engraved onto its surface, thus flattening the ribs of the fabric to this pattern.

- **Percale** is used for shirts and dresses in apparel and has a fine smooth handle due to the closely sett yarns. This fabric is produced in combed cotton or Egyptian cotton yarns and either piece dyed, printed or just bleached.
- **Poplin** was originally made for the pope and referred to a 'Papalino'. This is a lustrous and hardwearing fabric. A light cross-rib effect is formed through the use of plain weave combined with a warp sett that could have up to twice as many ends per centimetre as weft picks. Poplin is used for functional fabrics including jackets, dresses, shirts and linings.
- **Seersucker** fabrics are traditionally used in summer weight shirts or informal suits. The characteristic puckered stripe effect is either created through controlled tension differences in sections of the warp during weaving, through the use of yarns with differential shrinkage, or using chemical finishes (Fig. 11.2).
- **Terry** is produced on a loom fitted with a terry mechanism and an extra warp (pile warp) that forms uncut loops or pile on one or both sides of the fabric. Cotton fibres are usually used in the production of this cloth, where its inherent absorbency is suited to its use in dressing gowns.

11.2 Seersucker fabric.

11.3.4 Medium weight patterned

This category of fabrics is characterised by complex ornate patterns formed by jacquard weaving. Metallic threads, lustrous yarns and sumptuous textures create a beautiful range of fabrics used in exclusive and high-end garments. Fabrics include brocade, cloqué and damask. These are discussed below.

- **Brocade** is an ornamental fabric usually produced on a jacquard loom, to achieve highly patterned, multicoloured, richly figured patterns for use in evening dress fabrics and accessories. Traditionally figured with gold or silver thread, brocade was woven in silk, and the resulting cloth was of a heavy weight in comparison to damask. Brocade fabrics are usually designed so that only the face of the fabric shows the true pattern, combining simple ground weaves (twill, plain weave, satin, sateen) and supplementary figuring yarns in the warp or weft direction.
- **Cloqué** is characterised by the three-dimensional blister effect that is created in this apparel fabric, produced by weaving a double cloth fabric that combines warp and weft yarns with differential shrinkage. A blister effect results in the finishing process and may also be produced through a combination of weave structures that allow for more or less shrinkage (Fig. 11.3). These fabrics are used in a range of ladies apparel including lingerie, evening wear and blouses.

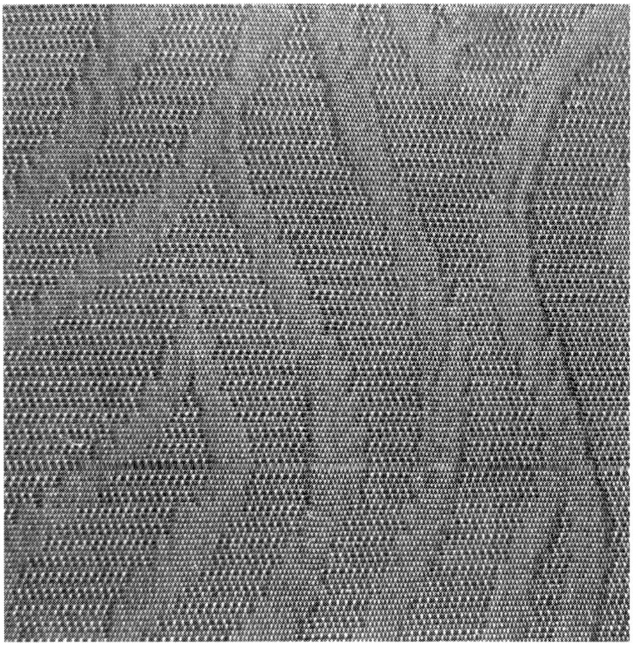

11.3 Cloqué fabric.

- **Damask** jacquard fabric's opulent appearance makes it most suitable for evening wear, and it is characterised by large-scale classical patterns, more frequently seen for interior and table-wear uses. Satin and sateen weaves are used to create this reversible fabric, which has a subtle sheen in the warp direction, contrasted with a more matt weft yarn.

11.3.5 Suiting fabrics

Smart, uniform, hard-wearing conservative and classic, the timeless elegance of the woven fabrics used in suiting is little changed in essence, but they have managed to keep their appeal in the uniform of the workplace. Suiting fabrics include worsted, barrathea, and serge. These are discussed below.

- **Worsted** fabrics are made from fine long staple fibres that have been carded and combed to align the fibres in parallel. Worsted yarns are tightly twisted and woven to form smooth, fine surface texture that has a crisp and firm hand. It is usually woven in a twill weave, the resulting fabric is more durable than woollen spun fabrics, and more expensive to produce. It is used predominantly for suiting and formal work-wear in part due to the clean-cut appearance of the fabric that has a flat and smooth finish. Wool is the most common fibre used in worsted fabrics, but at the high end of the market, blends that mix wool with cashmere, silk or the extremely rare Qivuik fibre are also used.

- **Barrathea** fabric is used for neckties, uniforms and trousers. This fabric has a smooth pebbly surface and is usually woven for worsted suitings. The weave is a twilled hopsack, but other broken-rib variations are used in the production of cotton and silk fabrics.
- **Serge** is a hard-wearing worsted-wool cloth; with a 2/2 twill weave construction. Compactly woven with high-twist yarns, it usually has a smooth face that becomes shiny with wear. End uses include coats, skirts, suiting and uniforms, where the fabric's natural ability to hold a crease is invaluable.

11.3.6 Jacketing fabrics

Warmth and protection from the elements are combined in these hard-wearing cloths. Medium weight jacket fabrics include flannel and gaberdine. These are discussed below.

- **Flannel** is medium weight, and is normally a plain or twill weave fabric that is typically made from cotton, a cotton blend or wool. This fabric is finished to give a very soft hand; by brushing on both sides the fibre is lifted to create a soft, slightly fuzzy surface.
- **Gaberdine** is a functional fabric with clearly defined diagonal twill lines on the face. This fabric is most suited to outerwear such as raincoats, jackets or trousers, skirts and sportswear. It can be made from cotton/polyester, polyester or viscose, but the best quality fabrics are made from wool (Fig 11.4). Tightly woven, the smooth uniform finish helps to make the fabric water repellent and hardwearing.

11.3.7 Work wear and casual wear

These fabrics are widely used both in the workplace and also extensively for casual wear. A mixture of traditional and conservative fabrics, including others that change each season following the latest whims of fashion, are all represented in this section. Fabrics include cavalry twill, chino, denim, drill and moleskin. These are discussed below.

- **Cavalry** twill fabric is formed by a steep pronounced double twill. This fabric is ideal for trousers, jackets, coats and, as it was originally, for uniforms (Fig. 11.5). Made from woollen or worsted yarns traditionally, it is now also produced in other natural and man-made fibres to give a strong wearing fabric.
- **Chino** cotton or cotton/polyester fabric also uses a warp-faced twill weave as its structure. Hard-wearing with a slight sheen, its name has

11.4 Gabardine fabric.

11.5 Cavalry twill fabric.

been adopted for a style of casual trousers in the United States but originates from China where it was first made.

- **Denim**, a hard-wearing utility cloth, has developed into one of the staples of the apparel industry. Denim is full to the touch and has diagonal ribbing formed with a twill weave (2/1 or 3/1 twill) and uses contrasting colours in the warp and weft (typically blue warp and white weft). Traditionally made from cotton, but now available in polyester/cotton blends, lyocell and combined with an elastomeric to give added comfort, this fabric is produced in a vast number of variations for both men's and women's wear.
- **Drill** fabric covers a variety of materials that are used for shirts, suitings and uniforms. A 3/1 twill is often used in combination with good quality, high-twist, cotton yarn to form very durable woven fabric, where the twill lines run in the opposite direction (bottom right to top left) to other fabrics. Drill may be undyed, warp striped or piece-dyed as in the case of khaki, a colour that gave its name to this army uniform cloth.
- **Moleskin** has a fine-brushed nap on the surface of this cloth, giving it the appearance of fine suede. This medium-weight fabric is usually made from cotton and woven using a satin-based weave with short weft floats on the surface, which are raised in finishing to give the suede-like handle. Trousers, work wear and jackets are all produced in this cloth.

11.3.8 Pile and texture

Surface textures including exaggerated ribs, cut pile surfaces, and sumptuous soft hand are all created through the use of weave structure, special yarns, and in the case of velvet and corduroy, cut brushed fibres aligned in finishing. Ranging from the casual look of cord through to the luxurious opulence of velvet, these fabrics can suffer from the trends of fashion. Fabrics of pile construction are velvet and velveteen and corduroy, whereas Bedford cord and Pique create ridges or cords using the weave structure alone.

- **Bedford cord** can be seen in use in coats, trousers, uniforms and other outerwear. A specific weave construction is used to create cords in the fabric that are formed by the tightly sett warp, in contrast to weft floats on the back of the cloth that causes the fabric to buckle. The name of this firm hard-wearing cloth comes from the Duke of Bedford who used the cloth to clothe his troops in the fifteenth century.
- **Corduroy** fabric is formed from raised and cut pile ribs of pile running down the length of the fabric. Usually woven in cotton, and of varying width (expressed in wale's per inch) this fabric is suitable for trousers, jackets and skirts.
- **Pique** derives it name from the French word meaning to quilt, and has an embossed look. Double cloth weaves are used in conjunction

with one fine and one thicker warp yarn that stitch together at intervals forming cords or ribs on the surface. Used in sportswear, dress shirts and waistcoats, the fabric is usually made of cotton and piece dyed.

- **Velvet** is a medium weight cut pile fabric in which the cut pile stands up very straight to give an even, uniform surface. It is woven using two sets of warp yarns; the extra set creates the pile. Velvet, a luxurious fabric used for evening wear, accessories and jackets, is commonly made with a filament yarn for high lustre and a smooth, soft hand. There are many variations of velvet that are defined by the finish applied, fibre content or length of the pile. Crushed velvet, for example, is a solid-coloured velvet that is passed through heated rollers in order to press the pile flat in an irregular pattern. Façonné velvet has a pattern 'burnt-out' using a devoré process to selectively remove the pile to reveal the ground weave composed of yarns resistant to the chemicals applied.
- **Velveteen** fabric is made in a similar way to velvet, but an extra set of weft yarns are used to form the pile instead of an extra set of warp yarns. The resulting fabric has more body than velvet but with less drape.

11.3.9 Heavy weight

These protective outerwear fabrics with textural, hairy surfaces have served as coating fabrics for many generations. The soft colour of the heathered tweed fabrics is influenced by the surrounding landscape and is dyed in fibre form to create complex colour blends woven into simple weave structures. The main fabric types in this category are loden, tweed and Harris Tweed. These are discussed below.

- **Loden** is traditionally woven in the Tyrol region of Austria. This fabric has a brushed, raised surface giving it a warm, tactile surface, usually dyed to a specific green colour. The felted effect on the surface of this woollen fabric, imparts a natural water repellence, ideal as a coating fabric.
- **Tweed** is traditionally a term that describes fabrics that are made from wool and are heavy weight and have a rougher, hairy handle. The yarns used to produce tweed fabrics are fibre-dyed and blended to produce colours that have a melange effect (Fig. 11.6). The rougher surface texture of this fabric type, originally made on the banks of the river Tweed in the Borders, means that it is most suitable for outerwear such as coats and jackets that can be lined.
- **Harris Tweed** is the best known of all the tweeds and is exclusively woven in the Outer Hebrides, on either the Isle of Harris or Lewis. Still produced in the homes of the islanders with Scottish wool, dyed locally the colours reflect the landscape of the area, and the fabric is

11.6 Tweed fabric.

finished in the region and can be accredited with the 'Harris Tweed' trademark. Twill, herringbone and plain structures are used to produce these expensive yet exclusive fabrics.

11.3.10 Performance apparel fabrics

Protective, functional, breathable, and durable fabrics are the mainstay of the outdoor pursuits and performance sportswear business. Engineered to protect the wearer from the elements or to control the climate of the wearer, garments that use woven fabrics in this category are highly specialised and tested to high standards.

- **Goretex®** fabric was developed by W. L. Gore and Associates, Inc., and was one of the first fabrics to fill the demand for a waterproof breathable fabric suitable for outdoor pursuits. Goretex® is essentially a sandwich of a woven face fabric (such as polyester) and a membrane of fluorocarbon and a backing fabric. Many outdoor brands use this fabric and have developed their own propriety fabrics to perform in a similar way, including Lowe Alpine, North Face and Bergaus.
- **Ripstop**, as the name suggests, has a resistance to being torn or ripped. A base weave such as plain weave is intersected at regular intervals in

both the warp and weft directions to form a check of double-lifting (two ends/picks together) ends or picks. As in the case of a hopsack, these areas are stronger than the plain weave ground. Usually woven in nylon, this lightweight fabric can also be back-coated with silicone to create a more waterproof material.

11.4 Practical design applications

Matching the fabric to the end use requires some knowledge of fabric properties and characteristics, characteristics that can be used to enhance the look, structure or aesthetic of a garment. Fabric shows such as Premiere Vision and Interstoff are vital in the fabric selection and development activities of fashion designers. Apparel fabrics at Premiere Vision are promoted through trend direction forums based around such categories as Relax-Distinction (wool suiting, shirting, jeans and cotton), Seduction (fluid and fancy, silks, fancy woollens and prints), Pulsation (technical and performance fabrics, extreme sports fabrics and lingerie), and Atelier Denim. Designers and fabric selectors need to be assured that the performance of the material meets their parameters for performance in a given garment. New and novel fabrics offered by suppliers may not have gone through such rigorous testing or field trials, requiring a leap of faith from the buyers. Conversations between the designers and producers at these fairs can lead to a valuable exchange of information regarding the performance of previous products and possibly the development of exclusive fabric ranges.

11.4.1 Surface interest: felted appearance

A level of milling or brushing applied to a fabric through finishing them with a combination of heat, pressure and friction gives rise to thick dense fabrics with a felted appearance that do not fray easily. A felted wool garment can be left with unfinished edges, abutted seams or top stitched hems. Fabrics such as Loden and Melton are ideal candidates for this approach.

11.4.2 Surface interest: texture

The texture of a fabric can alter depending on how we perceive the colour, especially in the case of velvets where the angle at which the fabric is viewed changes the depth of shade. Pile fabrics absorb light and can also have a greater thickness that imparts a rich luxurious look to a garment. Novelty fabrics, where texture is created using fancy, boucle, tape and slub yarns woven into colour and weave structures like dogtooth checks, are a staple of jackets in the Chanel's collections where cost is not such an issue.

11.4.3 Stretch

Stretch may be imparted into a fabric through finishing. It may also be inherent in the fibre type or be added through the use of stretch yarns that use fibres such as Spandex or Lycra. The comfort of denim and corduroy are improved through the use of cotton/lycra fibre blends which impart a degree of stretch to the fabric, so that not only is comfort improved, but also wrinkle resistance and fit. Stretch woven fabrics have also contributed to the development of fabrics that perform for sportswear and outdoor pursuits as well as redefining garment shaping for snug streamlined silhouettes.

11.4.4 Transparency

Fabrics used to produce garments with transparency have beautiful qualities but are also more difficult to work with due to their fineness, slippery surface and tendency to mark. Organza and organdie are examples of semi-transparent or transparent crisp fabrics. Softer fabrics in this category are chiffon, georgette and crepe chiffon, whereas voile, shirtings and gauze sit somewhere between the two. Considerations when pattern cutting and choosing fabrics are the weight, the thickness, shear (distortion in the warp and weft directions), drape and stretch of the fabric.

11.4.5 Pattern

Directional or large-scale patterns in traditional weave designs like dogtooth and Prince of Wales check require additional care when matching up seams compared with an equivalent semi-plain all over design. Fashion trends influence the popularity and scale of patterned fabrics, as well as the proportions of the garment they are aimed at. Large-scale jacquard motifs need to be carefully placed or used in garments with a greater volume of fabric to fold and drape, and smaller scale patterns are suited to closer-fitting garments. Fabric utilisation may well be affected by the matching of these larger scale patterns, with more wastage at the pattern cutting stage.

11.4.6 Tailored garments

The right choice of fabrics used in combination is significant in creating a well-fitted tailored garment such as a suit. Woollen fabrics for suits may be constructed from yarns produced on the worsted or woollen systems. Worsted yarns are composed of fibres that are longer and finely combed, with a resulting firm flat fabric used for the majority of business suits. Woollen yarns are composed of shorter, uncombed fibres, which are loosely twisted, creating softer, bulkier, hairier fabrics, of which Harris Tweed is a well-known fabric. In these types of garments, fabrics needed for hidden support include Melton for the collar, and pockets with canvas for other lighter weight interfacings.

11.4.7 Support fabrics

Corsetry requires fabrics to fulfil a role not unlike that of an engineered material, where the body is controlled, or distorted to fit a desired shape. Lustrous and opulent fabrics like satin and silks are supported by carefully chosen control fabrics that impart strength, stability and body to the garment. Canvas is used in corsetry to control the deformation or stretch of the face fabric in the garment, whereas a cotton drill woven in twill weave provides a strong, natural flexibility suitable for the foundation of the corset. Other support fabrics include organdie, muslin and organza. Cotton pocketing uses a closely woven twill or plain weave and has good durability due to the two-fold yarns used. Interlining fabrics such as a stiff organza must have the same shrinkage resistance as the other fabrics in the garment, or be fully shrunk already.

11.5 Application examples

11.5.1 Chanel and Linton Tweeds

The long-established relationship between Chanel and Linton Tweeds is a rare example in today's fast fashion, globally sourced, price sensitive world. Originally a producer of 100% woollen cloth, Linton Tweeds faced stiff competition from all the other woollen manufacturers in Scotland and Yorkshire. This Carlisle-based weaver, established in 1912 and synonymous with good quality, took steps to differentiate itself from the competition in the late 1960s. In order to make themselves unique, they developed fabrics that used more exotic and man-made yarns, as well as developing machinery to manufacture their own unique fancy yarns.

A long association with the couture houses of Paris had brought the company a high profile but little financial reward, but this was helped when the introduction of the ready-to-wear collections took off for their haute couture customers such as Chanel. Lintons expanded their horizons to supply markets in Japan and the United States, and they now promote their ranges at Premiere Vision in Paris, as well as in Milan, Japan and New York.

The novelty woven fabrics that Lintons create are elegant, fancy tweeds suitable for ladies jackets, skirts and coats. The fabrics that they produce for Chanel and other couture names are woven in short runs and are usually non-repeatable. Typical fabric constructions are based on hopsack, herringbone, basket weave, leno weaves and colour and weave effects including the four-point star and the hounds-tooth check.

The Chanel suit became the status symbol of the 1950s generation, and was updated by Karl Lagerfeld when he took over design at the label in the 1980s. Using inspiration from past collections, lighter tweed fabrics with a softer look using new yarns were developed for use in the iconic collarless style jacket now synonymous with Chanel. The tweed fabric from Linton Tweeds continues to be re-invented, updated and transcends fashion having established itself as one of the signature looks of Chanel.

11.5.2 Denim

'Denim is more than just a cotton fabric; it inspires strong opinion within the hearts of historians, designers, teenagers, movie stars, reporters and writers' (Downey, 2007). First produced in Nimes, France, the name of this cloth originated from the town that it was originally woven in and was described as 'de Nimes' (from Nimes), a name that evolved into the denim that we know today. This hard-wearing serge cloth was originally imported in the 1850s into the United States as a canvas for tents, and was adapted for work wear during the gold rush by Loeb (Levi) Strauss.

Denim jeans did not start their rise into mainstream fashion until the 1950s with the teenage rebellion around James Dean and his film *Rebel without a Cause*. In the 1960s and 1970s, denim became a staple of fashion trends, adapting its look to fit the style of those eras. Denim has elevated its status even further since the 1980s, as it has been styled, finished and labelled to command high prices in the collections of the top fashion brands. The art of finishing denim has created a whole new area of specialist knowledge around the colouration and fit of jeans as well as the destruction, manipulation and embellishment of this seemingly simple fabric.

The blue colour synonymous with denim originally came from the natural Indigo dye, and then moved to a synthetic version in the nineteenth century. Cotton is still the dominant fibre type used in the manufacture of jeans, with other synthetic fibres such as Lyocell, a cotton-like cellulosic, taking a small market share. The comfort and stretch of denim is enhanced where cotton and lycra are blended, and other properties have been developed by mixing cotton with polyamide or polyester.

One of the first companies to introduce a 'distressed denim' look in the 1980s was the Italian company, Diesel. Their 'aged/already worn' look was slow to take off, but this aesthetic is now at the forefront of trends for denim worldwide. The art of finishing denim, now ranges from sandblasting, acid-washing, bleaching, resin treatments, stone-washing with pumice and whiskering, a process that simulates the creases formed through years of normal wear. Throughout all these changes, the one constant factor that characterises denim has been the 3/1 twill structure.

11.5.3 The future of worsted suiting

Woven worsted fabrics, produced over the centuries for the top names of London's Savile Row, appear little changed on the face of it, but have in reality evolved into a diverse collection of luxury fabrics. Worsted suiting traditionally produced in wool and using weave structures that include plain-weave, twill, herringbone and satin has moved forward to embrace the needs of the modern day consumer as well as the bespoke international market for these fabrics.

Product developments led by the Woolmark company and Australian Wool Innovation Ltd (AWI) have continued the International Wool Secretariat 'Cool Wool' campaign which was first introduced in the early 1990s. The promotion of wool as a trans-seasonal fibre, supported by its natural properties, durability and ability to respond to changes in both environmental and body temperatures has led to Merino Cool™ which uses the finest Merino fibres and results in fabrics that weigh less than 165 gm^2. These lightweight breathable fabrics have excellent drape, making them suitable for use in less structured garments and for warmer climates.

The concept of the machine washable suit has also been taken one step further through the application of a polymeric shrink-resistant finish to plain woven, wool/polyester blend fabric. The sustainable qualities of wool fabrics have been promoted by the designer Paul Smith and improved efficiency and care for natural resources are at the heart of the latest campaign to promote the benefits of this fibre in woven and knitted fabrics. 'MerinoFresh™ technology has been developed with the environment in mind; a MerinoFresh™ garment typically requires less water and energy for after care than conventional woven products', according to Australian Wool Innovation. This push for development between the fibre suppliers, weavers and finishers and promotional activities like the Campaign for Wool is ensuring a bright future for one of the longest established and traditional worsted weave industries.

11.6 Sources of further information and advice

11.6.1 Literature

Aldrich W, *Fabric, Form and Flat Pattern Cutting*, 2nd edn, Oxford, Blackwell Publishing, 2008.

Fisher A, *Construction*, Lausanne, AVA Publishing, 2008.

Gioello D, *Understanding Fabrics; from Fibre to Finished Cloth*, 3rd edn, New York, Fairchild Publications, 2002.

Hallett C and Johnston A, *Fabric for Fashion; a Comprehensive Guide to Natural Fibres*, 1st edn, London, Laurence King Publishing, 2010.

Hardingham M, *Illustrated Dictionary of Fabrics*, London, Studio Vista, 1978.

Keiser S and Garner M, *Beyond Design, the Synergy of Apparel Product Development*, New York, Fairchild Publications, 2008.

Taylor M, *Technology of Textile Properties*, 3rd edn, Odiham, Hampshire, Forbes Publications, 2007.

Wilson J, *Classic and Modern Fabrics; the Complete Illustrated Sourcebook*, London, Thames & Hudson, 2010.

11.6.2 Websites

Worth Global Style Network: www.wgsn.com
Trendstop: www.trendstop.com
MPDclick: www.mpdclick.com
Design Boom®: www.designboom.com/eng/education/denim2.html
The Woolmark Company & Australian Wool Innovation: www.wool.com

11.6.3 Trade shows

Interstoff Asia Essential: www.interstoff.messefrankfurt.com
Intertextile Beijing: www.intertextile.com
Premiere Vision: www.premierevision.com
Techtextil: www.techtextil.messesfrankfurt.com
Texworld: www.texworldusa.com
Tissu Premier: www.tissu-premier.com

11.7 Reference

Downey, L., *A Short History of Denim*, Levi Strauss & Co., 2007.

12
Woven fabrics for geotextiles

A. RAWAL, Indian Institute of Technology Delhi, India

Abstract: Woven geotextiles have been extensively used in civil engineering applications. This chapter discusses the selection of fibres, most common production methods and the structure–property relationships of woven geotextiles. Case studies of woven fabrics used as geotextiles are also discussed.

Key words: design, durability, geotextile, woven fabrics.

12.1 Introduction

The term 'geotextile' has been defined by the American Society for Testing and Materials as: 'Any permeable textile material used with foundation, soil, rock, earth or any geotechnical engineering related material as an integral part of a manmade project, structure or system'. This definition of geotextile suggests that a textile material has to be designed and engineered so that it can be used for geotechnical or civil engineering applications. These applications are summarised below (Rawal *et al.*, 2010).

- *Filtration.* When the geotextile acts as a barrier to the majority of solid particles while allowing liquid to pass through, whether it is in contact with a liquid carrying fine particles or placed in contact with soil with water seeping through it.
- *Drainage.* When the transmission of fluid is in the direction of in-plane flow of the fabric without any loss of soil particles. A fabric with good filtration characteristics can be used for drainage applications.
- *Separation.* When the geotextile segregates different materials, e.g. fine soil from coarse material placed at applied loads.
- *Reinforcement.* When the stability of a material, e.g. soil needs to be complemented by the superior tensile strength of a geotextile fabric. It is well known that soil particles have good shear and compression resistance but are weak against tension. A tensile element such as fabric can be used within the soil to improve its strength.

In addition, other functions of geotextiles have been identified by Giroud (1984) as shown below.

- *Surfacing.* When a planar ground surface is required along with prevention of soil particle removal from the soil surface.

- *Solid barrier*. When the motion of solids is to be prevented or stopped.
- *Container*. Retention or protection of materials such as sand, rocks, fresh concrete, etc.
- *Tensioned membrane*. The principle of using a geotextile fabric to even out the pressure difference, when placed between two materials of differing pressure, by using the tension of the fabric to distribute the pressure.
- *Tie*. When it joins various pieces of a structure that are capable of moving apart.
- *Slip surface*. A geotextile fabric is placed between two materials to minimise the frictional properties of the structure.
- *Absorber*. When it shares in carrying the stresses and strains transmitted to the material that requires protection.

Depending on manufacturing techniques, textile materials can exhibit the above mentioned functions if they are appropriately designed and engineered.

12.2 Production and classification of geotextiles

Generally, most geotextiles are produced using traditional or conventional fabric production techniques. According to Giroud (1984), the manufacturing processes for the production of geotextiles can be classified in two ways: classical and special geotextile processes. Classical geotextiles include typical products of the textile industry such as woven, knitted and nonwoven fabrics. Special geotextiles (webbing, mats, nets, etc.) are similar in appearance to classical geotextiles, but they are not produced through conventional textile production techniques. Different production techniques for producing geotextile materials are shown in Fig. 12.1 (Giroud, 1984).
Classical geotextiles are produced in two stages:

- production of fibres/filaments/slit films (tapes)/yarns as constituent materials
- converting these constituent materials into a fabric.

The constituent materials required for making a fabric are produced using various techniques as discussed below (Giroud, 1984):

- *Filaments*. In a typical melt extrusion process, the filaments are produced by extruding molten polymer through spinnerets which is then drawn to improve molecular orientation along the filament, resulting in high tensile strength and modulus. The spinneret can contain more than one slot, so that numerous filaments may be extruded to produce *multifilament* yarns. These filaments can be in circular or non-circular shapes.
- *Short (staple) fibres*. Continuous filaments are cut into short lengths, ranging from 2 to 10 cm.

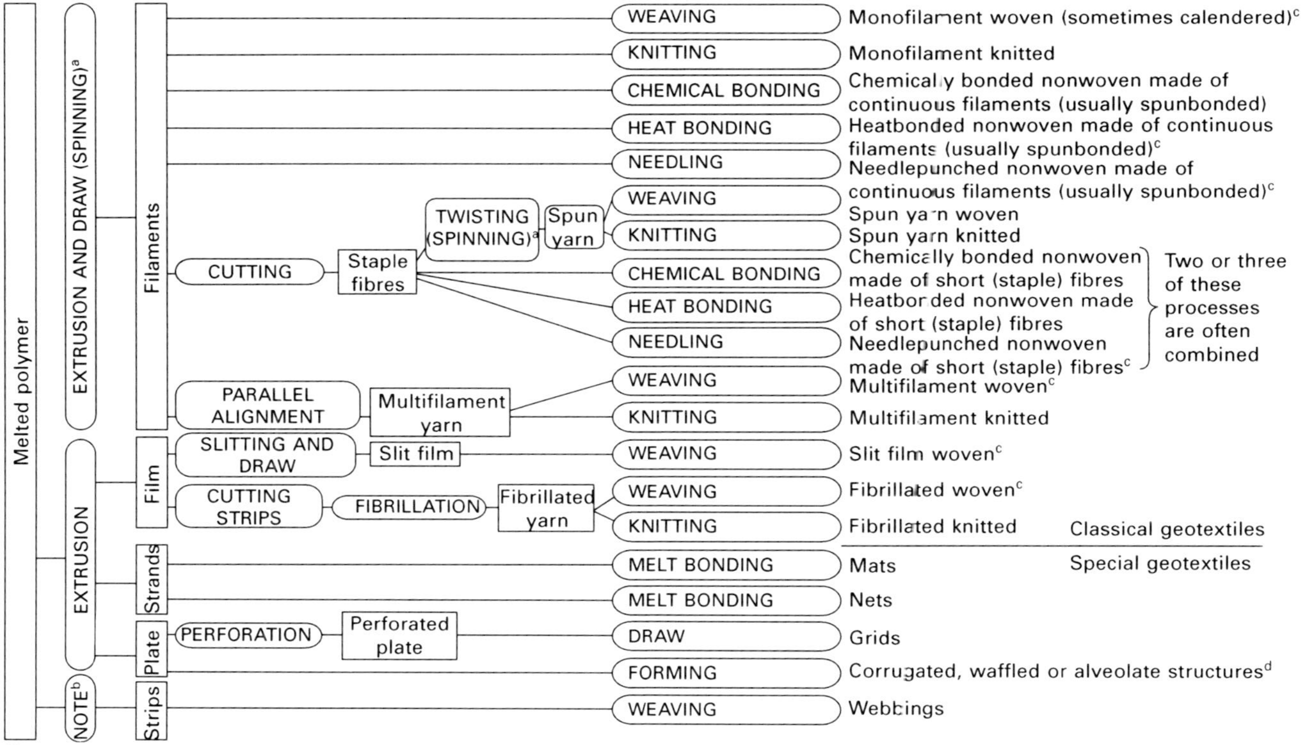

Notes:
[a] The word spinning has two meanings: extrusion through a spinneret to make a filament, and fabrication of yarns from staple fibres.
[b] Strips can be made using any appropriate process such as extrusion, calendering, weaving, yarn or fabric coating, etc.
[c] Some classical geotextiles (those indicated) are used more than others.
[d] Corrugated, waffled or alveolate structures are generally not used alone, they are used to make composite geotextiles.

12.1 Production of geotextiles (Giroud, 1984). Products are in lower case letters, processes are in capital letters. All special geotextiles are currently used.

- *Slit films*. The films are produced through a melt extrusion process using slit dies that are fitted with sharp blades. These films can be further fibrillated and broken into fibrous strands known as ***fibrillated yarn***.

The constituent materials (filaments, fibres, slit films or yarns) are converted into different types of classical and special geotextiles. Some of the most commonly used geotextiles are briefly discussed below.

- *Nonwoven geotextiles*. Nonwoven fabrics are defined as a sheet, web or batt of directionally or randomly oriented fibres/filaments, bonded by some combination of friction, cohesion and adhesion. In general, nonwoven fabric formation is a two-step process, i.e. web formation (layers of fibres having predefined orientation characteristics) and bonding the fibres by mechanical, thermal or chemical means (Rawal *et al.*, 2007). This two-step process forms the basis for classification of nonwoven fabrics, i.e. carded, air-laid, spun-bonded, melt-blown, needle-punched, hydro-entangled, adhesive-bonded, thermal-bonded, stitch-bonded, etc.
- *Knitted geotextiles*. These are produced by the interlocking of a series of loops of filaments or yarns to form planar structures. The loops in the knitted structure can be interlocked in various ways, similar to woven fabrics.
- *Braided geotextiles*. These are narrow, rope-like structures, formed by diagonally interlacing three or more strands of filaments or yarns. In general, the bundles of filaments/yarns in a braid are interlaced similarly to interlacements of ribbons formed in a maypole dance. This results in a tubular woven structure in which the constituent filaments/yarns follow helical paths, simultaneously forming the interlacements between them.
- *Woven geotextiles*. A woven fabric consists of two sets of interlaced filaments/films/yarns that are generally orthogonal to each other. The weave design or pattern is determined by the manner in which yarns or filaments are interlaced. Filaments or yarns in the longitudinal direction and transverse directions are known as warp and weft, respectively. Woven geotextiles can also be made from slit films that are thinner in comparison to multifilament, spun and fibrillated woven geotextiles.

In this chapter, our main focus is on the production and properties of woven fabrics that are used for geotextile applications. In addition, we will discuss some of the case studies that showcase the applications of woven fabrics in geotextiles.

12.3 Selection of fibres for woven geotextiles

Both synthetic and natural fibres are used in the production of woven geotextiles. Natural fibres offer high strength, high modulus and low breaking extension in addition to low levels of creep during use. Since, tensile strength is an important property for a woven geotextile, especially for reinforcement applications, natural fibres show the greatest potential for use in woven geotextiles. Moreover, natural fibre-based geotextiles are biodegradable and therefore suited to fulfil temporary functions. In summary, the main advantages of using natural fibres in geotextiles are (Rawal *et al.*, 2010):

- low cost, robustness
- strength/durability
- availability
- good drapeability
- biodegradability/sustainability.

Synthetic fibres are also used in the manufacture of woven geotextiles. The four predominant polymer families used as raw materials for geotextiles are: polypropylene, polyester, polyamide and polyethylene. It should be noted that within these polymer groups there are many subgroups and variants, each with their own set of characteristics. Polypropylene has been amongst the most widely used filament/fibre/film in the production of woven geotextiles, due to its low cost, reasonably good tensile properties and chemical inertness. This polymer has low density, which results in very low cost per unit volume. The disadvantages of polypropylene are its poor resistance to ultraviolet radiation and high temperature, and its poor creep properties. Polypropylene-based woven geotextiles should therefore be used only under suitable environmental conditions.

Another important synthetic polymer used in the manufacture of geotextiles is polyethylene terephthalate (PET), commonly known as polyester. It offers excellent creep resistance with high strength, and is used in applications where the geotextile is subjected to high stresses and elevated temperatures. The main disadvantage of polyester is its susceptibility to hydrolytic degradation in some soils. It is worth mentioning that the polypropylene, polyethylene and polyester filaments/fibres/films are produced from a common melt extrusion process. Other synthetic fibres such as polyamides (nylon 6,6 and nylon 6) are also used in small quantities as conventional geotextiles. The polymer selected can have a significant influence on the mechanical properties of geotextiles but other factors such as fabric structure, finishing treatments applied to the fabric and the properties of surrounding soil can also influence the properties of the geotextile (John, 1987).

12.4 Production of woven geotextiles

Generally, monofilament, slit film and multifilament-based synthetic materials have dominated the geotextile market. The production stages of woven geotextiles of these materials are thus of significant importance. Figures 12.2 and 12.3 show

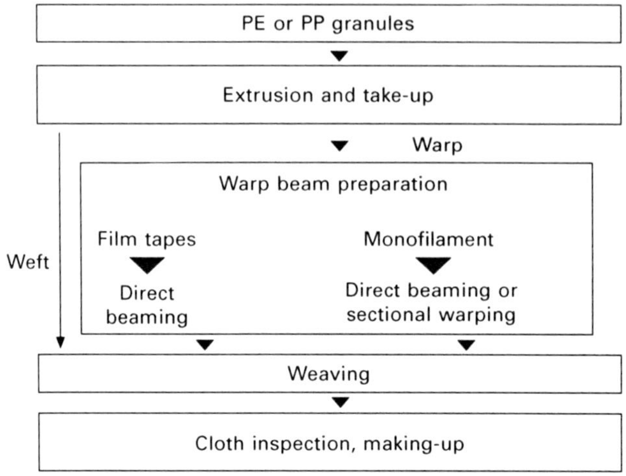

12.2 Flowchart for manufacture of woven polyethylene (PE) or polypropylene (PP) monofilament or tape fabrics (Achermann, 1985).

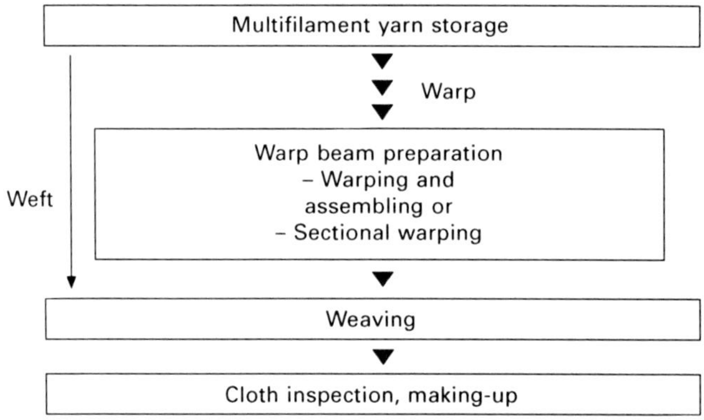

12.3 Flowchart for manufacture of woven polyester or polyamide multifilament fabrics (Achermann, 1985).

the flowchart of production of woven geotextiles made from monofilaments/films and multifilaments, respectively (Achermann, 1985). Woven geotextiles made from monofilaments/films are produced in the following stages:

- extrusion of a flat film/filament
- cooling of the film/filament in a water bath
- cutting the film into individual tapes
- drawing out the tapes/filaments and winding them onto packages.

The manufacture of multifilaments is similar except that the slit dies are used in place of spinnerets with slots in the melt extrusion process.

These filaments/films subsequently form an important constituent of 'beaming' in the warping process. In general, direct beaming (also known as direct warping) is commonly used for films and for circular cross-section monofilaments. Warps prepared from multifilament yarns generally have a higher number of ends and are produced on assembly warpers and subsequently assembled onto a weaver's beam. Following warping, the weaving of monofilament or multifilament geotextiles is commonly carried out either on projectile or rapier looms (Achermann, 1985). The projectile loom has been more popular than the rapier loom, due to its versatility in producing a range of dense and heavy fabrics as well as its flexibility in permitting a wide range of production widths or simultaneously weaving several different widths of fabrics on one machine in an economical way.

12.5 Specifications of woven geotextiles and their essential properties

Geosynthetic engineers or designers will include geotextiles in their designs when they have reliable technical data and design methodology for these materials (Leflaive, 1985). The selection of woven geotextiles should be made using the 'design by function' route to specify the properties of geotextiles (Koerner, 1984).

Woven geotextiles comprise two yarn systems that are orthogonal to each other, also known as warp and weft. These fabrics are porous in nature and the hydraulic (in-plane and cross-plane permeability) and mechanical (pull-out) properties are significantly influenced by the presence of predefined voids or pores available in the woven geotextile. In general, two types of pores are present in the woven geotextile, i.e. between and within the yarns. The pores between the yarns help to determine how the geotextile filters soil particles, and how liquid flows through the fabric. Assuming the soil particle to be spherical in shape, a simple relationship between the diameter of sphere that can pass between the yarns in a plain woven fabric is given by the following equation (see Fig. 12.4, Harten, 1986):

$$D_{sph} = S - d_y \quad [12.1]$$

where S is distance between the yarns, D_{sph} and d_y are the diameter of soil particle and yarn diameter, respectively.

In the case of woven geotextiles having maximum cover, the maximum diameter of a sphere that fits into the pore is given by Harten (1986):

$$D_{sph} = 0.732 d_y \quad [12.2]$$

While the above relationship is valid for plain woven fabric, the relationships for complex weaves can be determined by studying the geometry of woven fabrics (Peirce, 1937). The other important pore parameters in woven fabric are

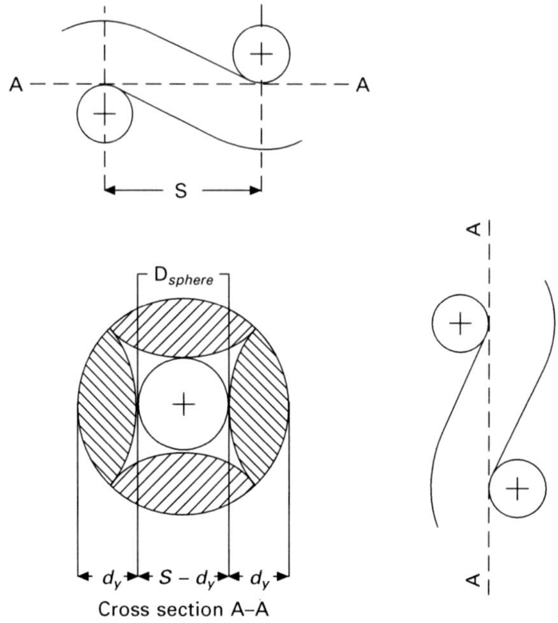

12.4 Model cross-section with sphere (Harten, 1986).

openness and porosity. Openness is applied in two dimensions (2D), whereas porosity is generally characterised in three dimensions (3D). Porosity (η) is defined as fraction of bulk volume of a material that is occupied by void space whereas the fabric openness (O_{Fab}) is defined as the proportion of area occupied by the pore in relation to the fabric area. Mathematically, fabric openness and porosity are given by Eqs [12.3] and [12.4], respectively:

$$O_{Fab} = \frac{(S - d_y)^2}{S^2} \tag{12.3}$$

$$\eta = 1 - \frac{W}{T_g \rho_f} \tag{12.4}$$

where W is the geotextile mass per unit area, T_g is the geotextile thickness and ρ_f is the density of constituent fibres. The permittivity of a woven fabric can thus be calculated by defining the unit cell of model fabric and using the well-known Kozeny–Carman equation, as shown below (Harten, 1986):

$$V_W = \frac{1}{K_{ng}} \frac{(h_1 - h_2)\gamma_w}{\eta_w} \frac{d_y}{32} \frac{\left(1 - \frac{\pi d_y l_y}{4S^2}\right)^3}{\left(\frac{\pi d_y l_y}{4S^2}\right)^2} \tag{12.5}$$

where V_w is the discharge velocity of water, K_{ng} is the coefficient of normal permeability of a geotextile, γ_w is the unit weight of water, η_w is the dynamic viscosity of water, l_y is the length of yarn in unit cell of woven fabric, and h_1 and h_2 are hydraulic head above and under the geotextile, respectively.

It is worth mentioning that the pore size of a woven fabric is critical in the design of geotextiles specifically for filtration and separation applications. A typical design criterion for a geotextile filter is defined as:

$$\frac{O_{95}}{d_{85}} \leq x \qquad [12.6]$$

where x is an arbitary ratio based on experimentally determined criteria (typical values between 1 and 3), O_{95} represents that 95% of the pores are smaller than this size, and d_{85} is the diameter such that 85% of the soil particles are smaller than this diameter.

The pore size of a woven geotextile is significantly affected by mechanical deformation. As an example, compression loading decreases the pore size of a woven geotextile primarily due to flattening of yarns in the fabric. Similarly, the effect of in-plane tensile load is likely to reduce the geotextile pore size and can cause clogging problems. Clogging is a result of saturation of fine particles in the geotextile leading to blocking of pores or caking up on the upstream side of the geotextile. Clogging of soil particles can cause progressive increase in the water head loss in the geotextile (Christopher and Fischer, 1992; Faure *et al.*, 2006). Interestingly, the application of unequal biaxial tensile loads in two orthogonal in-plane directions can significantly decrease the pore size of thick woven slit-film geotextile as a consequence of unequal in-plane strains (Fourie and Addis, 1999; Fischer *et al.*, 1992). An increase in tensile load causes closing of the smallest pores and the gradual closing of larger pores with increasing tensile strain, resulting in lower pore sizes when unequal biaxial loading has been applied to a woven geotextile (Fischer *et al.*, 1992). Interestingly, thinner and lighter woven slit-film geotextile shows the opposite behaviour, as there was increase in the pore size due to the application of biaxial tensile load (Fourie and Addis, 1999).

Generally, soil particles have high compression resistance but are weak in tension. When a geotextile, acting as a tensile element, is added to soil particles, a composite material is formed. This composite material is a result of intimate contact between the soil and the geotextile such that these two materials experience the same level of strain under operational conditions. Subsequently, the lateral movement of soil particles is prevented by the geotextile due to the development of lateral shear stress (Pritchard *et al.*, 2000). The tensile element, i.e. the woven geotextile, is suitable for reinforcement applications. Generally, a woven geotextile of the same mass per unit area is much stiffer in comparison to the equivalent nonwoven geotextile.

It is important to note that testing a narrow strip of a woven fabric in a uniaxial tensile test is not suitable for geotextile applications, since the lateral contraction is not able to fully develop in the part of the strip that is near to the clamps gripping the fabric. However, large widths of geotextiles with fully developed lateral contractions can be used in practice in appropriate applications. They demonstrate significantly different tensile behaviour than would be expected from narrow width fabric tensile tests. Lateral contraction can be prevented by applying biaxial stress, resulting in straightening of warp yarns and a simultaneous increase in crimp in the weft yarn. In this case, the 'crimp interchange' is limited and the resulting modulus of elasticity is much higher than that from the uniaxial tensile test. Alternatively, the multi-axial tensile test can be performed to avoid the lateral contraction effect in geotextiles (Andrejack and Wartman, 2010).

In the past, it has been reported that the woven geotextiles are highly susceptible to the changes in strain rate and temperature; however, the intrinsic properties of the constituent fibres play an important role in determining the deformation behaviour of woven geotextiles (Andrawes *et al.*, 1984). Ideally, the tensile characteristics of woven geotextiles should be determined under confined conditions, i.e. with the geotextile sandwiched between layers of soil particles under the desired level of stresses. It is important to note that the shapes of stress–strain curves are significantly affected at higher confined stresses which cannot be simulated even with a wider specimen size (Leflaive *et al.*, 1982).

Nevertheless, Poisson's ratio (ratio of lateral contraction to the longitudinal strain) is a key parameter defining the amount of lateral contraction occurring in the geotextile under a defined level of strain. Poisson's ratio (v) can be computed based on the assumption that the ratio of volume (V) of geotextile corresponding to strain (ε) to volume (V_0) of geotextile at the beginning of the tensile test is a linear function of applied strain (ε), as shown in Eq. [12.7] (Giroud, 2004):

$$v = \frac{1}{\varepsilon}\left(1 - \sqrt{\frac{1 + \varepsilon(1 - 2v_0)}{1 + \varepsilon}}\right) \qquad [12.7]$$

where v_0 is the Poisson's ratio for zero strain.

The tensile properties of woven geotextiles significantly influence other mechanical and frictional properties. Puncture failure of woven geotextile, for example, is caused by subgrade surface irregularities leading to concentrated forces normal to the plane of the fabric while the geotextile is already under in-plane tension. Hence, the puncture resistance of a woven fabric is significantly dependent upon its in-plane characteristics, as shown below (Ghosh, 1998; Cazzuffi and Venesia, 1986):

$$T_f = \frac{F_p}{2\pi r} \quad [12.8]$$

where T_f is the tensile force per unit width, r is the radius of plunger and F_p is the puncture force. Interestingly, it has been reported in the literature that the tensile strength of woven geotextiles obtained from wide width tensile tests was found to be higher than the strength determined from puncture tests (Murphy and Koerner, 1988).

As mentioned earlier, woven geotextiles are considerably stiffer materials than other types of geotextiles and have been extensively used for reinforcement applications. However, the question arises how shear stresses are transferred from soil to the reinforcement. There have been two mechanisms proposed for transfer of shear stresses from soil to the reinforcement:

- The geotextiles used as tensile elements transfer most of the shear stresses from the soil to the reinforcement by surface friction and any remaining stress by interlock (John, 1987; Koerner, 1986).
- The passive resistance of the soil against the reinforcing elements (geotextiles), when placed normal to the direction of shear displacement, can be another mechanism (Kabeya *et al.*, 1993).

The former mechanism is governed by the surface roughness of geotextiles, whereas the latter is dictated by pore size and deformation capability. The efficiency of geotextiles in generating shear resistance at the soil–geotextile interface is known as the contact efficiency or coefficient of interaction (α). This is defined as the ratio of the tangent of the friction angle (δ) for the soil–geotextile to the tangent of the friction angle of the soil alone (ϕ), i.e. $\tan \delta/\tan \phi$. The soil–geotextile interaction can be classified in two ways: shearing (bond) and pull-out (anchorage) interactions (Fourie and Fabian, 1993; Collios *et al.*, 1980). In shearing, the soil slides over a geotextile, whereas in pull-out tests, the strain is applied to the fabric to mobilise the shear resistance in the direction of the applied load. The pull-out test is difficult to interpret and is more closely related to the failure mechanism of the reinforced soil structure, as shown in Eq. [12.9] (Karmokar *et al.*, 1996):

$$P = S + F \quad [12.9]$$

where P = pull-out force, S = shearing force, and F = frictional force.

On the other hand, the shear strength at the soil–geotextile interface is a function of the stress normal to the interface, as illustrated in Eq. [12.10]:

$$\tau = a + \sigma \tan \delta \quad [12.10]$$

where τ is the soil–geotextile or geotextile–geotextile interface shear strength, a is the interface adhesion, σ is the stress normal to the interface and δ is the interface friction angle.

When the stress on the geotextile is applied constantly, there is an increase in the extension of the geotextile with time, a phenomenon known as creep. The performance characteristics of geotextiles are significantly affected by creep behaviour. As an example, if the function of a geotextile is separation, the fabric should be capable of separating soil layers. However, when it is elongated, the pore size significantly changes, affecting the separation function of the geotextile. Hoedt (1986) has reported that constituent material properties, construction of geotextiles, loading conditions, temperature and time play an important role in determining the creep behaviour of geotextiles.

Creep is one of the main parameters affecting the durability of woven geotextiles. Durability of a geotextile is defined as the ability of a material to remain stable and fulfil its prescribed function(s) effectively during the predefined life of the project. In general, the durability of a geotextile is determined primarily by the resistance offered by the constituent fibres to environmental conditions. A geotextile should exhibit sufficient strength and ability to resist rupture, puncture, cutting, compression, abrasion and silting when stretched. It must also have the required filtration characteristics, sufficient hydraulic resistance, low clogging and moisture absorption (Rawal *et al.*, 2010). Moreover, a geotextile must be biologically and chemically stable and resistant to UV radiation. The properties of woven fabrics should also be selected to ensure that the geotextile maintains its performance during transportation, storage and installation as well as performing its function during service.

The use of recycled polymers for the manufacture of geotextiles has been increasing, primarily due to the significant availability of low-cost recycled fibre (Akkapeddi *et al.*, 1995; Rebeiz *et al.*, 1993; Davies and Horrocks, 2000). However, the long-term durability of geotextiles using recycled materials needs to be established. Geotextiles are sometimes employed in a wide variety of chemical environments. The engineering of a geotextile should include the biochemical resistance of its constituent fibres/filaments/films. The use of geotextiles in environmental applications requires good biochemical resistance as they are exposed to contaminants leaching from landfill. Contamination comes in the form of high-strength wastewater, characterised by extremes of pH, chemical oxygen demand (COD), biochemical oxygen demand (BOD), inorganic salts and toxicity (Keenan *et al.*, 1984). These wastes are complex materials that can cause physical, chemical and biological damage to geotextile filters as they are a mixture of various chemical compounds from different types of waste generated from biotic and abiotic processes in the system (Oman and Hynning, 1993).

Table 12.1 illustrates the essential physical, mechanical and chemical properties required for the desired functions of geotextiles (Horrocks and Anand, 2000). In the past, these essential properties of geotextiles have formulated the basis for successful deployment of woven fabrics in geotextile

Table 12.1 Functional requirements for geotextiles (Horrocks and Anand, 2000)

	Geotextile function				
	Reinforcement	Filtration	Separation	Drainage	Erosion control
Tensile strength	iii	i-ii	ii	na	ii
Elongation	iii	i-ii	iii	i-ii	ii-iii
Chemical resistance	ii-iii	iii	iii	iii	i
Biodegradability	iii	iii	iii	iii	iii
Flexibility	i	i-ii	iii	i-ii	iii
Friction property	iii	i-ii	i	na	ii
Interlock	iii	iii	ii	ii	i
Tear resistance	i	iii	iii	ii-iii	ii
Penetration	i	ii	iii	iii	ii
Puncture resistance	i	ii	ii	iii	i-ii
Creep	iii	na	na	na	na
Permeability	na-i	ii-iii	ii-iii	iii	ii
Resistance of flow	i	i	i	i	iii
Property of soil	iii	ii	na	na	na
Water	iii	iii	iii	iii	iii
Burial	iii	iii	iii	iii	na
UV light	ii	na	na	na	iii
Climate	na	iii	i	iii	iii
Quality assurance and control	iii	iii	ii	iii	i
Costs	iii	iii	iii	iii	iii

iii = Highly important, ii = important, i = moderately important, na = not applicable applications. In the next section we include case studies where woven geotextiles have been engineered and deployed for specific civil engineering applications.

12.6 Applications of woven geotextiles

Some of the major applications where woven geotextiles have been successfully employed are given in the following examples.

12.6.1 Jute-blended woven geotextile for unpaved rural road

As mentioned earlier, natural fibres have distinct advantages compared to synthetic fibres. Despite the fact that natural fibres such as jute, flax, coir, etc., are low cost, they have certain functional characteristics including high tensile strength, modulus, good dimensional stability, high moisture absorption, availability, sustainability and biodegradability. Nevertheless, the advantages of synthetic fibres cannot be ignored and the combination of synthetic and natural blends is ideal for a range of geotextile applications. One such example is the use of high density polyethylene (HDPE) film of 111 tex in the warp direction woven on a circular weaving machine with jute yarns (2 × 360 tex) used in the weft direction (Basu *et al.*, 2009). A range of blended plain woven fabrics were produced and optimised to obtain a sample of mass per unit area of 316 g/m^2 for a field trial in road construction. This field trial involved constructing a medium traffic, unpaved rural road 3 m wide in Deochasaranda village in West Bengal, India. The soil consisted of sand (59%), slit (5.6–7.2%) and clay (33.6–35.6%). A schematic diagram of the cross-section of the road in the field trial is shown in Fig. 12.5 (Basu *et al.*, 2009). Interestingly, it was pointed out that, in spite of heavy rainfall (216 cm) and a flash flood in the surrounding area, no significant changes were observed in the apparent condition of the road. The results of

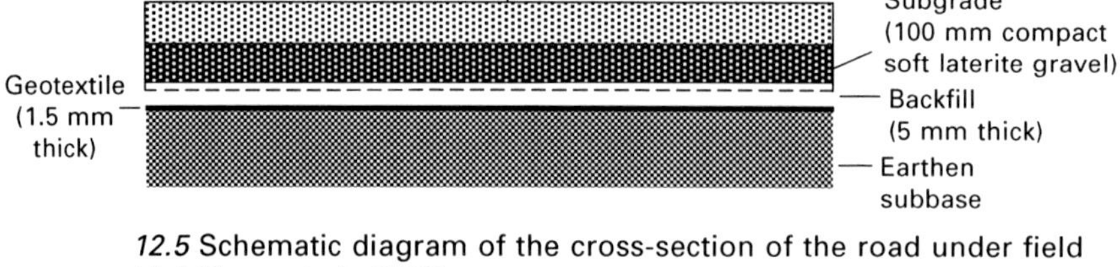

12.5 Schematic diagram of the cross-section of the road under field trial (Basu *et al.*, 2009).

California bearing ratio* (CBR) tests confirmed a considerable increase of 67 and 73% after the deployment of blended plain woven geotextile after 11 and 18 months, respectively. Digging up certain sections of the road showed that the jute element was either partly or fully degraded, leaving behind the HDPE tape as sub-base strainer. This case study clearly showed the successful deployment of low cost jute-based woven geotextile for an unpaved rural road.

12.6.2 Woven geotextile soilbags for protection and reinforcement applications

Woven bags consisting of concrete or recyclable construction waste such as asphalt, tyres, tiles, etc., can be used for protection, including flood emergency protection in dams and dykes, reinforcing elements in erosion control and many other similar applications. Soilbags are rarely used for constructing permanent structures because of poor stability (Matsuoka and Liu, 2003). The bearing capacity of soft ground can be increased by 5–10 fold using soilbags. Secondly, the soilbag has a high compressive strength and can absorb vibrations. Finally, they can be easily constructed and are environmentally friendly.

Recently, two case studies have reported the use of woven soilbags in pond filling and soil protection (Xu *et al.*, 2008). The first case study is that of the construction of a highway in Jiangsu Province in China, where a new reinforcement method – using polyethylene woven soilbags containing natural soils of optimum water content – was selected to fill up the foundations in an especially waterlogged and unstable pond site. These soil bags of 40 × 40 × 10 cm dimensions were connected using high strength ropes and compacted using a vibro-roller. A comparison of the pond before and after filling using soil bags is shown in Fig. 12.6 (Xu *et al.*, 2008). In another application, the soilbags were used in the construction of retaining walls to protect a soil slope. Groups of four soilbags of the same dimensions were connected and compacted in the lower part of the slope to create a slope angle of 30° (see Fig. 12.7; Xu *et al.*, 2008). Since, these soilbags were made up of polyethylene material, they were sensitive to sunlight so a thin layer of grass was planted on the outside surface of the wall. Overall, 2,000 soilbags were successfully employed in the construction of retaining slope walls. This case study has clearly demonstrated the successful use of woven geotextile soilbags in permanent structures.

*California bearing ratio test signifies the ratio of the force per unit area required to penetrate a soil mass with a circular piston to that required for corresponding penetration of a standard material.

(a) (b)

12.6 Construction of the pond by filling up of soilbags: (a) initial condition of pond bottom and (b) compaction of soilbag reinforced foundation using vibro-roller (Xu *et al.*, 2008).

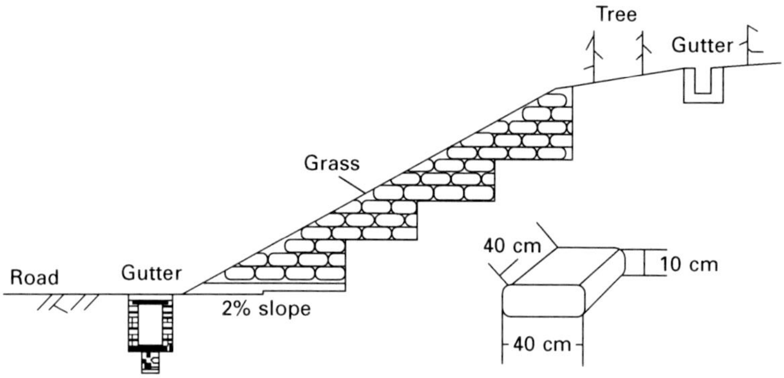

12.7 Schematic design of a retaining wall using soilbags (Xu *et al.*, 2008).

12.6.3 Woven geotextile tubes for fluid discharge applications

Large diameter geotextile tubes are often used for dewatering various types of slurries ranging from contaminated dredged material to chemical plant wastes. The principle of filtering solid waste from liquid is fairly simple, i.e. the slurry is pumped into the tube and the liquid under internal pressure and, through gravity, it migrates to the surrounding geotextile and passes through it. Subsequently, the solids are retained and form a cake inside the tube (Weggel *et al.*, 2011). The easiest method to evaluate a geotextile is by filling and suspending with a known quantity of slurry and then measuring the rate at which the slurry liquid seeps from the tube. This method clearly indicates that pore size and permittivity characteristics are of significant importance, but that the tensile properties of the hanging tube also play an important role in determining the performance characteristics of the geotextile tube. This is the reason that woven geotextiles exhibiting high tensile strength are used in combination with nonwovens for geotextile tube applications.

A further case study demonstrates the use of woven geotextile tubes for dewatering fine-grained contaminated dredged material from the Miami River and the Port of Oakland, CA (Fowler *et al.*, 1997). Two small geotextile tubes, of 122 cm and 178 cm circumference, respectively, were supported vertically on a wooden frame prior to filling. The outer bag consisted of a commercial product, Nicolon Geolon GT 500, which is a woven polypropylene having an ultimate tensile strength of approximately 222 kgf per unit cm width. The inner bag consisted of polyester and polypropylene based non-woven materials. Samples of suspended solids, heavy metals and bacterial count were collected after the contaminated dredged material was made to pass through the geotextile tube. It was found that there was significant consolidation or reduction in the volume of the sludge in the bag in addition to the significant reduction in heavy metals and bacterial count in the effluent water. The quality of pore water or effluent passing through the geotextile has met environmental standards and could be discharged into the Mississippi River and/or return to the treatment plant. This was claimed to be the first successful use of geotextile tubes for dewatering sewage sludge in the United States.

12.6.4 Other case studies of woven geotextiles

In addition to the aforementioned applications of woven geotextiles, there are other case studies which have successfully employed woven geotextiles for reinforcement, filtration, separation and drainage. In most cases, the woven geotextile was expected to perform two or more functions simultaneously. Achermann (1985) has summarised some of the European case studies where woven geotextiles have been used. These include a shopping centre built in a lake delta zone outside Rennaz, at the south-eastern end of Lake Geneva, where the ground had a limited bearing capacity; the Dottikon railway siding built by Swiss Federal Railways, vulnerable to subsidence and landslide, that was stabilised by a woven fabric; the combination of woven and nonwoven geotextiles to improve the ground for construction of Sargans Station in Switzerland by constructing a drainage system; and the lining of water-facing surfaces near Saint Pierre du Vauvray in France with a woven polypropylene tape geotextile.

12.7 Future trends

Geotextiles are one of the fastest growing sectors of the technical textile market, which is expected to keep growing in the future. Synthetic fibres have dominated the woven geotextile market, but the use of natural fibres should not be ignored especially in developing countries. Furthermore, the standardisation of certain properties and test methods will deliver more reliable technical data and design methodology for woven geotextiles. This will eventually lead to design of woven geotextile by 'design by function' route. As a result, woven geotextiles will perform their prescribed functions more effectively during the lifetime of the project. The combination of woven fabric mechanics and soil mechanics can effectively solve innumerable construction and design problems. However, this requires a clear understanding between textile professionals and civil engineers.

12.8 Sources of further information and advice

Faure Y H, Gourc J P and Gendrin P (1990), Structural study of porometry and filtration opening size of geotextiles. In Peggs I D (ed), *Geosynthetics: Microstructure and Performance*, ASTM STP 1076, Philadelphia, PA.

Hamid S H, Amin M B and Maadhah A G (1992), *Handbook of Polymer Degradation*, Marcel Dekker, New York.

Ingold T S (1994), *The Geotextiles and Geomembranes Manual*, Vol. 1, Elsevier, Oxford.

Koerner R M (1989), *Durability and Aging of Geosynthetics*, Vol. 3, Elsevier, New York.

Koerner R M (1990), *Biological activity and potential remediation involving geotextile landfill leachate filters*. In: ASTM STP 1081, Philadelphia, PA.

Shukla S K, Yin J H (2006), *Fundamentals of Geosynthetic Engineering*, Taylor and Francis, London.

Zanten R V V (1986), *Geotextiles and Geomembranes in Civil Engineering*, A.A. Balkema Publishers, Accord MA 02018, Netherlands.

12.9 References

Achermann A (1985), 'The application and manufacture of woven geotextiles', *Geotextiles and Geomembranes*, 2, 151–168.

Akkapeddi M, Biskirk B V, Mason B, Chung C D, SwamiKannu X (1995), 'Performance blends based on recycled polymers', *Polymer Engineering and Science*, 35, 72–78.

Andrawes K Z, McGown A, Kabir M H (1984), 'Uniaxial strength testing of woven and nonwoven geotextiles', *Geotextiles and Geomembranes*, 1, 41–56.

Andrejack T L, Wartman J (2010), 'Development and interpretation of a multi-axial tension test for geotextiles', *Geoetextiles and Geomembranes*, 28, 559–569.

Basu G, Roy A N, Bhattacharyya S K, Ghosh S K (2009), 'Construction of unpaved rural road using jute–synthetic blended woven geotextile – a case study', *Geotextiles and Geomembranes*, 27, 506–512.

Cazzuffi D, Venesia S (1986), 'The mechanical properties of geotextiles: Italian standard and inter-laboratory test comparison', In *Proceedings of 3rd International Conference on Geotextiles, Vienna, Austria*, 695–700.

Christopher B R, Fischer G R (1992), 'Geotextile filtration principles, practices and problems', *Geotextiles and Geomembranes*, 11, 337–353.

Collios A, Delmas P, Gourc J-P, Giroud J-P (1980), 'The use of geotextiles for soil improvement', *ASCE National Convention, Portland, Oregon.*

Davies P J, Horrocks A R (2000), 'Effect of recycled polyolefin inclusion on the properties of PP filaments, Part 1: Physiochemical properties', *Textile Research Journal*, 70, 363–372.

Faure Y H, Baudoin A, Pierson P, Ple O (2006), 'A contribution for predicting geotextile clogging during filtration of suspended solids', *Geotextiles and Geomembranes*, 24, 11–20.

Fischer G R, Holtz R D, Christopher B R (1992), 'A critical review of geotextile pore size measurement methods', in *Proceeding of the First International Conference 'Geofilters' Karlsruhe, Germany.*

Fourie A B, Addis P C (1999), 'Changes in filtration opening size of woven geotextiles subjected to tensile loads', *Geotextiles and Geomembranes*, 17, 331–340.

Fourie A B, Fabian K J (1993), 'Laboratory determination of clay-geotextile interaction', *Geotextile and Geomembranes*, 6, 275.

Fowler J, Bagby R M, Trainer E (1997), 'Dewatering sewage sludge with geotextile tubes', *Geotechnical Fabrics Report*, September, 26–30.

Ghosh T K (1998), 'Puncture resistance of pre-strained geotextiles and its relation to uniaxial tensile strain at failure', *Geotextiles and Geomembranes*, 16, 293–302.

Giroud J P (1984), 'Geotextiles and geomembranes', *Geotextiles and Geomembranes*, 1, 5–40.

Giroud J P (2004), 'Poisson's ratio of unreinforced geomembranes and nonwoven geotextiles subjected to large strains', *Geotextiles and Geomembranes*, 22, 297–305.

Harten K V (1986), 'The relation between specifications of geotextiles and their essential properties', *Geotextiles and Geomembranes*, 3, 53–76.

Hoedt G D (1986), 'Creep and relaxation of geotextile fabrics', *Geotextiles and Geomembranes*, 4, 83–92.

Horrocks A R, Anand S C (2000), *Handbook of Technical Textiles*, Cambridge, Woodhead, 372–406.

John N W M (1987), *Geotextiles*, Blackie and Sons, Glasgow.

Kabeya H, Karmokor A K, Kamata Y (1993), 'Influence of surface roughness of woven geotextiles on interfacial frictional behaviour – evaluation through model experiments', *Textile Research Journal*, 63, 604–610.

Karmokar A K, Kabeya H, Tanaka Y (1996), 'Shearing and friction in pullout behaviour of woven geotextiles for reinforcement applications', *Journal of the Textile Institute*, 87, 586–594.

Keenan J D, Steiner R L, Fungaroli A A (1984), 'Landfill leachate treatment', *Journal of Waste Pollution Control Federation*, 56, 27–33.

Koerner R (1984), 'Should I specify a woven or nonwoven?' *Geotechnical Fabrics Report*, IFAS, Minnesota.

Koerner R M (1986), *Designing with Geosynthetics*, Prentice Hall, Englewood Cliffs, NJ.

Leflaive E (1985), 'Geotextiles: Their rationale and future', *Geotextiles and Geomembranes*, 2, 23–30.

Leflaive E, Paute J L, Segouin N (1982), 'La mesure des caracteristiques detraction en vue des applications pratiques', *Proc. Second Int. Conf. on Geotextiles, Las Vegas.*
Matsuoka H, Liu S H (2003), 'A new earth reinforcement method by bags', *Soils Found*, 43, 173–188.
Murphy V P, Koerner R M (1988), 'CBR strength (puncture) of geosynthetics', *Geotechnical Testing Journal*, 3, 167–172.
Oman C, Hynning P A (1993), 'Identification of organic compounds in municipal landfill leachates', *Environmental Pollution*, 80, 265–271.
Peirce F T (1937), 'Geometry of cloth structure', *Journal of the Textile Institute*, 28, T45–T96.
Pritchard M, Sarasby R W, Anand S C (2000), 'Textiles in civil engineering. Part 2 – Natural fibre geotextiles', in Horrocks A R and Anand S C (eds), *Handbook of Technical Textiles*, Cambridge, Woodhead, 372–406.
Rawal A, Lomov S V, Ngo T, Verpoest I, Vankerrebrouck J (2007), 'Mechanical behavior of thru-air bonded nonwoven structures', *Textile Research Journal*, 77, 417–431.
Rawal A, Shah T, Anand S (2010), 'Geotextiles: production, properties and performance', *Textile Progress*, 42, 181–226.
Rebeiz K, Fowler D W, Paul D R (1993), 'High performance polymer composites using recycled plastic', *Trends in Polymer Science*, 1, 315–321.
Weggel J R, Dortch J, Gaffney D (2011), 'Analysis of fluid discharge from a hanging geotextile bag', *Geotextiles and Geomembranes*, 29, 65–73.
Xu Y, Huang J, Du Y, Sun D (2008), 'Earth reinforcement using soilbags', *Geotextiles and Geomembranes*, 26, 279–289.

13
Hollow woven fabrics

X. CHEN, University of Manchester, UK

Abstract: Hollow woven fabrics are a variety of woven structures that can be produced based on the conventional weaving principle. Hollow fabrics started to attract attention as lightweight high performance materials, demand for them has increased for various applications. This chapter introduces the classification of hollow woven fabrics and their structural attributes, describes the possible routes of manufacture for different types of hollow fabrics, explains the most attractive properties and potential applications, and discusses the geometrical modelling and techniques used in finite element analysis of this type of structures. Hollow woven fabrics are usually used as reinforcement to hollow composites, some of which are described in this chapter. The ease of manufacture and the unique properties such as lightweight, energy absorption, and bulkiness are the most attractive properties and it is reasonable to believe that hollow woven fabrics will find more applications as technical and engineering materials.

Key words: hollow woven fabrics, weaving principles, energy absorption, lightweight, force attenuation, modelling.

13.1 Introduction: overview and potential applications

Weaving as a mature technology of interlacing threads to form fabrics is able to manufacture fabrics with different features. The most commonly seen woven fabrics are single layer fabrics made from natural and synthetic fibres, mainly used in clothing and other domestic applications. Woven fabrics can also be made to have substantial thickness (3D fabrics) to satisfy requirements mainly for technical applications (Chen *et al.*, 2011). The increase in thickness can be achieved by binding multiple layers of warp and weft yarns (such as orthogonal and angle interlock structures), making one set of yarn to weave multiple sets of yarns in the other direction (such as the backed cloth structures), or linking distinctive layers together during the weaving process (such as multilayer fabrics). Based on the multilayer fabric principle, the so-called hollow woven fabrics can be created by connecting adjacent distinctive layers of fabrics according to certain rules.

The term hollow fabric was originated at the University of Manchester, UK, because the cross-sections of such fabrics are porous. This type of 3D fabric is also known as hollow fabrics and spacer fabrics for obvious reasons. Despite the intrinsic capability of the weaving technology in making hollow fabrics, only a limited amount of work has been reported on the design, manufacturing and application. Depending on the cell shape adopted, a hollow fabric can be either one with flat surfaces or one with uneven surfaces. Chen *et al.* (2004) worked on the mathematical modelling of hollow woven fabrics with uneven surfaces and on the CAD of such fabrics. Chen and Wang (2006) reported on mathematical modelling of hollow fabrics with flat surfaces and on the establishment of CAD software for such fabrics. Chen *et al.* (2008a) worked on the design, manufacture and evaluation of hollow fabrics and composites, and Tan and Chen (2005) and Tan *et al.* (2007) carried out characterisation of composites made from hollow fabrics. Chen and Zhang (2006) further developed and patented multi-level multi-directional hollow woven fabric. There have been many potential applications. Yu and Chen (2004) explored the application for protection against trauma impact. Kunz and Chen (2005) attempted to use it as a conductive layer in body armour for heat and moisture. Gong (2011) tested and simulated the mechanical performance such as impact energy absorption of the composites using hollow fabrics as reinforcing structure. Work has also been carried out to investigate the possibility of using hollow fabrics as insulating materials for oil pipelines (Kaddar and Ibrahim, 2011). Eriksson *et al.* (2011) reported work on using hollow fabrics as capacitors for interactive textiles.

13.2 Principles of hollow woven fabrics

13.2.1 Classification

In general, a hollow woven fabric refers to one whose cross-sections are porous. The materials between the two surfaces could be interlaced fabrics or simply yarns and fibres. Under this definition, hollow fabrics can be classified into hollow fabrics with uneven surfaces and those with flat surfaces.

13.2.2 Hollow fabrics with uneven surfaces

The cells in this type of hollow woven fabrics, formed by fabric layers are basically hexagonal, and a cell could have a symmetrical shape so that they can be folded down to make the entire fabric flattened. The formation of such a fabric is based on the regular linkage between the adjacent fabric layers in a multilayer fabric assembly. The fabric must be opened to assume the 3D shape. Figure 13.1(a) depicts how this type of hollow fabric is formed and Fig. 13.1(b) is an illustration of the hollow fabric with uneven surfaces viewed at the cross-section, where each line in the diagram represents a section of fabric. Because the cells in this type of fabric are of a hexagonal shape, they are also called the hexagonal hollow fabrics.

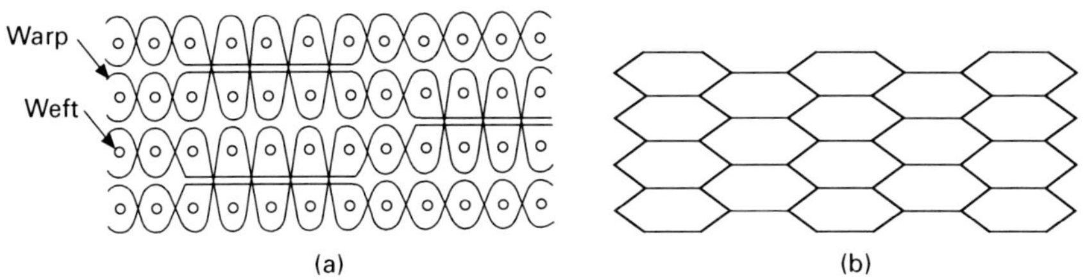

13.1 Illustration of a hollow fabric with uneven surfaces: (a) fabric construction, (b) cross-section of the opened fabric.

13.2.3 Hollow fabrics with flat surfaces

As indicated by the name, this type of fabric has flat top and bottom surfaces with the supporting sections in between either yarns or fabric layers. In general, this type of fabric cannot be flattened into one level as in the previous case. In the situation where the middle sections are in the form fabric, the longer length of fabric in the middle is achieved by inserting more weft yarn. In this structure, the cells in the cross-section are usually four-sided cells, such as rectangles and trapezoids. Figure 13.2(a) describes the construction of flat surfaced hollow fabric and Fig. 13.2(b) depicts the cross-section of this type of hollow fabric. Because the cell in this type of hollow fabric has a four-sided geometry, hollow fabrics with flat surfaces are also known as the quadratic hollow fabrics.

Figure 13.3 shows resin-treated hollow fabrics with flat and uneven surfaces. The hollow fabric with uneven surfaces can be used in curved and flat forms, whereas the hollow fabric with flat surfaces can only be used as a board structure.

Hollow fabrics with flat surfaces can also be made with yarns travelling between the face and back fabric layers based on the face-to-face principle (Grocicki, 1977). Such fabrics are more commonly known as spacer fabrics if made using warp knitting technology. Figure 13.4 displays a face-to-face hollow fabric based on the weaving technology. The face-to-face weaving technology has been used for manufacturing carpet, by cutting the connecting yarns between the face and back.

13.2.4 Variations of hollow fabrics

Usually, the hollow woven fabrics are woven with the tunnels in the weft direction. However, the hollow fabric with uneven surfaces can be made with the tunnels in the warp direction for applications where long tunnel length is necessary, whilst the hollow fabric with flat surfaces can only be made with tunnels in the warp direction. Chen and Zhang (2006) produced

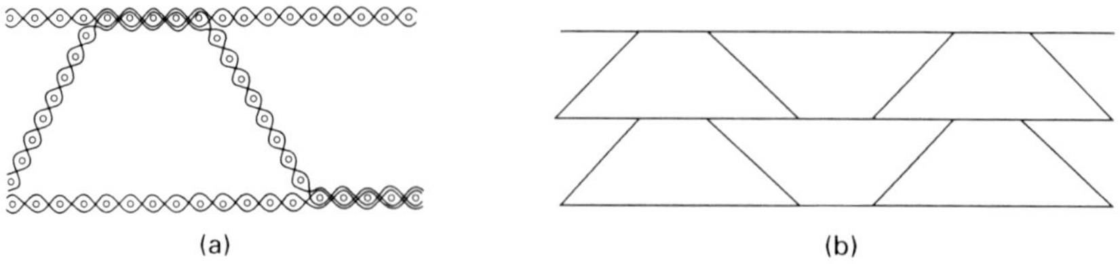

13.2 Illustration of a hollow fabric with flat surfaces: (a) fabric construction, (b) cross-section of the fabric.

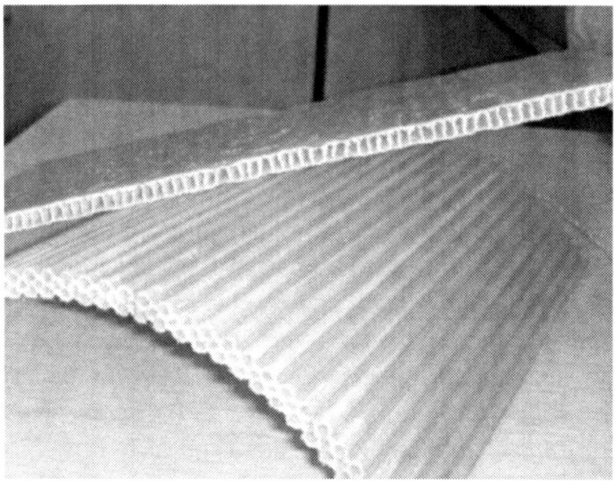

13.3 Resin-treated hollow fabrics with flat surfaces (top) and uneven surfaces (bottom).

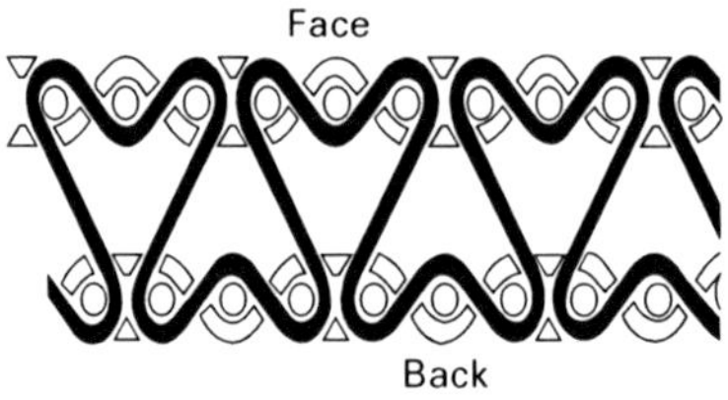

13.4 Hollow fabric structure based on the face-to-face weaving technology.

a the computerised design of hollow fabrics with tunnels running in both warp and weft directions.

By nature, the conventional technology is also capable of creating hollow fabrics with tunnels running in any direction parallel to the fabric plane. The tunnels can be situated on different levels, one running about the other; and they can be arranged on the same level in the fabric thickness, intersecting with one another (Chen and Zhang, 2006). Figure 13.5 shows a hollow fabric with multi-tunnels running in different relations to one another, which could

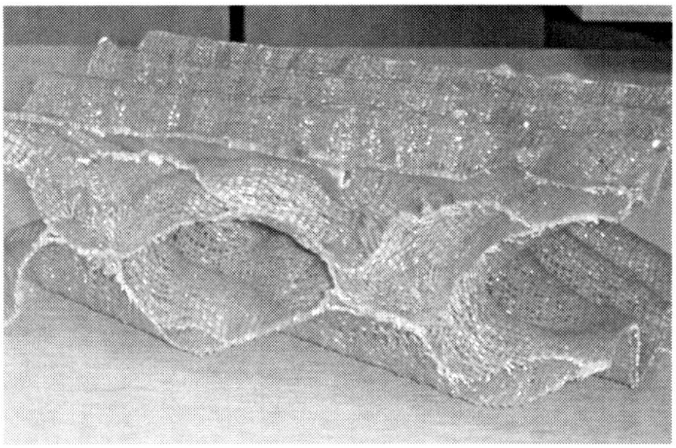

13.5 A hollow fabric with multi-tunnels in multi-directions.

be one over the other in an angle, one intersecting into the other, and tunnels running in any direction in the fabric plane.

13.3 Properties and performance of structures and materials based on hollow woven fabrics

13.3.1 Structural features

Hollow woven fabrics possess attractive unique features, which include bulkiness, lightweight, possibility for embedding, strength, energy absorption and force attenuation. Because of the existence of voids within the structure and possibility of opening and closing, such fabrics could also be used for developing temperature regulating materials.

13.3.2 Possible loading directions

In general terms, woven hollow fabrics are a special type of hollow solids. Such structures have found wide use in many areas including thermal insulation, packaging, structural, buoyancy as well as some other applications (Gibson and Ashby, 1999). As a structural material, they have been notably used in civil engineering and aerospace engineering.

Hollow solids can be loaded in different directions, providing different properties. Figure 13.6 illustrates a hollow solid in three dimensions, with the principal directions being, x, y, and z. If load is applied in the direction of either x or y, then the loading is known as in-plane loading. In such cases, cell deformation will be the major mechanism affecting material behaviour. It is usually true that hollow solids under in-plane loading manifest as a soft material which serves as a damping medium as it is easier to absorb impact energy.

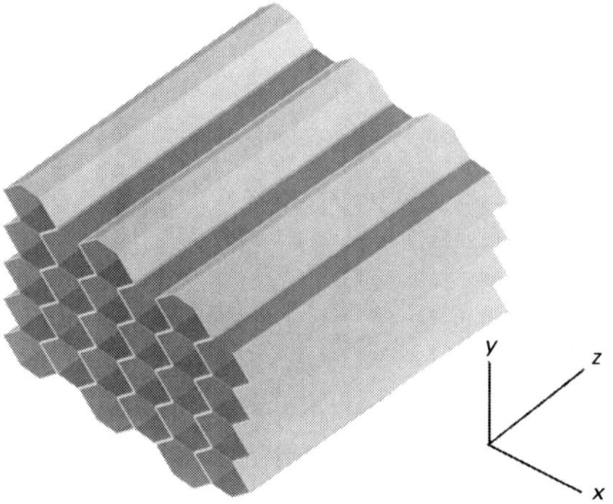

13.6 Illustration of a hollow solid in three dimensions.

Hollow solids can also be loaded in the z direction, and this is called out-of-plane loading. Hollow solids are popularly used with out-of-plane loading for their high compression rigidity and lightweight.

13.3.3 Lightweight of hollow composites

Due to the hollow construction, the volume density of the hollow composites reinforced by hollow fabrics can be very low, resulting in a type of lightweight composite material which is desirable for many applications.

Figure 13.7 displays the cross-section of a hollow composite. To evaluate the volume density of hollow composites, let us assume the following:

W = width of one repeat of hollow composite;
H = height of the hollow composite;
D = depth of the hollow composite;
n = number of cells in the longer column of cells;
a = length of inclined cell walls;
b = length of horizontal cell walls;
θ = open angle of the cell;
t = thickness of the cell walls; and
ρ = volume density of the solid composite.

From Fig. 13.7(a), it can be obtained that

$$W = 2(a\cos\theta + b) \tag{13.1}$$

$$H = 2n\,a\sin\theta \tag{13.2}$$

So, the volume, V_c, of one repeat of the hollow composite is

$$V_c = 2Dna^2\sin 2\theta + 4Dnab\cos\theta \tag{13.3}$$

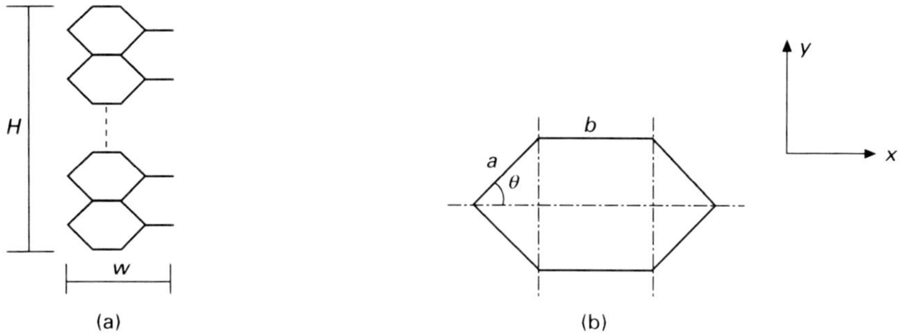

13.7 Geometric description of cross-section of a hollow composite: (a) one repeat of the hollow composite, (b) cell geometry.

The mass of the hollow composite, M_c, can be calculated from Fig. 13.7(b), taking into account the composite density, ρ.

$$M_c = 2nt\rho\left(2a + \frac{t}{2n} + 1\right) + (2n + 1)b\rho \tag{13.4}$$

Therefore, the volume density of this type of hollow composite, ρ_c, can be expressed as:

$$\rho c = \frac{M_c}{V_c} = \frac{2nt\rho\left(2a + \frac{t}{2n} + 1\right) + (2n + 1)b\rho}{2Dna^2\sin 2\theta + 4Dnab\cos\theta} \tag{13.5}$$

It is obvious that the volume density of the hollow composite is affected by the cell size, cell shape, cell network in the cross-section as well as the open angle of the cells. The volume density of the hollow fabrics with flat surfaces can be evaluated in a similar fashion.

13.3.4 Compression modulus of hollow composites

Assume that the cell wall is a composite of homogeneous and isotropic materials with Young's modulus being E_s for the sake of simplicity in the analysis. According to previous researchers (Abd El-Sayed *et al.*, 1979; Gibson *et al.*, 1982), the Young's moduli, E_y and E_x, in the compression directions, x and y, can be expressed as follows:

$$E_y = \left(\frac{t}{a}\right)^3 \frac{\sin\theta}{\left(\frac{b}{a} + \cos\theta\right)\cos^2\theta} E_s \tag{13.6}$$

$$E_x = \left(\frac{t}{a}\right)^3 \frac{\left(\frac{b}{a} + \cos\theta\right)}{\sin^3\theta} E_s \qquad [13.7]$$

When the cell shape assumes a regular hexagon, where the length of cell walls is the same (i.e., $b = a$) and θ is 60°, both E_y and E_x assume the same expression, which is shown as Eq. [13.8]:

$$E_y = E_x = 2.3\left(\frac{t}{a}\right)^3 E_s \qquad [13.8]$$

It is clear from Eq. [13.8] that for regular shaped hexagonal cells, the compression modulus is determined by the thickness of the cell wall, the length of the wall as well as the Young's modulus of the wall material. Thicker and shorter walls lead to higher modulus of the hollow composites under in-plane compression.

Gibson and Ashley (1999) also revealed that when the hollow structure under the same assumption is subject to out-of-plain compression, i.e. under loading in the z direction, the Young's modulus is expressed as in Eq. [13.9]. The term in the square brackets represents the cross-sectional area of the cell walls:

$$E_z = \frac{t}{a}\left[\frac{\frac{b}{a} + 2}{2\left(\frac{b}{a} + \cos\theta\right)\sin\theta}\right] \qquad [13.9]$$

For such a structure to be composed of regular hexagonal cells, Eq. [13.9] can be simplified as:

$$E_z = 1.15\frac{t}{a} \qquad [13.10]$$

Hence, for out-of-plane loading, the Young's modulus for the hollow composite increases when the cell wall becomes thicker and when the wall length is shorter. Figure 13.8 shows the schematic diagrams by Gibson and Ashby (1999) of the strain–stress curves for in-plane and out-of-plane compression, demonstrating the influence of the deformation processes and the influence of the t/a ratio.

13.3.5 Energy absorption

Due to the volume density of the hollow composites made from hollow fabric, one of the applications of this type of materials is energy absorption. Tan and Chen (2005) simulated the influence of the hollow structures on

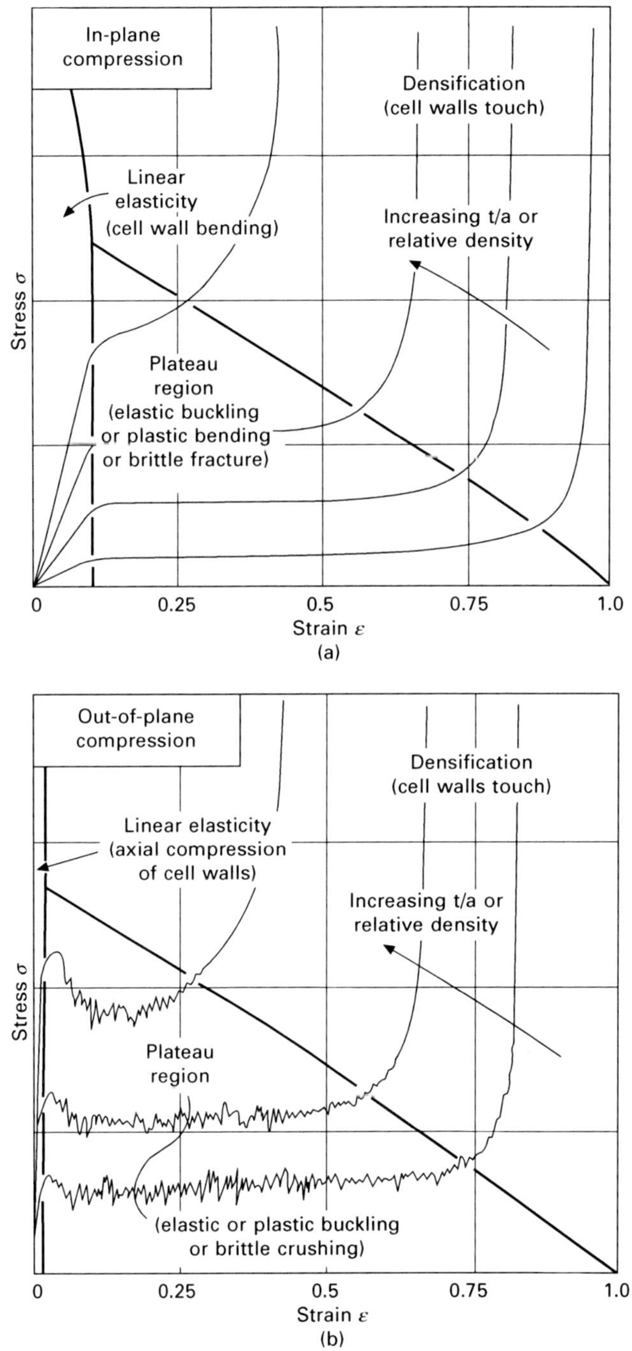

13.8 Schematic strain–stress curves for (a) in-plane and (b) out-of-plane compressions to a hollow structure (Gibson and Ashby, 1999).

mechanical performance including energy absorption, assuming quasi-static compression. The parameters considered range from the open angle, length of cell wall, free and bonded wall thickness, and length ratio of free and bonded walls. It revealed that energy absorption per unit material volume is quite sensitive to the change in the open angle, cell wall length, and the free wall thickness when loading the hollow structure in-plane and in the y direction as indicated in Fig. 13.7. Figure 13.9 illustrates the effect of open angle on hollow structures.

Chen *et al.* (2008a, 2008b) carried out experimental investigations on impact energy absorption and other mechanical performance. The structural parameters used in the investigation were the cell open angle, cell wall length, length ratio of cell walls between the free and bonded walls and the volume density of the hollow composites. This work conformed much of the simulated outcomes by Tan and Chen and also provided other interesting findings. For example, when the cell shape is not a regular hexagon, although the amount of absorbed energy is about the same, the mode of energy absorption can be quite different. Figure 13.10 illustrates such a situation. The open angle of all these samples is 60°. The group of hollow composites with wall length < a, e.g. 8L(3+6)P60 where the first figure in the bracket indicates the length of b and the second a, demonstrated absorption of similar energy to that of 8L(6+3)P60, a case of $b > a$. Figure 13.10(a) shows that the energy absorption of these two groups is about the same, and Fig. 13.10(b) shows that the $b < a$ group absorbs energy mainly because of large composite deformation and the $b > a$ group due to large force applied to the material. In other words, the $b < a$ group are soft composites, and the $b > a$ group are hard ones.

13.3.6 Force attenuation

As a result of impact energy absorption, the transmitted force through the hollow composites can be significantly reduced. The transmitted force can be measured by a load cell embedded in the anvil through experiment. The force attenuation factor, f_{att}, is used to demonstrate the force-blocking effectiveness of the sample and it is defined as follows (Dionne *et al.*, 2003):

$$f_{att} = \left(1 - \frac{F_{trans}}{F}\right) \times 100\,(\%) \qquad [13.11]$$

where F_{trans} is the transmitted force through the specimen and F is the impact force acting directly on the anvil. Table 13.1 lists the transmitted force through various hollow composites, the direct impact form and the calculated force attenuation factor, when the hollow composites were subject to the same impact energy, i.e. 8.5 J.

It is evident from Table 13.1 that the hollow structure plays an important role in influencing the performance in force attenuation. Figure 13.11(a)

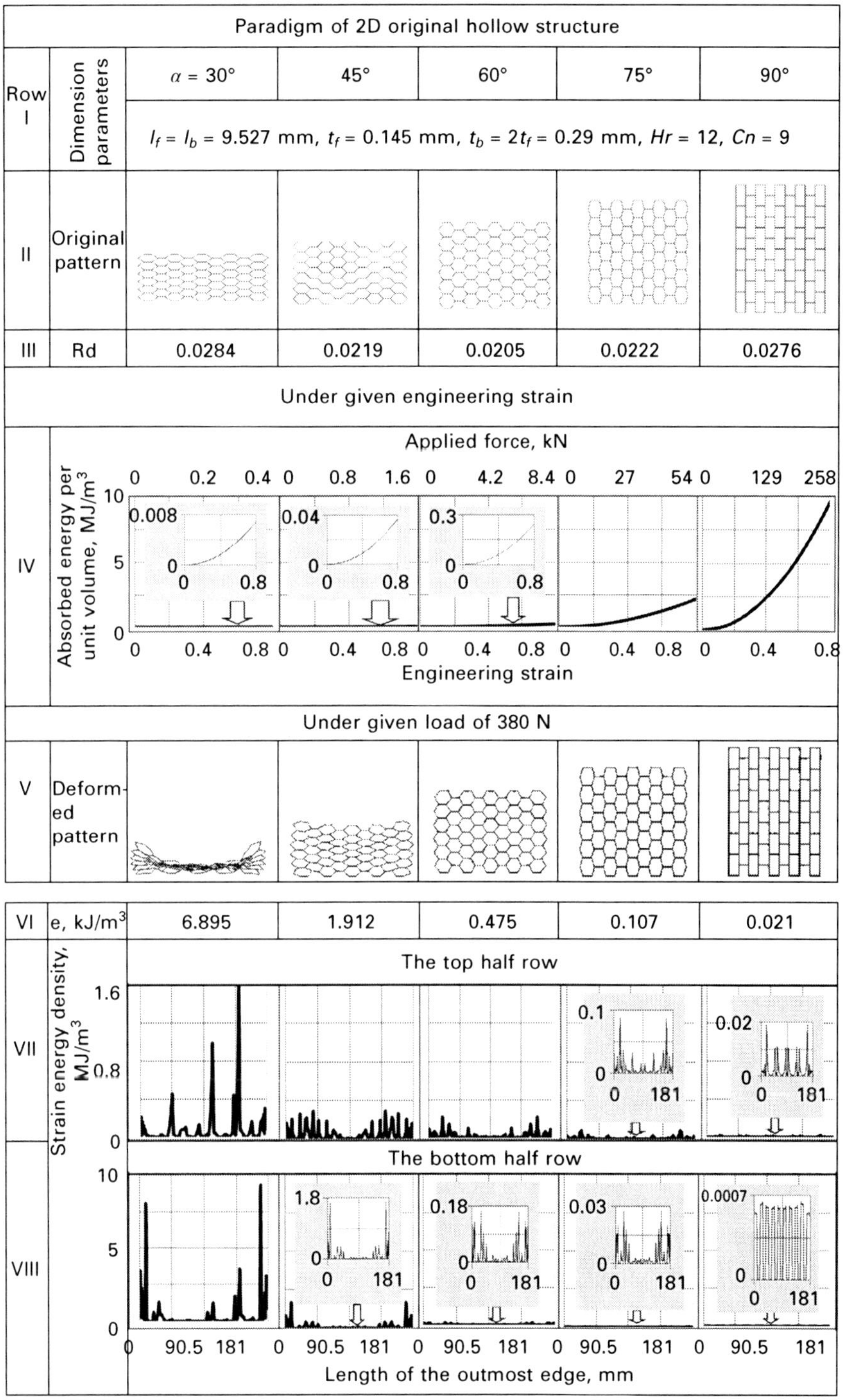

13.9 Effect of open angle on energy absorption and structure deformation.

Table 13.1 Experimental data of various hollow composites on force attenuation (Gong, 2011)

Sample	Peak time (ms)	F_{trans} (kN)	F (kN)	f_{att} (%)
4L6P60	6.94	1.27	17.5	92.3
6L4P60	3.05	0.50	17.5	97.0
8L3P60	3.52	0.95	17.5	94.2
8L4P60	3.15	0.60	17.5	96.3
8L5P60	7.18	0.55	17.5	96.7
8L6P30	4.24	0.49	17.5	97.0
8L6P45	7.35	0.42	17.5	97.5
8L6P60	4.91	0.35	17.5	97.9
8L6P75	7.69	0.29	17.5	98.3
8L6P90	6.48	0.27	17.5	98.4
8L(4+3)P60	3.97	0.77	17.5	95.3
8L(6+3)P60	4.65	0.66	17.5	96.0
8L(3+6)P60	4.47	0.40	17.5	97.6
8L(4+6)P60	7.46	0.40	17.5	97.6

illustrates the effect of cell size on the force attenuation factor. For hollow composites reinforced by eight-layer hollow fabrics, the force attenuation factor increases as the cell wall length changes from three picks to six picks. Figure 13.11(b) demonstrates the influence of the open angle on the force attenuation. Basically, the larger the open angle, the more force attenuating is the hollow composite. Table 13.1 also reveals the influence of ratio of wall length between the free and bonded walls.

13.4 Modelling of hollow woven fabrics

3D hollow fabrics can be manufactured using the existing weaving machinery as well as some specially created weaving devices. However, the design of such fabrics is more complicated than the fabrics that are used for clothing and other domestic use. Modelling of 3D hollow woven structures has been carried out for creating computerised design and manufacture of such fabrics.

13.4.1 Hexagonal hollow fabrics

Chen *et al.* (2004) have pointed out that a hexagonal hollow structure has two columns of cells containing n and $(n-1)$ cells, respectively. The whole structure is made out of $2n$ layers of fabrics. It is assumed in this context that the tunnels formed by the cells run in the weft direction, although they can be arranged to go in the warp direction too. One repeat of the hollow structure can be divided into four regions, i.e., regions I, II, III and IV, as shown in Fig. 13.12. Region I corresponds to the section of the 3D hollow

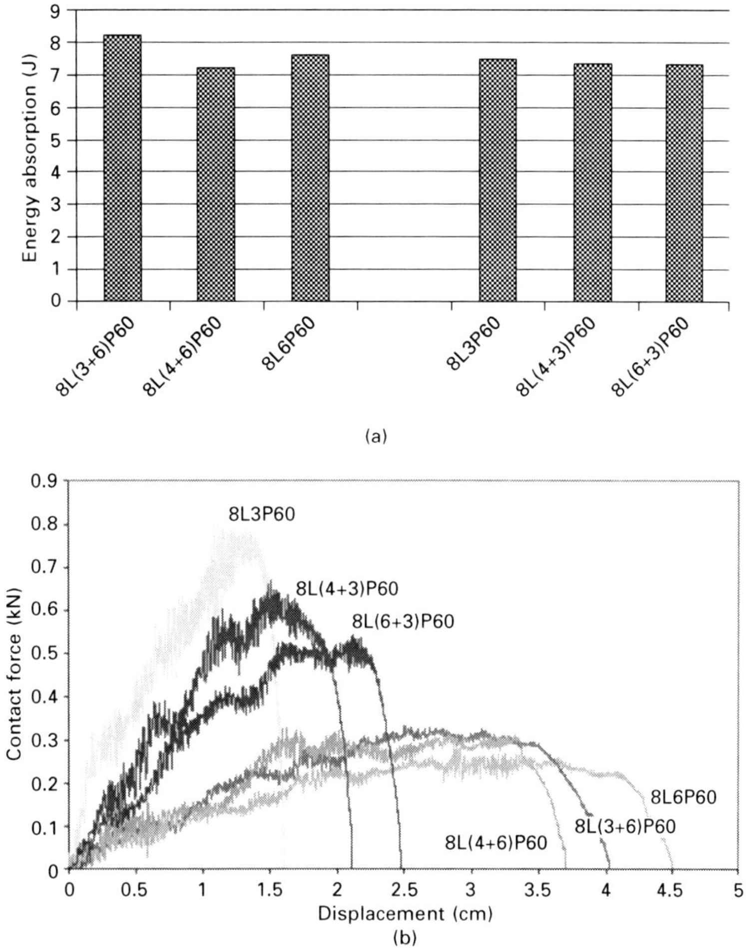

13.10 Influence of cell wall ratio on energy absorption: (a) energy absorption, (b) impact force–displacement curves.

structure where the fabric layers are all separated from each other; region II is where the adjacent layers join together at an alternate interval; region III is the same as region I; and region IV is again the joining section but the joining layers are different from those in region II. Because of the nature of weaving, a hollow fabric of this type is woven with all cells flattened as indicated in Fig. 13.12(a), and the hollow structure is achieved when the fabric is opened up and consolidated as illustrated in Fig. 13.12(b). According to the cell definition, regions II and IV correspond to the bonded walls with l_b being the length and regions I and III the free walls with l_f being the length. It needs pointing out that Fig. 13.12(b) shows only part of the hollow structure that would open up to form Fig. 13.12(a).

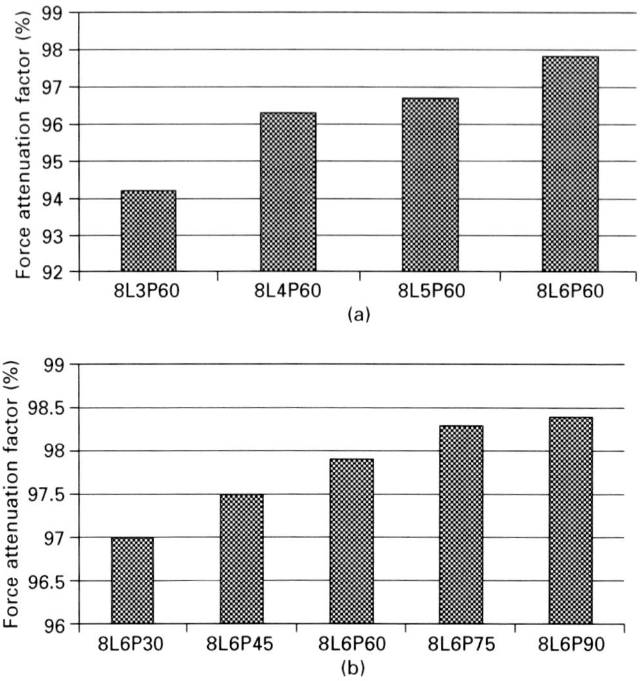

13.11 Influence of hollow structure on force attenuation: (a) effect of cell size, (b) effect of cell open angle.

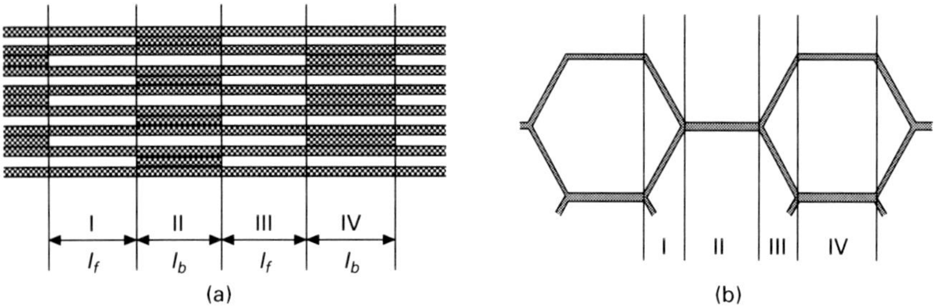

13.12 Regional division of a hollow structure: (a) hollow fabric woven before opening, (b) hollow fabric after opening.

A woven hollow structure can be defined by the specification of a group of structural parameters. The following general coding format is used to denote a particular hollow structure:

$$xL(l_b + l_f)P\theta$$

where x is the number of fabric layers used to form the hollow structure; l_b is the length of bonded wall measured in the number of picks; l_f is the length of free wall measured in the number of picks; θ is the open angle between the free cell walls and the horizontal line; L is used to denote 'layer'; and P is used to denote 'pick'.

There are situations when the lengths of free and bonded walls are the same, i.e. $l_b = l_f = y$, say. In such a case, the coding format can be reduced to:

$$xLyP\theta$$

Further, when the opening angle of the cell is 60°, the coding format becomes:

$$xLyP$$

In all the above coding expressions, x, l_b, and l_f are integers and $x \geq 2$, $l_b \geq 1$ and $l_f \geq 1$.

According to the format, a 4L6P hollow structure stands for a hollow structure comprising four layers of fabric, where the length of both free and bonded walls is six picks and the opening angle of the cells is 60°. On the other hand, code name 8L(4+3)P refers to a hollow structure made from eight layers of fabric, where the lengths of the bonded and free walls are four picks and three picks, respectively, with the cell opening angle being 60°.

13.4.2 Quadratic hollow fabrics

The cross-section of a quadratic hollow structure can be created based on the shape of the cell and the arrangement of cells. A cell can be fully defined either by the coordinates of its nodes, or by lengths and angles of fabric sections (Chen and Wang, 2006). The arrangement of cells can be described by the number of cell levels and the shift between the levels. A quadratic hollow structure becomes more complicated when the cells are of different geometrical features and when the cells are arranged differently in the cross-section. Figure 13.13 illustrates a quadratic hollow structure with

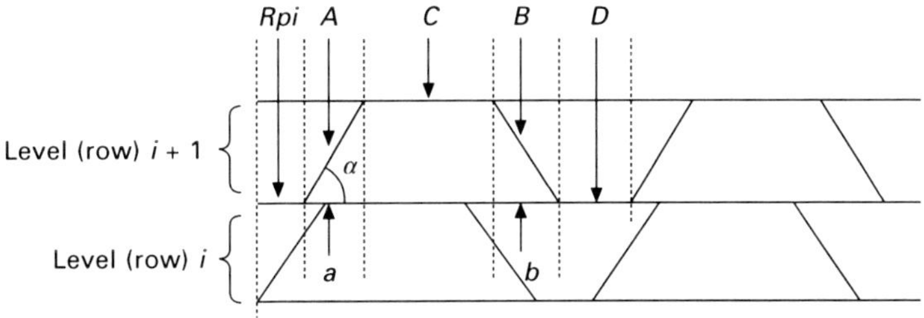

13.13 Geometrical description of a hollow structure with trapezoidal cells.

trapezoidal cells. In general, the geometrical model of a quadratic hollow structure with a trapezoidal cross-section can be specified using the following parameters: *A*, *B*, *C* and *D* are the lengths of trapezoidal sides, measured in number of picks; Rp_i is the shift between levels *i* and *i* + 1; *a*/*A* or *b*/*B* is the ratio between long and short fabric sections and α is the expansion angle. Obviously, $\cos \alpha = a/A$.

13.4.3 Generation of weaves for hexagonal hollow fabrics

It has been a common practice that a 2D matrix is used to record the data of a weave (Li *et al.*, 1988; Milasius and Reklaitis, 1988; Chen *et al.*, 1996). In the case of a single-layer fabric, a 2D binary matrix is used to represent the weave, whose element values are either 0 or 1. '1' indicates a warp-over-weft crossover, and '0' means a weft-over-warp crossover. The position of each element in the matrix is located by coordinates (x, y) where x indicates the xth column from the left and y the yth row from the bottom. This approach is adopted in generating the weaves for hexagonal hollow structures (Chen *et al.*, 2004).

For hexagonal hollow structures, if there are n cells in the thickness direction, then $2n$ layers of single layers of fabric must be used. The adjacent layers of the $2n$ fabrics will be joined with and separated from one another in a predesigned pattern. Based on such specifications, a cross-sectional view of a hexagonal hollow fabric can be defined. Weaves will be assigned to the single layers as well as the joined layers. Once all these layers have received weaves respectively, all these weaves will be combined to form the overall weave for the hexagonal hollow structure based on the layer relationship.

Consider the combination of weave of two single-layer fabrics. Let M_1 and M_2 be the matrices for two single-layer fabric sections. In order to unify the repeat size of the two layers, the lowest common multiple (*lcm*) are calculated based on the dimensions of these two matrices. Assume that d_1 is the dimension of M_1 and d_2 that of M_2. If $lcm(d_1, d_2)$ is equal to d_1 (or d_2), the dimension of M_2 (or M_1) will be expanded to be of the same size of d_1 (or d_2). If $lcm(d_1, d_2)$ is larger than both d_1 and d_2, then M_1 and M_2 will have to be expanded to the new dimension $lcm(d_1, d_2)$. Let the enlarged matrix be *ME*. The elements of this matrix are assigned as follows:

$$ME_{k(j,k)} = M_{k(i\%d_k, j\%d_k)} \quad i, j = 1, 2, \ldots, lcm(d_1, d_2), k = 1, 2 \qquad [13.12]$$

The final task in this procedure is to combine the two enlarged matrices into an overall matrix *MF*. Equation [13.13] describes the algorithm from which *MF* is enumerated:

$$MF_{(i,j)} = \begin{cases} ME_{1\left(\frac{i}{2},\frac{j}{2}\right)} & \text{if } i\%2 = 0 \text{ and } j\%2 = 0 \\ 0 & \text{if } i\%2 = 0 \text{ and } j\%2 = 1 \\ 0 & \text{if } i\%2 = 1 \text{ and } j\%2 = 0 \\ ME_{2\left(\frac{i}{2},\frac{f}{2}\right)} & \text{if } i\%2 = 1 \text{ and } j\%2 = 1 \end{cases} \qquad [13.13]$$

Note that $i\%2$ and $j\%2$ are integer divisions, returning only the value of the integer part of the result. For instance, $4\%3 = 1$.

For hollow structures, two columns of cells represent one structural repeat. One repeat can be further divided into four areas as illustrated in Fig. 13.12, with areas I and III identical to each other. As can be seen, area I contains only single layers, area II has two single layers (top and bottom) and several double layers, and area III is the same as area I, and area IV has only two-layer fabrics.

A base matrix is a rectangular matrix which is prepared to contain the weave for an individual area. If all component weave matrices are assigned to weaves with the same weave repeat, the width and height of the base matrix will be equal to the product of the number of layers and the dimension of each layer. Otherwise, the dimension of the base matrix will be the lowest common multiple of all the dimensions of the component weave matrices. One repeat in the hollow structure has three different areas (areas I and III being the same); therefore there are three base matrices involved in a weaving diagram for the hollow structure. Each base matrix is further divided along the warp direction by the dimension of the weave matrix. Take Fig. 13.14 for example, where 2/1 twill weave is used for all four layers in area I. Counting from the left-hand side, the first three columns correspond to fabric layer 1, the second three layers 2, the third three layers 3, and the last three layers 4. Each section in the base matrix is constructed by superimposing the weave matrix for each layer and by insertion of 'additional lifting points' (Chen *et al.*, 1996), which is added to reflect the multi-layer weaving principle that when a pick for a lower layer is to be inserted, all warp ends for layers above must be lifted. Such finished base matrix will represent the overall weave for the area concerned.

Area II includes two single layers and several two-layer fabrics and area III has only two-layer fabrics. In a similar way, the base matrices for areas II and IV can be generated. Figure 13.15 lists the base matrices for areas I, II and IV.

The final weave for the whole hexagonal hollow structure is the combination of these base matrices for the three areas. Figure 13.16 is the overall weave responsible for weaving one repeat of the hexagonal hollow structure with four layers of fabric, in the sequence of areas I, II, III and IV.

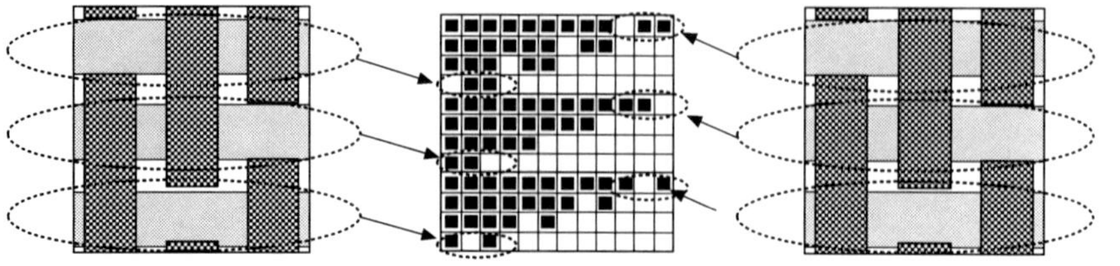

13.14 Creation of an example of base matrix (2/1 twill weave, Area I).

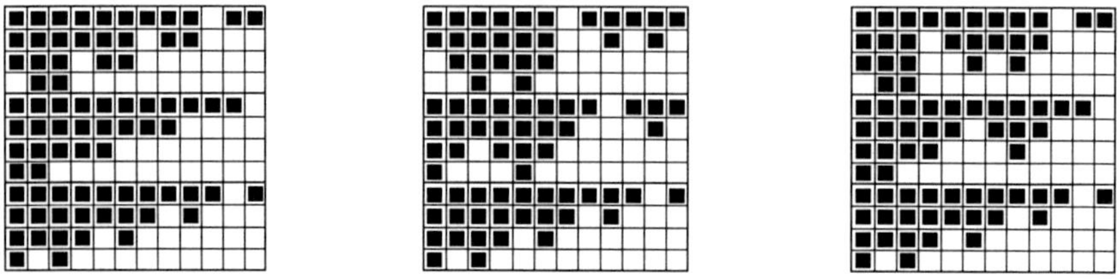

(a) (b) (c)

13.15 Base matrices.

13.4.4 Generation of weaves for quadratic hollow fabrics

Figure 13.17 shows an example of a three-layer single level quadratic fabric with subdivision of the repeat unit into four different 'areas' (Chen and Wang, 2006). Area 1 includes three layers of fabric, which have different lengths and possibly different weaves. All layers in this area are single-layer fabrics. The length ratio of these three fabric layers in area 1 is defined as $l_1 : l_2 : l_1$. However, picks of weft yarns must be distributed evenly into the three fabric layers. It is necessary to divide the number of picks for each section into the same number of groups.

Let us denote the number of such groups by k. For a quadratic structure with m layers, group i of weft yarn distribution is recorded as $w_{1i} : w_{2i} : w_{3i} : \ldots : w_{m\,i}$, where all w_{qi} ($q = 1, 2\ldots, m$ and $i = 1, 2, \ldots, k$) are integers. When the weft density is taken into consideration, the following relation holds:

$$\sum_{i=1}^{k} w_{1i} : \sum_{i=1}^{k} w_{2i} : \ldots : \sum_{i=1}^{k} w_{qi} : \ldots : \sum_{i=1}^{k} w_{mi} = l_1 : l_2 : \ldots : l_q : \ldots : l_m$$

Area 2 includes two sections of fabrics, each having the same lengths and possibly different weaves. However, the top fabric section is a combined layer made from top and middle fabric layers from area 1, and the bottom section is a single-layer fabric, the same as layer III in area 1. The weave for area 3 is the same as that for area 1, and area 4 is similar to area 2 except that the top section is a single layer and the bottom section is a stitched layer in this case.

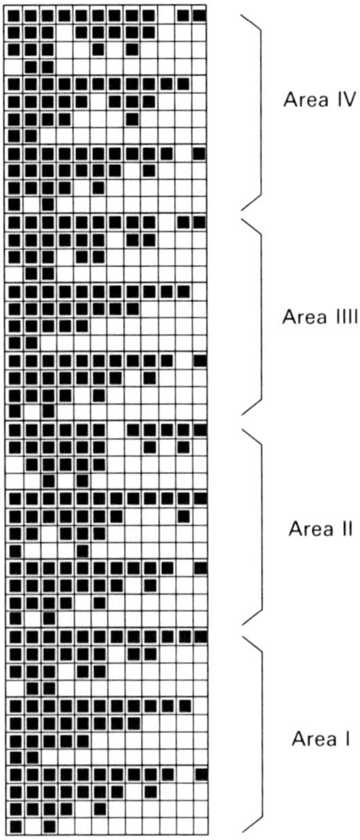

13.16 Final weave for the hexagonal hollow fabric.

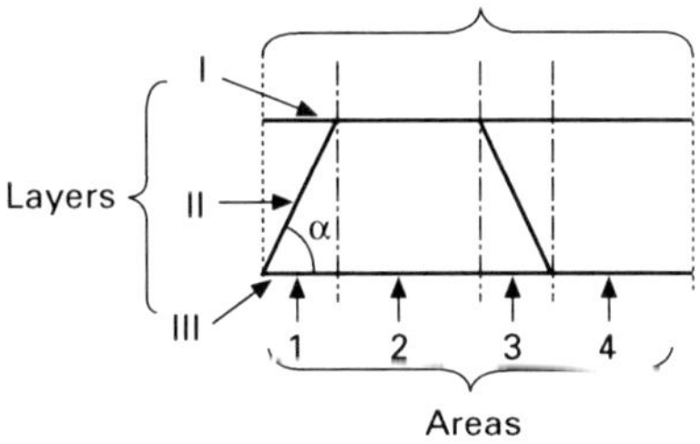

13.17 Illustration of cross-section of a quadratic hollow fabric.

The weave for each fabric layer is recorded in a 2D matrix. The 2D weave matrix representing the overall weave for each area can be generated by combining the 2D matrices of all fabric sections in the area concerned. Suppose there are n fabric layers in the area of concern. The weave for layer i is recorded in matrix M_i ($i = 1,2, \ldots, n$), and the element of this matrix at the xth warp and the yth weft is $M_i(x,y)$ ($1 \leq x \leq r_{ie}$, $1 \leq y \leq r_{ip}$) where r_{ie} is the number of warp ends in M_i, and r_{ip} the number of weft picks in M_i. l_i is the length of layer i.

The warp dimension of all constituent matrices, r_{lcm}, can be found by calculating the lowest common multiple (LCM) of the warp repeats of all constituent weave matrices, i.e.,

$$r_{lcm} = LCM\,(r_{1e}, r_{2e}, \ldots, r_{ne}) \qquad [13.14]$$

Accordingly, the warp dimensions of the constituent weave matrices are expanded to r_{lcm}. The expanded weave matrix for the constituent layer i is denoted by M_i' and its elements can be obtained by repeating the weave of a single layer r_{lcm}/r_{ie} times, namely,

$$M_i'(x, y) = M_i\,(x \bmod r_{ie}, y) \qquad [13.15]$$

The warp repeats r'_{ie} of the expanded matrix M_i' are now r_{lcm}, while its weft repeats r'_{ip} remain r_{ip}.

When a similar treatment is applied, the weave matrix for fabric layer i will be changed from M_i' to M_i'' by repeating the rows of M_i' until its weft dimension is equal to the layer length l_i, i.e.,

$$M_i''\,(x, y) = M_i'\,(x, y \bmod r_{ip}) \qquad [13.16]$$

Based on what has been achieved, all constituent weave matrices M_i'' in this area will be combined to generate the overall weave matrix W for this area. The elements of matrix W, $W(x,y)$, are assigned by the following expression:

$$W(x,y) = \begin{cases} 0, & \text{when } x > \sum_{i=1}^{m} r''_{ie},\ \sum_{i=1}^{m-1} r''_{ip} < y < \sum_{i=1}^{m} r''_{ip} \\ 1, & \text{when } \sum_{i=1}^{m-1} r''_{ie} < x < \sum_{i=1}^{m} r''_{ie},\ y > \sum_{i=1}^{m-1} r''_{ip}, \\ M_i''\left(x - \sum_{i=1}^{m-1} r''_{ie}, y - \sum_{i=1}^{m-1} r''_{ip}\right), & \text{when } \sum_{i=1}^{m-1} r''_{ie} < x < \sum_{i=1}^{m} r''_{ie}, \\ & \sum_{i=1}^{m-1} r''_{ip} < y < \sum_{i=1}^{m} r''_{ip} \end{cases} \qquad [13.17]$$

where $0 < m < n + 1$.

Based on the aforementioned, the weave for the entire repeat of quadratic structure can be obtained by combining the 2D matrices for the different areas in the corresponding sequence. The procedure of combining all the areas may involve the enlargement of the matrix for each area in order to make them have the same warp ends.

Consider a three-layer quadratic structure, where $A = B = 6$ picks, $C = D = 4$ picks, and $A/a = 3:2$ ($\alpha \approx 48.2°$). The weaves used for the top, middle, and bottom fabric layers are $\frac{1}{1}$, $\frac{2}{1}$, and $\frac{1}{1}$ and respectively. The weave for one repeat unit of this quadratic hollow fabric is created and is shown in Fig. 13.18.

13.4.5 Finite element (FE) modelling of low velocity impact on hollow composites

This section reports on some work on FE modelling of composites reinforced by hexagonal hollow fabrics under low velocity impact (Yu and Chen, 2004). The hexagonal cells interact with each other during the impact process, causing propagation of the strain wave. Figure 13.19 illustrates the geometric models of the hollow composite and the impact object. The impact velocity used for FE modelling is 15 m/s. The low velocity impact is loaded in the in-plane direction y, as indicated in Fig. 13.6. Tables 13.2, 13.3 and 13.4 list the cell geometry, material properties and attributes of the impact objects, respectively.

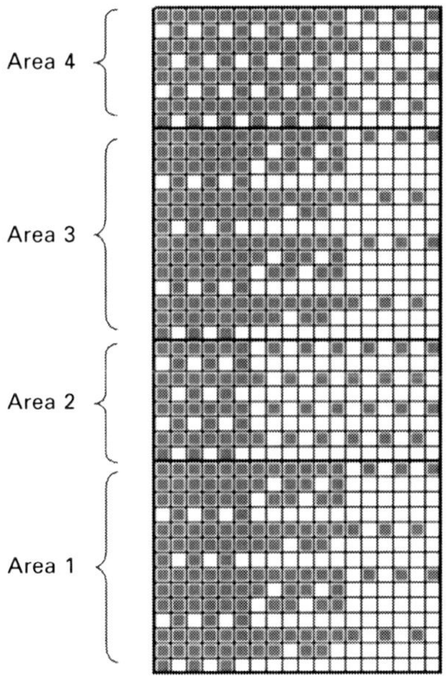

13.18 The weave diagram created for a three-layer quadratic hollow fabric.

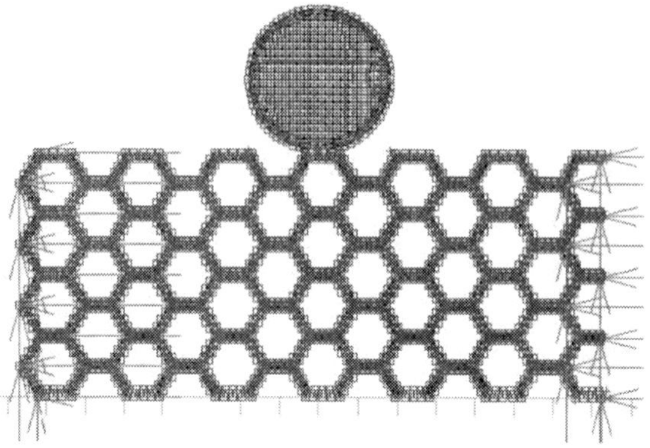

13.19 Geometric model of composite based on a hollow fabric.

The essence of protection against trauma impact is the material's ability to absorb energy and to keep the transmitted force on the protected object below a limit, over which damage will be caused to the protected object. Energy is absorbed as the cell walls bend, buckle or fracture depending on stage of the impact. The total strain energy absorbed by a deformed structure can indicate the degree of damage caused by the dynamic impact from different objects at a given speed. Typical total strain energy curves for the hollow composites under the impact of four different objects at a speed of 15 m/s are shown in Fig. 13.20. It can be seen that objects with sharp corners cause deformation and damage to materials more easily. For instance, FE simulations show that the rectangular shaped impacting object of iron weighing 0.998 kg caused almost the same level of damage as that caused by a spherically shaped iron ball with a mass of 2.091 kg.

The shape and size of impacting objects were selected from those found in a riot situation, therefore the objects used are with different masses and dimensions. When they impact the hollow composites in the perpendicular direction, heavier objects and objects with sharp corners will cause more damage which leads to high levels of strain energy as well as high transmitted force. Figure 13.21 shows that regardless of the shape of the objects, the total strain energy and the transmitted force increase when the object mass is increased.

The FE simulation also reveals the deformation process when the hollow composites were impacted by different impacting objects. Figure 13.22 illustrates the densification of the hollow composites when impacted by a rectangular concrete object. The cells at the outmost surface started to deform when impact started. This mainly involves the deformation in cell walls at the initial stage. As impact goes on, the inner cells begin to take up impact energy. The propagation of the deformation continues until the end of the impact process. As is evident from the illustration, this process involves

Table 13.2 Structural parameters of cells

Cell parameters	l_b	l_f	t_b	t_f	α
	5 mm	5 mm	1 mm	1 mm	60°

Table 13.3 Material properties used for FE modelling

Materials	Young's modulus (N/mm^2)	Mass density ($tonne/mm^3$)
Glass/epoxy composite	13,200	2.0 E-09
Iron	200,000	7.8 E-09
Wood	25,000	1.5 E-09
Concrete	27,000	2.5 E-09
Glass	94,000	2.6 E-09

Table 13.4 Details of impact objects

Shape of impact object	Dimension (mm)	Mass (kg)			
		Iron	Wood	Concrete	Glass
Spherical	ø 80	2.091	0.402	0.67	0.697
Cylindrical (end impacts)	ø 70 × 60	1.801	0.346	0.577	0.6
Rectangular	40 × 60 × 53.33	0.998	0.192	0.32	0.333
Cylindrical (side impact)	ø 40 × 53.33	0.519	0.0999	0.167	0.173

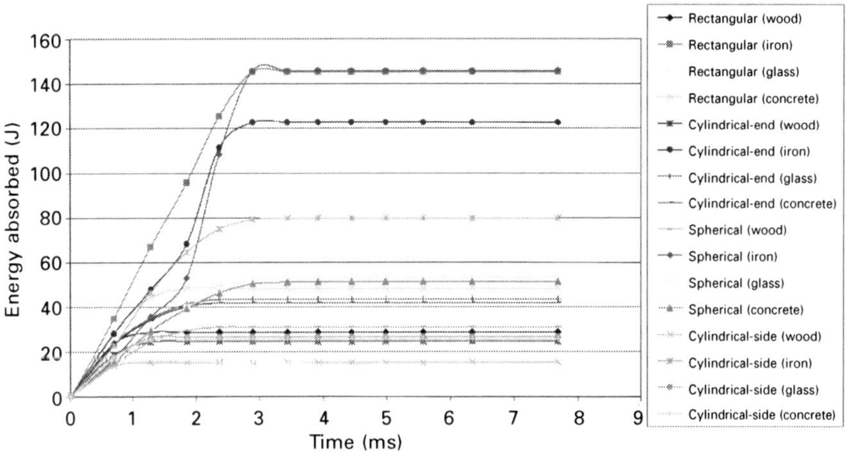

13.20 Total strain energy when hollow composites impacted by different objects at 15 m/s.

elastic deformation, plastic deformation and collapse of cell walls. Figure 13.23 records the indentation of the material against time. It indicates that during the first 1 ms, the hollow composite received an indentation of more than 10 mm, whereas for the remaining 2 ms the indention is only less than 2 mm.

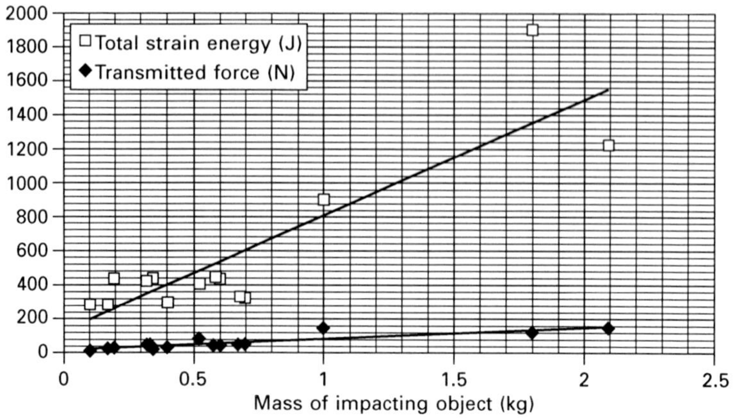

13.21 Impacting object mass, the total strain energy and the transmitted force.

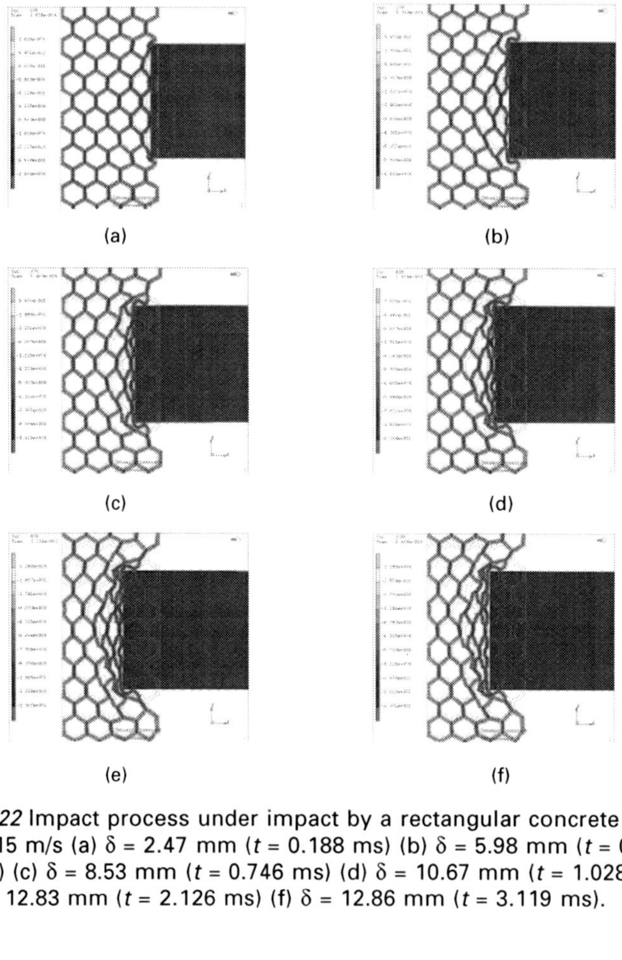

13.22 Impact process under impact by a rectangular concrete object at 15 m/s (a) δ = 2.47 mm (t = 0.188 ms) (b) δ = 5.98 mm (t = 0.491 ms) (c) δ = 8.53 mm (t = 0.746 ms) (d) δ = 10.67 mm (t = 1.028 ms) (e) δ = 12.83 mm (t = 2.126 ms) (f) δ = 12.86 mm (t = 3.119 ms).

13.5 Possible applications of hollow fabrics and future trends

13.5.1 Possible applications

Obviously, the 3D hollow woven fabric is an interesting fabric construction and the composites made from it offer unique properties including lightweight, volumetric, internal tunnelling, force attenuating, and energy absorbent. In addition, the hollow composites offer different mechanical behaviours when loaded in different directions. As indicated earlier, the chief deformation mode is buckling when the hollow structure is in the out-of-plane direction and therefore the structure will offer high rigidity. This would be of particular interest to areas where lightweight, high strength and high compression rigidity are the primary requirements. Sandwiched board materials have been used in aircraft applications.

Use of the hollow structure in the in-plane directions offers high capacities in impact energy absorption and in impact force attenuation. This is suitable for areas where damping and protection against impact are of importance. Packaging and personal and property protection are possible areas for application. There has been work on using hollow composites reinforced by 3D hollow fabrics with uneven surfaces for limb protection against trauma impact (Yu and Chen, 2004).

The attributes in volume and ventilation could also be utilised. A company tried to create office walls and partitions from 3D hollow woven fabrics.

13.5.2 Future trends

Hollow woven fabrics are regarded as a special branch of multi-layer fabrics, where the adjacent layers are stitched together at periodical internals during the weaving process. Demands for strong and lightweight materials for various applications, most notably the aerospace industry, and for energy absorbent materials, e.g. protection and packaging, will make the hollow fabric more

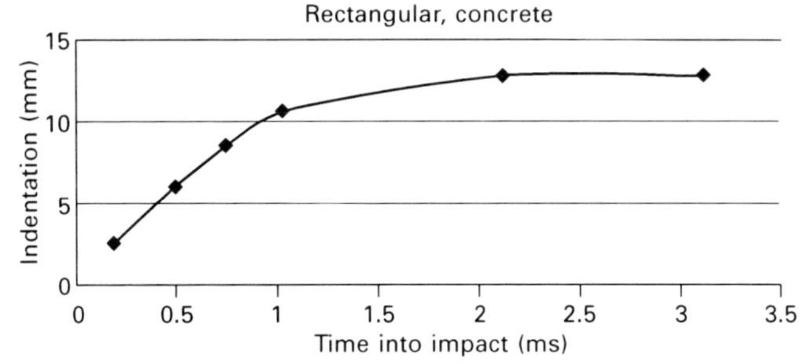

13.23 Rate of indentation during the impact.

interesting and attractive. The use of various fibre types will lead hollow fabrics and their final products to have different properties. The cell walls in the hollow fabrics may also be made using various weave structures including 3D structures so that the hollow fabric products can be applied to situations where the load varies from small to large.

13.6 Sources of further information and advice

Adanur, S. (2001), *Handbook of Weaving*, Technomic Publishing, Lancaster, PA.

Bitzer, T. (1997), *Honeycomb Technology*, Chapman & Hall, London.

Chen, X. (ed.) (2010), *Modelling and Predicting Textile Behaviour*, Woodhead Publishing, Cambridge.

Gibson, L.J., Ashby, M.F. (1999), *Cellular Solids: Structure and Properties*, 2nd edn, Cambridge University Press, Cambridge.

Grocicki, Z. (1977), *Watson's Advanced Textile Design: Compound Woven Structures*, 4th edn, Newnes-Butterworths, London.

Marks, R., Robinson, A.T.C. (1986), *Principles of Weaving*, The Textile Institute, Manchester.

13.7 References

Abd El-Sayed, F.K., Jones R., Burgens, I.W. (1979), 'A theoretical approach to the deformation of honeycomb based composite materials', *Composites*, **10**(4), 209–214.

Chen, X., Wang, H. (2006), 'Modelling and computer aided design of 3D hollow woven fabrics', *J. Text. Inst.*, **97**(1), 79–87.

Chen, X., Zhang H. (2006), 'Woven textile structures', Patent Number GB2404669.

Chen, X., Knox, R.T., McKenna, D.F., Mather, R.R. (1996), 'Automatic generation of weaves for the CAM of 2D and 3D woven textile structures', *J. Text. Inst.*, **87**, Part 1, No. 2, 356–370.

Chen, X., Ma, Y., Zhang, H. (2004), 'CAD/CAM for cellular woven structures', *J. Text. Inst.*, **95**, Nos.1–6, 229–241.

Chen, X., Sun, Y., Gong, X. (2008a), 'Design, manufacture, and experimental analysis of 3D honeycomb textile composites, Part I Design and manufacture', *Text. Res. J.*, **78**(9), 771–781.

Chen, X., Sun, Y., Gong, X. (2008b), 'Design, manufacture, and experimental analysis of 3D honeycomb textile composites, Part II: Experimental analysis', *Text. Res. J.*, **78**(10), 1011–1021.

Chen, X., Taylor, L.W., Tsai, L-J. (2011), 'An overview on fabrication of 3D woven textile preforms for composites', *Tex. Res. J.*, **81**(9), 932–944.

Dionne, J.P., El Maach, I., Shalabi, A., Madris, A. (2003), 'A method for assessing the overall impact performance of riot helmet', *Journal of Applied Biomechanics*, **19**, 246–254.

Eriksson, S., Berglin, L., Gunnarsson, E., Guo, L., Lindholm, H., Sandsjö, L. (2011), 'Three-dimensional multilayer fabric structures for interactive textiles', *Proc. 3rd World Conference on 3D Fabrics and Their Applications*, Wuhan, China, 63–67.

Gibson, L.J., Ashby, M.F. (1999), *Cellular Solids: Structure and Properties*, 2nd edn, Cambridge University Press, Cambridge.

Gibson, L.J., Ashby, M.F., Schajer, G.S., Robertson, C.I. (1982), 'Mechanics of two-dimensional cellular materials', *Proc. R. Soc. Lond.*, **A382**, 25–42.

Gong, X. (2011), Investigation of different geometric structural parameters for honeycomb textile composites on the mechanical performance, PhD Thesis, University of Manchester.

Grocicki, Z. (1977), *Watson's Advanced Textile Design: Compound Woven Structures*, 4th edn, Newnes-Butterworths, London.

Kaddar, T., Ibrahim, A. (2011), '3D spacer fabric for insulating oil pipelines', *Proc. 3rd World Conference on 3D Fabrics and Their Applications*, Wuhan, China, 200–205.

Kunz, E., Chen, X. (2005), 'Analysis of 3D woven structure as a device for improving thermal comfort of ballistic vests', *International Journal of Clothing Science and Technology*, **17**(3), 215–224.

Li, M, Chen, X, Liu, Z (1988), 'Mathematical models for fabric weaves and their application in fabric CAD', *J. Text. Res. China*, **9**, 319.

Milasius, V., Reklaitis, V. (1988), 'The principles of weave coding', *J. Text. Inst.*, **79**, 598.

Tan, X., Chen, X. (2005), 'Parameters affecting energy absorption and deformation in textile composite cellular structures', *Materials & Design*, **26**, 424–438.

Tan, X., Chen, X., Conway, P.P., Yan, X-T. (2007), 'Effect of plies assembling on textile cellular structures', *Materials & Design*, **28**, 857–870.

Yu, D.K.C., Chen, X. (2004), 'Simulation of trauma impact on textile reinforced cellular composites for personal protection', *Proc. Technical Textiles for Security and Defence (TTSD)*, Leeds.

14 Woven textiles for medical applications

S. RAJENDRAN and S. C. ANAND,
University of Bolton, UK

Abstract: The application of woven textiles in healthcare and hygiene is discussed. It is demonstrated in this chapter that the conventional woven structures are still playing a crucial role in designing and engineering hi-tech medical devices. A brief discussion on classification of wounds and their management, infection control, antimicrobial, implantable products, wound dressings and bandages makes this chapter interdisciplinary. Specification and manufacturing aspects of a few product developments are outlined. Commercial names of a few medical devices and their specific applications are highlighted.

Key words: woven structures, wound management, wound dressings, bandages, medical devices.

14.1 Introduction

With the increasing demand for novel textiles, structures and their associated products for various applications, it is worthwhile to mention that conventional woven textiles are still largely used to a significant extent in healthcare and hygiene sectors such as antimicrobial and associated textiles used in hospitals, although nonwoven, knitted and emerging spacer structures are increasingly dominant. Weaving technology is also employed in manufacturing various hi-tech medical devices such as vascular grafts, tendons and wound dressings. Woven structures are used for a wide spectrum of healthcare, medical and surgical products. Woven fabrics are commonly used for gauze dressings, capillary dressings, support and compression bandages, plasters, vascular prostheses, scaffolds for tissue engineering, surgical gowns, drapes, hospital textiles such as sheets, blankets, pillowcases, uniforms and a wide range of operating room garments, hospital bed linen and staff uniform garments. Woven fabrics are extremely versatile in terms of the type of yarns and structures that can be engineered on the loom, but suffer from the tendency of fraying when cut into shapes and the anisotropic nature of their properties. The woven structure tends to scissor when extended in the bias direction which could be detrimental for some applications such as mattresses for pressure sores and garments for hypertrophic scarring treatment.

The market potential for healthcare and medical textile devices is considerably increasing. In the EU alone the sales of medical textiles already account for 10% of the technical textile market. The US is the largest consumer of medical devices followed by Western Europe. With the rise in income levels and increasing adoption of health insurance, the demand for healthcare is expected to grow in Asian countries such as India and China. The Indian healthcare market is expected to grow to US$50.2 billion and US$78.6 billion by 2011 and 2016, respectively[1] and the predicted medical textile market is US$575 million by 2011.[2] The development of the medical textile industry in recent years in China is relatively rapid. The annual growth rate of medical textiles in 2009 was 10% of the total textile market. The US Health Industry Manufacturers Association (HIMA) acknowledges that China is the world's fastest growing market in fibre-based biomedical products that include wound dressings, surgical sutures, artificial lungs, artificial kidney dialysers, vascular grafts and artificial heart valves with an average annual increase of 28% since 1996. In the UK the medical device market is dominated by the National Health Service (NHS) accounting for approximately 80% of healthcare expenditure, even though there are fewer private sectors. It is forecast that the share of hygiene and medical textiles would be 12% of the global technical textiles market and would account for US$4.1 billion.

The healthcare and medical devices market is driven by various factors and these include population growth, rising standards of living and higher expectations of quality of life and ageing of the population. The relative change within the elderly population in the UK between 1985 and 2010 is represented in Fig. 14.1.[3] It is forecast that there will a significant shift in the population with a major increase in the number of people over 60.

It has been predicted that there is a substantial market potential for advanced wound dressings. The forecast for annual growth would be between 10% and 15% in 2012. In the US alone there are over 100,000 surgeries performed daily involving surgical wounds. An ageing population creates increased demand for all types of surgical intervention, particularly cardiovascular,

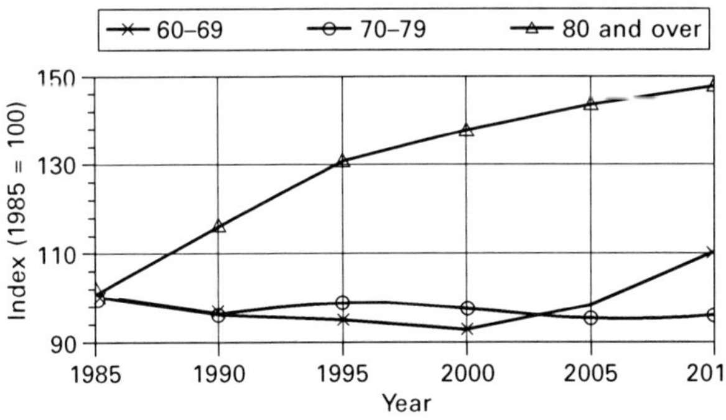

14.1 Relative changes of population aged over 60 in the UK (1985–2010).

orthopaedic, urological and dermatological procedures. The treatment of venous leg ulcers creates considerable demands upon healthcare professionals throughout the world. In the UK alone the treatment of this condition costs the NHS £650 million per year.[4]

14.2 Application of woven textiles in managing acute and chronic wounds

14.2.1 Medical devices

It should be stressed that specialised textile products, irrespective of their structures (woven, nonwoven, knitted and spacer) and production techniques, used in humans are classified as medical devices and are regulated by regulatory agencies. Generally it covers most of the textile medical devices such as simple wound dressings to life-saving implantable devices. The EU Commission has issued the following three CE marking directives related to medical devices;

- The Medical Devices Directive (MDD) (1992): Applies to all general medical devices not covered by the Active Implantable Medical Devices Directive or the In Vitro Diagnostics Directive;
- The Active Implantable Medical Devices Directive (AIMDD) (1990): Applies to all active devices and related accessories intended to be permanently implanted in humans; and
- In Vitro Diagnostics Directive (IVDD) (1998): Applies to all devices and kits used away from the patient to make a diagnosis of patient medical conditions.

Since 14 June 1998 no medical device covered by the MDD 93/42/EEC shall be placed on the market that does not carry a CE mark. The only devices not requiring a CE mark are 'custom-made devices' and 'devices intended for clinical investigations', where the manufacturer must keep documentation in accordance with MDD. In the UK, the Medicines and Healthcare Products Agency (MHRA) regulates medicine and medical devices. The agency implements the EC medical devices directives into UK law. In the United States, the US Food and Drug Administration (FDA) regulates via its MEDWATCH programme.

14.2.2 Classification of wounds

Wounds can be classified into acute wounds and chronic wounds. While acute wounds take only a few weeks to heal, chronic wounds require several months to heal completely. Chronic wounds include venous leg ulcers and pressure sores. Generally the wound healing process involves three phases:

- the inflammatory phase, which occurs immediately after injury to tissue and during which swelling takes place;
- the proliferation period, in which new tissues and blood vessels are formed; and
- the maturation phase, in which tissues laid down during the proliferation stage are remodelled.

14.2.3 Wound management

The care and treatment of wounds are a matter of clinical judgement. The wounds should first be examined to determine the amount of exudate and whether the symptom of infection exists. In addition, the surrounding tissue should be assessed to establish the extent of ulceration. The first stage of treatment (debridement) often requires the removal of foreign material and dead tissue from the wound in order to prevent infection and promote healing. Subsequent stages in wound management include control of exudate, stimulation of wound healing and wound protection. Briefly the various stages involved in wound management are:

- autolytic debridement using the body's own enzymes and moisture to dissolve and clean the dead tissues;
- mechanical debridement that consists of wet-to-dry technique utilising saline-moistened gauze and injecting pressurised saline solution (pulse lavage) on the wound bed;
- chemical debridement which involves the use of topical enzymatic gels and solutions to dissolve and remove dead tissues from the wound;
- surgical debridement by utilising surgical instruments such as scalpel, forceps, scissors, and laser; and
- bio-debridement using live maggots.

14.2.4 Woven wound dressings

The healing of wounds depends not only upon medication but also upon the use of proper dressing techniques and suitable dressing materials. Dressing materials include woven, nonwoven and knitted structures of which nonwoven dressings are increasingly used because they are soft and highly absorbent. Generally, the dressing is placed directly over the wound (primary dressing) and is covered with an absorbent pad (secondary dressing). The whole dressing is then retained with adhesive tape or a suitable bandage, depending on the location of the wound in the body. The primary dressing is expected to maintain the wound temperature and moisture level to permit respiration and to allow epithelial migration. The secondary dressing must not be too absorbent as it may cause the primary dressing to dry out too quickly. Different shapes are available which are suitable for dressing wounds in difficult positions such as heels, joints, digits and the perineal area.

An ideal dressing is normally expected to:

- provide a barrier against microorganisms, dirt and other foreign bodies;
- provide a humid environment at the wound surface;
- control exudates; and
- be capable of being removed without trauma.

Dressings are also used to protect against further injury and abrasion by acting as a cushion and also help to promote rather than interfere with the healing process. Traditionally, woven cotton gauze is used for dressings because of its good absorption properties and soft handle. A gauze dressing or bandage is a woven piece of material that covers an injury or a wound to prevent microorganisms from entering into the affected area. It also can hold a dressing in place and may contain ointment to enhance healing. However, it has been established that cotton gauze allows moisture to evaporate from the wound which means that cotton gauze dressings do not maintain the moist environment to facilitate faster wound healing. Furthermore, cotton gauze adheres to the wound bed. This means that patients suffer pain when the gauze is removed. Also cotton gauze requires frequent changes. In view of the above disadvantages, the woven cotton gauze is no longer in use in European hospitals. However, it is still considered a popular choice of dressing in the majority of countries in Asia, Africa and Middle East, mainly because it is affordable to treat different wounds both in hospital and home environments. Dry gauze dressings are used as a primary dressing for open wounds with heavy exudate and are also used to protect a closed wound from additional trauma and infection. In some situations, wet-to-dry woven gauze dressings can be used for mechanical debridement to remove dead tissues from the wound bed. Moist gauze dressings help to maintain a moist wound healing environment and are often considered as wet-to-moist or damp dressings. This type of dressing can be used on a granulating wound. Illustration of commercial wound dressings is published elsewhere.[5]

It should be stressed that the traditional gauze dressing consists of an open woven structure made from 100% cotton. Commonly available woven dressing comprises a number of layers of such gauze material and is of a predetermined shape. It absorbs fluids secreted from the wound and on drying up, the gauze tends to stick to the wound and loose cotton fibres often get caught in the wounds. This causes irritation, trauma and even further injury when the dressing is changed. In addition, the cotton dressings, once dry, cannot provide a moist environment which is essential for faster wound healing. There are a number of patents relating to non-adherent/low adherent wound dressings. Nonwoven structures made from gel forming fibres, for example alginate, are considered to be preferred dressings for managing exudating wounds. However, these dressings have less tensile strength in both wet and dry states than that of woven dressings. Moreover, the nonwoven dressings have little integrity, once used on the wound, to remain intact. Although alginate fibres are preferred for achieving non-adherence by gelling on contact with wound fluid, the fibres are expensive and have comparatively poor strength. The main drawback in using alginate dressing is that, if allowed to dry, the dressing may adhere to the wound[6] and, therefore, it is not suitable for dry wounds.

To overcome the above limitations, the University of Bolton, UK, has developed a non-adherent wound dressing by using a plain woven structure for managing both the exudating and dry wounds. The dressing uses a single-layered fabric woven using dense huckaback weave. The fabric is treated in a solution of a non ionic flocculent and a softening agent. On one side of the fabric a laminated hydrophobic film is coated to prevent seepage. The non-ionic flocculent reacts with the exudate to form a film of gel between the dressing and the wound and, thereby, prevents the dressing from sticking to the wound. The ratio of non-ionic flocculent to softening agent used is in the range of 1:3 to 1:6 per litre of water. The softening agent is a fatty acid amide derivative and the flocculent is a polyacrylamide.

The above gauze dressing (Table 14.1) is engineered using 100% cotton yarn. The fabric is woven using conventional two shafts with the warps and wefts following a set pattern and it possesses a greater absorption capability stipulated in BP method[7] (Table 14.2). In addition such fabric has greater strength and the tendency to shed loose fibres is rare. This unique weave for gauzes allows the use of a single layer in place of eight to twelve layers. A number of fabric layers are clubbed together to form a swab. These layers absorb any secretion from the wound. On drying up, the swab tends to stick to the wound and the area around it. The absorbed liquid, however, is dissipated from the gauze upon the application of pressure to the wound.

The above dressing meets the following requirements of an ideal wound dressing:

- produces gels on contact with wound fluid that facilitate non-traumatic dressing removal;

Table 14.1 Huckaback cotton wound dressing

Particulars	Area density (g/m^2)	Thickness (mm)	Bulk density (g/cm^3)
Grey fabric	149	1.2	0.12
Scoured and bleached	212	1.4	0.15
Scoured, bleached and non-adherent finished	202	1.3	0.16

Table 14.2 Absorbency of woven cotton gauze dressings

Number of ply	Initial weight (g)	Final weight (g)	Liquid absorbed (g)	Absorbency (g/g)	Absorbency (g/25cm^2)	Absorbency (g/100cm^2)
1	0.0425	0.2572	0.2147	5.1	0.2	0.9
2	0.0819	0.5586	0.4767	5.8	0.5	1.9
3	0.1257	0.9294	0.8037	6.4	0.8	3.2
4	0.1661	1.3660	1.1999	7.2	1.2	4.8
5	0.2113	1.8840	1.6727	7.9	1.7	6.7
6	0.2424	2.1988	1.9564	8.1	2.0	7.8
7	0.2924	2.8201	2.5277	8.6	2.5	10.1
8	0.3309	3.2381	2.9072	8.8	2.9	11.6

- soft and easy to handle, so no gelling in the dry state; they can also be used on dry wounds, unlike alginate dressings;
- maintains moist environment at the wound site that promotes healing;
- enhances the absorption of body fluid;
- softening agent serves to soften the fabric and also increases the absorption capacity of the wound dressing; and
- free from toxic, irritant and allergic characteristics.

In addition to the above woven gauze wound dressing, some research into wound management has already been carried out using woven gauze. A noteworthy piece of research includes the development of a novel cotton gauze that absorbs neutrophil elastase present in choronic wounds.[8] It is known that the success of post-operative treatment for split-thickness skin graft (STSG) depends on maintaining graft integrity, preventing graft and wound desiccation and minimising infections. Researchers applied one layer of woven mesh gauze dressing (Xeroform) impregnated with bismuth tribromophenate onto STSG followed by layers of dry gauze dressings, and observed the growth of the graft after five days. The results revealed that minimal post-operative nursing care is sufficient and this is not possible in the case of other dressings for skin grafts.

It should be noted that there are numerous types of wound dressings available for the management of different kinds of acute to chronic wounds (Table 14.3), but typically a modern wound dressing is composed of absorbent layers held between a wound contact layer and a base material (Fig. 14.2). The absorbent layers absorb blood, body fluids and exudate. The wound contact layer is low-adherent and can easily be removed without disturbing new tissue growth. Generally the material is placed directly over the wound (primary dressing) and covered with an absorbent pad (secondary dressing) and the whole dressing is retained with a tape or a suitable bandage depending on the location of the wound in the body. The primary dressing is expected to maintain the wound temperature, moisture level, permits respiration and

Table 14.3 Illustration of commercial wound dressings and their applications

Commercial name	Major application
Low-adherent dressing Adaptic, Release, Melolin, Telfa, Skintact, Mepore, Primapore, Tricotex, NA Ultra, Jelonet, Paranet, Paratulle, Unitulle, Vasaline gauze, Inadine, Sofra-Tulle, Silicone NA, Mepitel, Metalline	Reduces risk of adherence to wounds. Less trauma
Semipermeable film dressing Mefilm, Tegaderm, Bioclusive, Cutifilm, Epi View, Opsite Flexigrid	Lightly exuding wounds, superficial pressure sores, primary dressing and secondary dressing in combination with alginates and hydrogels
Odour adsorbing dressing Actisorb Plus, Carbonet, Kaltocarb, Metrotop, Cliniflex	Undesirable odour producing wounds
Hydrocolloid dressing Aquacel, Granuflex, Comfeel, Combiderm, Cutinova Foam, DuoDerm, Tegasorb, Hydrocoll	Light to medium exuding wounds. Not suitable for infected wounds
Hydrogel dressing AquaForm, Sterigel, Purilon Gel, Intrasite Gel, Granugel Hydrocolloid Gel	Dry and necrotic wounds. Lightly exuding wounds and granulating wounds. Not suitable for infected and heavily exuding wounds
Polyurethane foam dressing Allevyn, Tielle, Lyofoam, Sterigel, Nu-Gel, Purilon Gel, Intrasite Gel, Granugel, Flexipore, Spyrosorb	Light to medium exuding wounds. Not recommended for dry superficial wounds
Alginate dressing Sorbsan, Tegagel, Kaltostat, Algosteril, Algisite, Algoderm, Melgisorb, Kaltogel, Tegagel, Algosteril and Comfeelseasorb	Mainly primary dressing for medium to heavily exuding wounds and cavity wounds. Not suitable for dry wounds
Alginate-collagen dressing Fibracol	Suitable for dressing foot ulcers and heel pressure sore

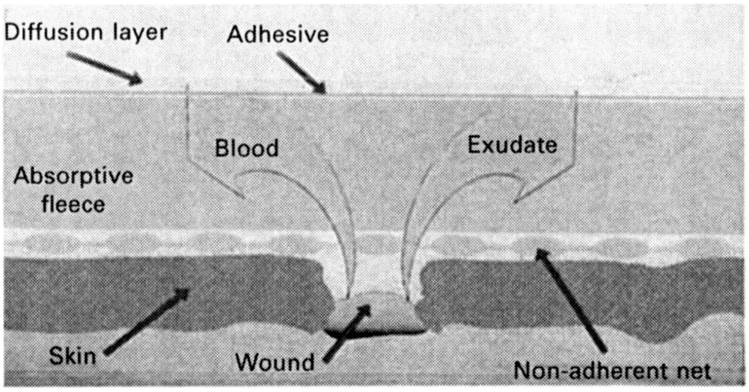

14.2 Wound dressing concept.

allows epithelial migration. The secondary dressing must not be too absorbent as it may cause the primary dressing to dry out too quickly.[9]

14.2.5 Woven wound bandages

Bandage types

Bandages can be used for many purposes including dressing retention, support and compression;

- retention bandages are used to retain dressings in the correct position;
- support bandages provide retention and prevent the development of a deformity or change in shape of a mass of tissue due to swelling or sagging; and
- compression bandages are employed mainly for the treatment of leg ulcers and varicose veins.

Modern bandages are either woven or knitted and are designed to provide prescribed levels of compression in accordance with specified performance-based standards[10] (Table 14.4).

Compression bandages are harmful if not applied properly. They provide high tension as well as high pressure. A thorough assessment involving several criteria is therefore essential before applying a compression bandage on a limb. For example, it is important to consider the magnitude of the pressure, the distribution of the pressure, the duration of the pressure, the radius of the limb and the number of bandage layers. Illustration of bandages used in compression therapy is published elsewhere.[11]

Bandages are selected according to their level of performance rather than their construction. An ideal compression bandage should:

- provide compression appropriate for the individual;
- provide pressure evenly distributed over the anatomical contours;
- provide gradient pressure diminishing from the angle to the upper calf;
- maintain pressure and remain in position until the next change of dressing;
- extend from the base of the toes to the tibial tuberosity without a gap;
- function in a complementary way with the dressing; and
- possess non-irritant and non-allergenic properties.

Woven structures have been in use for designing compression bandages for venous leg ulcer management for a long time. A patent describes a narrow woven compression bandage constructed from cotton weft yarns and textured stretch nylon multifilament warp yarns having total deniers in the range of

Table 14.4 Compression properties of bandages

Bandage type	Selected leading brand names	Remarks
Retention bandage	Easifix, Slinky, Stayform, Tensofix	Exerts very little pressure on a limb
Support bandage	Crepe BP, Elastocrepe	Prevents formation of oedema and support joints. Exerts various amounts of pressure, according to the type, on a limb
Light compression (3a)	J-Plus, K-Crepe	Gives sub-bandage pressures of 14–17 mm Hg at the ankle
Moderate compression (3b)	Granuflex Adhesive Compression, Veinopress	Gives sub-bandage pressures of 18–24 mm Hg
High compression (3c)	Setopress, SurePress, Tensopress	Gives sub-bandage pressures of 25–35 mm Hg
Extra high compression (3d)	Bilastic Forte, Blue Line Webbing	Gives sub-bandage pressures of up to 60 mm Hg
Paste bandage	Icthaband, Quinaband, Tarband, Zincaband	Woven cotton fabric impregnated with a medical cream or paste. Used for the treatment of eczema and dermatitis
Tubular bandage, elasticated	Tubifast, Tubigrip	Dressing on awkward sites
Foam padded	Netelast, Tubipad	Provides padding and protection against physical damage

210 to 560. The construction consists of from 10 to 16 picks per inch and from 30 to 40 warp yarns per inch. Spaced apart across the width of the fabric are pairs of warp yarns interwoven with the weft yarn in a leno construction that provides high frictional resistance to slippage between those warp yarns and the weft yarns. The ratio of leno pairs to non-leno warp yarns is in the range of from 1/6 to 1/14. The non-leno warp yarns and the weft yarn are interwoven in a 1 × 1 weave. The leno pairs are in a pretensioned state when the bandage is in an unstretched condition.[12] Elastomeric yarns such as spandex are used at intervals across the width of the woven bandage with inelastic cotton yarn.[13]

A woven short stretch crepe bandage made from 100% high twist cotton yarns is used for the treatment of varicose veins. It provides compressive post-surgery or light orthopaedic support. It is also used for support to joints and ligaments after mild strains. ProGuide (Smith & Nephew) is a two-layer bandage system in which the first layer is a nonwoven and the second layer is a woven elastic compression bandage coated with a pressure sensitive hot melt adhesive which is applied in a ladder pattern to reduce slippage and keep the system in place for extended periods.

As described above, support bandages provide retention and prevent the development of a deformity or change in shape of a mass of tissue due to swelling or sagging. These can be either flat, tubular or shaped bandages and can be used either on their own, or as a part of a series of bandages for the management of venous leg ulcers.

The specifications and manufacturing details of a woven support bandage Profore No. 2 Flow Wrapped Bandage (10 cm wide × 3 m long) are given below:

Loom specifications:	Bonas 2/110 or 2/175
Number of heald shafts:	2
Latch needle type:	7762G07
Front reed gauge:	7.2 dents per cm
Yarn specifications:	Warp: Beams 1 and 2; 2/78 dtex textured polyamide filament twisted with 20 tex ecru cotton, 125 turns per metre Weft: 36.9 tex (1/16 Ne) unbleached cotton yarns Catch thread: 1/44 dtex/13 filaments polyamide yarn
Weave:	Plain
Ends per bandage:	152
Picks per cm:	7.0 ± 2
Fabric width:	10 cm
Fabric area density:	80 g m^{-2}
Cut length:	3.0 m
Stretched length:	4.5 m
Bandage performance:	BS 7505: 1995
Bandage type:	Type 2
Tension ratio (%) S:	≥40
Working extenson (%) E:	≥20
Differential extension factor (%) F:	≥1.5

It has been highlighted that tightly woven fabric structure creates an ideal environment for the growth of microorganisms by maintaining a moist and warm environments.[14]

14.3 Woven vascular prostheses and meshes

Textile-based vascular grafts are widely accepted as one of the most important medical devices to manage vascular diseases. In 1952, Voorhess *et al.*[15] successfully deposited fibrin within the interstices of the fabric and found that ingrowth of fibroblast from adjacent soft tissue occurs first and then endothelium migration would follow. This finding led researchers to further research to develop vascular prostheses. The first commercial attempt was made by Edward[16] in 1955 using a braided polyamide and subsequently replaced polyamide with PTEE in the form of Teflon in 1957. PET was introduced in both warp and weft in 1961.[17] Today, implants are made from a variety of synthetic fibres. The majority include polyethylene terephthalate (PET), expanded PET (ePET), polytetrafluoroethylene (PTFE), expanded polytetrafluoroethylene (ePTFE) and polyurethane. However, PET, ePET and PTFE are the most commonly used fibres in commercial vascular prosthesis. The ePTFE graft is non-porous and popularly used for femoro-popliteal and axillo-femoral by-passes. The most common commercial grafts are listed[18] in Table 14.5.

An artificial vascular prosthesis was first experimented in dogs[19] and subsequently it was implanted in humans to replace an aneurysm (dilation) of the abdominal aorta.[20]

The essential criteria for an ideal vascular graft are:[21]

- mechanical strength and compliance to withstand long-term hemodynamic stresses;
- non-toxicity, non-immunogenicity, biocompatibility, 'off-the-shelf' availability in various sizes for emergency care;
- operative suturability and simplicity of surgical handling;
- resistance to *in vivo* thrombosis;
- ability to withstand infection; and
- complete incorporation into the host tissue with satisfactory graft healing and ability to grow when placed in children.

Table 14.5 Commercial vascular grafts

Manufacturer	Material
Bard	PET (woven, weft knit, velour weft and warp knit)
Meadox	PET (woven, warp knit, velour warp knit, umbilical vein)
Golaski	PET (weft knit, velour weft knit)
Sorin Biomedica	PET (carbon coated woven and knitted double velour)
Rhone Poulenc	PET (velour warp knit)
Coates-Paton Intermedia CS	PET
Gore	ePET
Impra	ePET
Johnson & Johnson	ePET, bovine carotid
St Jude Medical	Bovine carotid
Socol	Bovine carotid
Genetics Lab	Umbilical vein

In addition to the above, it is essential to consider parameters such as biocompatibility of fibres in the graft, non-fraying textile structure, flexibility, durability and resistance to sterilisation. The fibrous structure of the graft must have sufficient porosity to allow tissue growth as well as the formation of a thin fibrin-based blood clot resistance layer (non-thrombogenic) on the inner surface of the graft. The porosity of the commercial grafts is relatively constant and does not change with time because the grafts are constructed by using a plain non-absorbable polyester (Dacron).

It should be noted that Dacron (Du Pont) graft has been one of the preferred choices since it was introduced in 1950s. The original grafts were woven constructions but subsequent development of manufacturing technology led to the production of knitted and velour grafts. Vascular prostheses are either woven or knitted. Woven grafts exhibit a high degree of stability in both warp and weft directions, high strength and low porosity (50–200 ml/min/cm^2). Because of their low porosity, the prostheses are used without precoating in large-calibre, high-flow arteries.[22] Since woven structures provide greater strength and stability, they are therefore more suitable for high stress locations such as the thoracic aorta. Woven structures also provide high dimensional stability, low permeability to blood and less kinking. The main disadvantages of plain and associated woven structures for use as vascular fabrics are that they fray at cut edges, have low healing porosity, poor compliance and difficulty in suturing. However, velour woven fabrics, which are made by reducing the number of intersections along both warp and weft directions, are more flexible and produce loose surface structures which promote tissue attachment and ingrowth that conventional woven fabrics do not have.

Knitted prostheses are either single jersey or warp knitted velour structures. In knitted structures, the strength and porosity of the graft are influenced by loop configuration. Knitted grafts generally have a higher porosity (2,000 ml/min/cm^2) and lower strength than the woven ones. The average porosity of the graft is described by the internodal distance (IND) and the optimal IND of 60 μm (high porosity) is experimentally proposed for tissue ingrowth and endothelialisation of 4 mm ePTFE grafts.[23] High pore grafts are pre-coated with collagen or albumin or gelatine to decrease the porosity/permeability and to avoid the need for blood pre-clotting prior to implantation. In the early years of development, it was a common practice to pre-coat the graft with human blood before implantation to prevent blood leakage. However, the development of a new technique has eliminated the need for pre-coating the prosthetic tube. The technique involves:

- the immersion of the tube in water;
- draining the excess water from the inside surface; and
- coating the inner walls, making use of a suitable biocompatible elastomer.

The coating material is usually a copolyester-based polymer. However, collagen, albumin, chitosan and elastane derivatives are also used to coat the graft.[22] Carbon is used for coating to improve the thermogenic features of Dacron prostheses.[22]

Unlike knitted grafts, woven grafts do not rely on looping of the yarn around a needle and the grafts are therefore more compact. However, the stiffness of the woven structure makes the graft more difficult to handle and suture during implantation. The intraoperative handling qualities of knitted grafts are preferred by surgeons although they possess larger pores than the woven grafts. Velour construction increases the growth of tissues and closely-woven double-velour synthetic vascular grafts are less porous and facilitate rapid tissue ingrowth. Gelweave Valsalva (Vascutek) is the woven polyester root graft that is used for repair and replacement of damaged and diseased thoracic aorta (Fig. 14.3).[24]

The performance of woven and collagen-impregnated knitted arterial prosthetic grafts has been directly compared. A randomised trial was conducted using 141 and 126 patients who received woven and knitted grafts, respectively. The graft handling qualities, operative blood loss, associated complications and potency rates were assessed on discharge and at one year by duplex imaging technique. The study reveals that both woven and knitted grafts have similar clinical performance but less blood

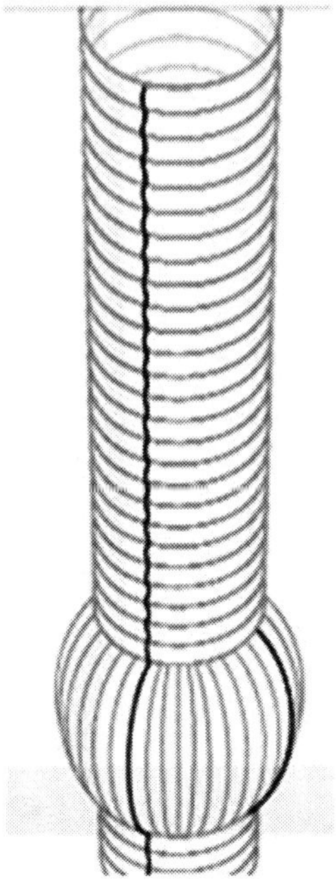

14.3 Typical example of a woven polyester graft.

loss is observed in knitted graft perhaps due to the collagen coating. After one year of implantation, a reduction in diameter of the woven grafts and an increase in diameter of the knitted grafts are observed. The study further suggests that less expensive woven graft may be the choice of graft unless haemorrhagic complications are anticipated. It does not support the view that bifurcated prostheses made of knitted and impregnated Dacron graft have better patency equivalents.[25]

It is established in another study involving biophysical, histopathological and ultrastructural aspects that the fibre structure and resistance to degradation are better in woven grafts than the knitted grafts.[26] The stability of the structure of woven graft helps to withstand more resistance to stress than the knitted graft where the loose structure is more prone to deterioration. Scott *et al.*[27] studied *in vivo* the dilatation of 74 woven and 69 knitted aortic grafts and found that knitted and woven grafts suffered a dilation of 49% and 28%, respectively. CT scan evaluation also reaffirmed the finding that the dilation of knitted grafts is higher than the woven grafts.[28] A two-year CT scan study involving woven polyester prosthetic grafts (Gelweave) and knitted grafts (Gelseal) concluded that, in contrast to the woven prostheses, the knitted prostheses show an early and slow late dilation.[29] The merits and demerits of PET woven and knitted grafts, and also the ePTFE and polyurethane based grafts (Table 14.6) are published elsewhere.[30]

In order to enhance the functional properties of vascular prostheses in the *in vivo* environment, woven vascular fabrics were developed using both the non-absorbable (polyester) and absorbable (polyglycolic acid; PGA) yarns.[31] Table 14.7 shows the fabric characteristics and properties of such bicomponent vascular fabrics.

Yu and Chu[31] found that the incorporation of absorbable yarns in the weft of a woven bicomponent vascular graft promotes faster tissue growth by being loose and porous. On the other hand, the absorbable yarns in the warp do not exhibit this unique surface. The following advantages are derived from the bicomponent vascular fabrics containing non-absorbabale and absorbable yarns:

- soft;
- adequate strength and stability;
- high integrity;
- increase in porosity; and
- enhanced tissue ingrowth.

A woven graft for the repair of an abdominal and aortic aneurysm, weakened wall of the aorta between the renal arteries and the bifurcation of the iliac arteries has been developed.[32] The thin polyester graft fabric has a double wall thickness of less than 0.51 mm and can be woven into a seamless tube or a woven fabric sheet can be formed into a cylinder and sewn to provide

Table 14.6 Merits and demerits of woven and knitted grafts

	PET (Dacron, Terylen)		ePTFE (Teflou, Gore-Tex)		Polyurethane	
	Woven	Knitted	Low-porosity (<30 μm IND)	High-porosity (>45 μm IND)	Fibrillar	Foamy
Advantages	Better stability, lower permeability and less bleeding	Greater prorosity, tissue ingrowth and radial distensibility	Biostability, no dilation over time	Biostability, better cell ingrowth	Compliance, good hemo-and biocompatibility, less thrombogenicity	
Disadvantages	Reduced compliance and tissue incorporation, low porosity, fraying at edges, infection risk	Dilation over time, infection risk	Stitch bleeding, limited incorporation infection risk, perigraft seroma formation	Late neointimal desquamation in 90 μm IND, infection risk	Biodegradation in first generation, infection risk, carcinogenic?	
Healing	Inner fibrinous capsule, outer collagenous capsule, scarce endothelial islands	Fibrin luminal coverage, very sporadic endothelium, transanastomotic endothelialisation in animals	Luminal fibrin and platelet capet, connective tissue capsule with foreign body giant cells, no transmural tissue ingrowth	Macrophages and polymorphonuclear invasion, capillary sprouting, fibroblast migration, certain angiogenesis, thicker neointima, endothelialisation in animals	Thin inner fbrin layer, outside foreign body cells, limited ingrowth	Better ingrowth with bigger pores

Table 14.7 Fabric specifications and properties of woven vascular fabrics

Fabrics	Ends/inch	Picks/inch	Thickness (cm)	Weight (g/cm^2)	Fabric density (g/cm^2)	Bending stiffness (mg-cm)	Bursting strength (kg/cm^2)	Water permeability (mL/cm^2/min/mm Hg)
W1	168	104	0.034 ± 0.001	0.0205 ± 0.0026	0.6029	G_{x1} = 2640 G_{x1} = 1430	15.0	300 ± 25
W2	78	128	0.050 ± 0.002	0.0300 ± 0.0022	0.6000	G_{x1} = 823 G_{x1} = 670	11.61	650 ± 70
W3	78	136	0.053 ± 0.002	0.0360 ± 0.0017	0.6792	G_{x1} = 420 G_{x1} = 810	13.51	880 ± 20
W4	156	128	0.052 ± 0.008	0.0300 ± 0.0022	0.5769	G_{x1} = 1045	22.3	400 ± 10
DeBakey woven	135	94	0.025	0.017	0.68	G_{x1} = 1895 G_{x1} = 1345	26.3	250 ± 3

the desired shape. It is claimed that the ends of the graft minimise the formation of gaps between the implant and the vessel wall that enhances tissue in-growth and provides a smooth transition for blood flow.

14.4 Application of woven structures in hospitals

14.4.1 Infection control

Woven textiles, specifically plain and twill weaves, are largely used in hospitals as reusable materials, such as patient's bed sheets and curtains. Their contribution in spreading infection is enormous. Textiles in any form used in hospitals are susceptible for bacterial growth under appropriate moisture and temperature conditions. Patients shed bacteria and contaminate their pyjamas and sheets. The temperature and humidity between the patients and the bed are appropriate conditions allowing for effective bacterial proliferation. Several studies have found that human contact with contaminated textiles is the source of transmission of microorganisms to susceptible patients.

It must be pointed out that infection control is a growing problem in places where hygiene is required and most particularly in hospitals. It is known that microorganisms which include bacteria, virus and fungi create and aggravate problems in hospitals and other environments by transmitting diseases and infections through clothing, bedding, etc. Patients in hospitals are more prone to infection because of their illness and the hospital acquired infections are among the top ten leading causes of death. In spite of a great deal research, clinicians are still facing problems with 'super-bugs' such as Methicillin resistant *Staphylococcus aureus* (MRSA) and Methicillin susceptible *Staphylococcus aureus* (MRSSA) bacteria in hospitals. When antibiotics are used incorrectly (for example, to treat colds and flu) and too frequently, bacteria will adapt and become resistance to antibiotics. It is well established that microorganisms create and aggravate problems in hospitals and other environments by transmitting diseases and infections through clothing, bedding, etc. (cross-infection). In this situation, it is essential that textile materials used in hospitals have the capability to eradicate/minimise cross-infection

14.4.2 Antimicrobial activity

With a view to develop antimicrobial textile materials, considerable research has already been carried out making use of organic and inorganic compounds, antibiotics, heterocyclics, quaternary ammonium compounds, etc. The biocidal properties of silver compounds have been known for thousands of years, and have been increasingly used nowadays to impart antibacterial properties to textile materials for hospital use. It is interesting to note that the crosslinking agent, dimethylol-5,5-dimethylhydantoin (DMDMH), commonly used in the wet processing industry to improve functional properties of textile fabrics, possesses a certain level of antibacterial effect on plain woven cotton and polyester/cotton (65/35) plain woven fabric[33] (Table 14.8).

The use of natural products as potential antimicrobial agents on textiles has currently received much attention due to the awareness of environmental issues. It is stressed that, although the synthetic antimicrobial agents are effective against a range of microorganisms and provide a durable effect on textiles, they possess limitations in use such as associated side effects, action on non-target microorganisms and water pollution. Longer durability of antimicrobial effects can be achieved by imbuing antibacterial agents into the structure of fibres rather than depositing them on the surface. It is important that antimicrobial textiles provide protection against a wide range of Gram-positive and Gram-negative bacteria that include 'super-bugs'.

Research on 100% cotton plain weave fabric finished with a synergistic system of formulation comprising inorganic chemicals involving a metal salt of a monocarboxylic acid, a carbamic acid derivative, a chelating agent, a boron compound, a dimethylene siloxane derivative and an alkane polymer has shown that the treated fabrics imparted antimicrobial activity in arresting the growth of several bacteria (Gram-positive and Gram-negative), fungi and mildew.[34]

Table 14.8 Antibacterial activity of woven fabrics treated with dimethylol-5,5-dimethylhydantoin (DMDMH)

Fabric	Microorganism	Log reduction of bacterial challenge	
		2% DMDMH	6% DMDMH
Cotton	*E. coli*	6	6
Cotton/PET		6	6
Cotton	*Staph. aureus*	6	6
Cotton/PET		6	6
Cotton	*Salmonell*	6	7
Cotton/PET	*choleraesuis*	7	6
Cotton	*Shigella*	6	6
Cotton/PET		6	7
Cotton	*Candida albicans*	2	6
Cotton/PET		6	6
Cotton	*Brevibacterium*	8	8
Cotton/PET		8	8
Cotton	*Pseudomonas*	6	6
Cotton/PET	*aeruginosa*	6	6
Cotton	*Methicillin-resis.*		3
Cotton/PET	*Staph. aureus*		6
Cotton	*Vancomycin resis.*		6
Cotton/PET	*Enterococcus*		6

Bed linens, patients' gowns and staff aprons were tailored using both the treated and untreated fabrics and were put to use in post-operative, gynaecology and labour wards of a reputed hospital. The presence of bacteria on treated and untreated items was tested after several use-wash-use cycles. It is observed from the results after 50 cycles (Table 14.9) that while the untreated samples are rich in some types of bacteria, the treated ones are almost devoid of them. It should be mentioned that the users did not experience any discomfort such as skin irritation, disagreeable odour and unpleasantness during the trial. The treatment also prevents the deterioration of fabrics by microorganisms (Figs 14.4 and 14.5).

Plain and twill weave woven fabrics (Table 14.10) are used as reusable operating room (OR) surgical gowns.[35] Microbial barrier effect has been achieved by using microfilament polyester as well as controlling the pore size in the woven structure. It is established that the barrier efficiency of the woven fabrics depends directly on the arrangement of the filaments in the yarn and the construction of the woven fabric.

Recent research indicates that a desized, scoured and bleached plain weave cotton fabric weighing 130 g m^{-2} finished with natural antimicrobial agent, neem seed, showed 99.5% antibacterial activity against *Staphylococcus aureus*.[36] It should be mentioned that neem seed is obtained from neem tree, *Azadirachta indica*, which is abundantly found in the Indian subcontinent. It has an excellent potential as an antimicrobial agent and its main constituents such as azadirachtin, salannin and meliantriol are proven insect growth regulator and antifeedent.[37] It is observed that particle size of the natural polymers influences the antibacterial activity.[38] The antibacterial activity of normal chitosan, nanochitosan and silver loaded nanochitosan applied on woven polyester fabrics shows that chitosan gets much enhanced antibacterial activity in nanoparticle form, as indicated by reduction in minimum inhibitory concentration (MIC) from 0.5% to 0.01%. The silver loaded chitosan nanoparticle shows a further increase in activity (MIC

Table 14.9 Antibacterial activity of 50 times washed woven fabrics: hospital trial

Wards	Organisms isolated from aprons, gowns and linen	Observation (growth of organisms)	
		Untreated	Treated
Post-operative	*Escherichia coli*	Moderate	Nil
	Klebsiella aerosens	Moderate	Nil
	Staphylococcus pyogens	Heavy	Insignificant
Gynaecology	*Escherichia coli*	Moderate	Nil
	Staphylococcus pyogens	Moderate	Nil
	Pseudomonas pyocynans	Moderate	Nil
Labour	*Escherichia coli*	Heavy	Heavy
	Klebsiella pneumoniae	Heavy	Heavy
	Staphylococcus pyogens	Heavy	Insignificant

14.4 Untreated woven fabric on storage.

14.5 Antimicrobial treated woven fabric on storage.

0.001%) due to synergistic effect of Ag^+ and chitosan nanoparticles. These particles additionally show a release mechanism as evident from a clear zone of inhibition (Figs 14.6(a)–(d)).

Recent research described the antimicrobial activity of bioactive-treated fabric (BTF) that contains silver for use in the hospital environment.[39] Unlike other biocides used in hospital fabrics, the prolonged use of silver has not been related to the appearance of resistant bacteria or cross-resistance

Table 14.10 Specifications of plain and twill weave operating room (OR) surgical gowns

Sample	Type of weave	Fineness of filament in dtex		Cross-section of filament		Number of filaments in the yarn		Fineness of filament yarn in tex		Yarn density /10cm		Fabric density (g cm^{-3})
		Warp	Weft	Warp	Weft	Warp	Weft	Warp	Weft	Warp	Weft	
P 4	Plain	0.85	0.85	Deformed	Deformed	112	102	9.5	8.5	456	370	0.55
P 5	Plain	2.60	1.25	Triangular	Round	48	198	13.0	25.0	572	313	0.98
P 6	Twill 2/1	0.60	1.35	Round	Round	206	69	9.5	12.5	458	362	0.37

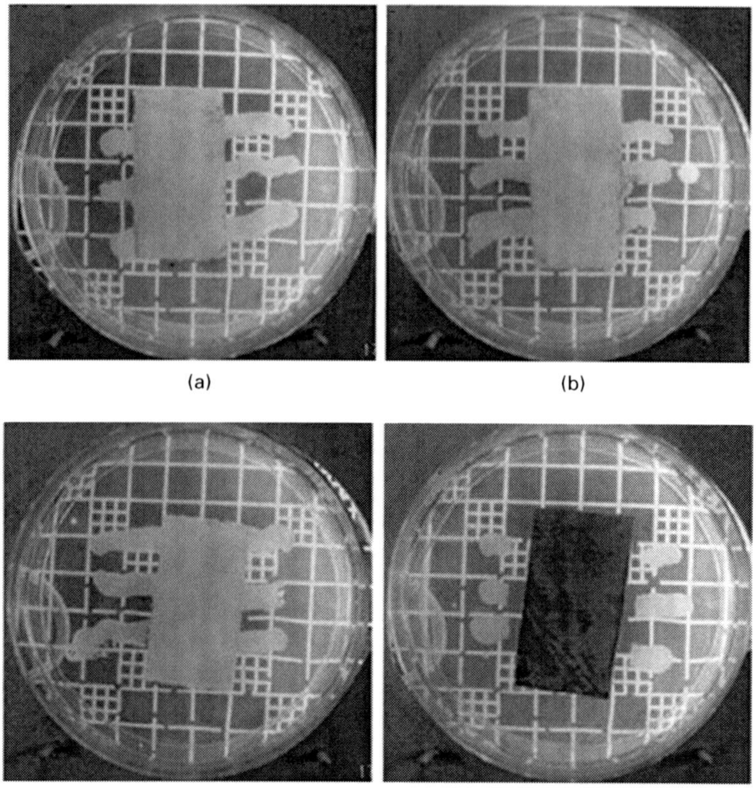

(a) (b) (c) (d)

14.6 (a) Untreated polyester woven fabric; (b) chitosan treated polyester woven fabric; (c) nanochitosan treated polyester woven fabric; and (d) nano silver-chitosan treated polyester woven fabric.

to antibiotics, in spite of being extensively used in some treatments. The antibacterial activity of the treated fabrics was tested against 33 hospital strains and a significant reduction was found in the number of bacteria present on the BTF. The physical, mechanical, moisture and vapours transmission and water repellence properties of antibacterial cotton/amicor woven fabrics containing different woven structures and that are intended for use as antimicrobial hospital sheets are published elsewhere.[40] It will be borne in mind that amicor is an acrylic based fibre into which organic antibacterial and antifungal additives are imbued.

14.5 Other medical applications of woven structures

The University of Bolton has developed novel reusable incontinence products for the benefit of the elderly community. The product specifications are:

Face fabric:	Comfortable, dry, stable, good even appearance.
Soaker material:	Highly absorbent, stable, washable at 95°C up to 200 washes/tumbling cycles.
Barrier material:	Non-leaking, light, stable, non-shrinking.
Non-slip material:	Good surface friction, good even appearance.
1. Wings:	Plain woven polyester.
2. Face fabric:	Loop raised warp knit polyester (125 g m^{-2}).
3. Soaker fabric:	Neddle-punched 60% polyester/40% viscose.
4. Barrier fabric:	PVC coating (500 g m^{-2}).
5. Non-slip fabric:	Weft knitted polyester (60 g m^{-2}).

A number of woven structures are being used as electronic conductive materials by integrating sensors for biomedical applications. Sensors integrated in clothing ensure a non-invasive measurement method without interfering in the human body. A company in Italy, Soliani SML, has recently developed an electroconductive woven structure by using silver for ECG measurement.[41] Meshes are implanted mainly to reinforce soft tissues. The application includes:

- hernia repair (abdominal, inguinal, diaphragmatic, epigastric, gastroesophageal, hiatal, intermuscular, mesenteric, paraperitoneal, rectovaginal, retrocecal, uterine and vesical);
- abdominal wall repair;
- prolapsed tissue support/repair;
- perforated tissue repair; and
- general tissue repair (pelvic floor, bladder and thoracic wall).

In addition, the use of mesh grafts on humans also includes protection of esophagogastrostomy, treating a damaged kidney by external splinting or encapsulation, repairing pelvic peritorium and replacing the membrane covering the brain. In repairing/reinforcing abdominal wall defects, polyester (Mersilene), polypropylene (Marlex), ePTFE (Gore-Tex), PGA (Dexon) and polyglactin (Vicryl) have been used for many years. However, none of these materials has proved entirely satisfactory for repairing large or difficult defects. No single mesh graft has gained universal acceptance.

Woven, nonwoven and knitted structures are used in surgery for the repair, reconstruction or substitution of tissues. The utilisation of mesh grafts in humans is based on the fact that, during the absorption period, a neomembrane is formed in the site where the mesh has been implanted. Initially animals were employed for *in vivo* studies. A polyglactin (Vicryl) mesh placed in the thoracic aorta of a pig as a patch graft facilitated the early growth of muscle cells around the mesh and subsequently the mesh was completely absorbed in 40 days. It has been recently reported that the tensile strength of Vicryl woven mesh has increased at 30 days after being implanted in rats.[42] An illustration of a mesh implant is shown[43] in Fig. 14.7.

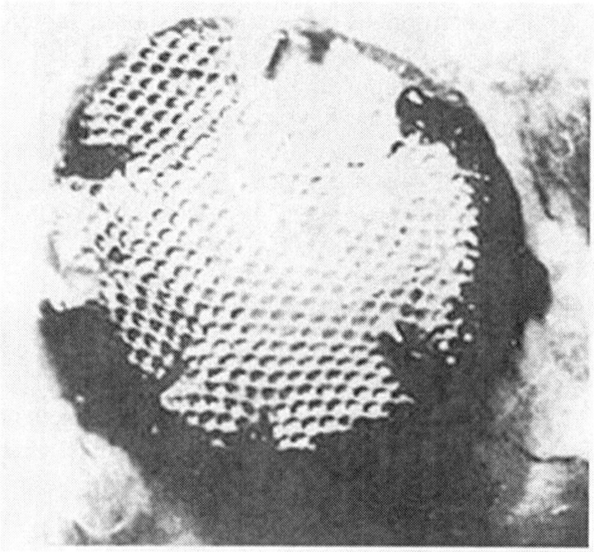

14.7 Polyester-carbon composite mesh implant after 5 weeks of implantation.

The combination of non-absorbable PTFE, a copolymer of hexafluoroethylene and tetrafluoroethylene fibres encased in a bioabsorbable polymer composite mesh has been utilised for the repair of defects in the abdominal wall.[44] It was demonstrated that the non-absorbable component made of knitted, woven or nonwoven structures serves as a reinforcement material in preventing hernia formation.

Heim and Gupta[45] have discussed the application of woven fabric for designing textile heart valve prostheses. They studied the long-term fatigue behaviour of polyester woven fabric for engineering heat valve prosthesis. The findings of the study may pave the way to develop criteria for designing a highly suited woven fabric for heart valve application. It is demonstrated during the development of a special woven polyester fabric for heart valve prosthesis that the woven structure gives the heart valve tissue a macroscopic behaviour that closely resembles the collagen-elastin heart valve tissue structure.[46] The effect of cyclic mechanical stimulation on a composite structure that consists of a woven fabric scaffold and tendon fibroblasts has been studied *in vitro* to overcome the complications associated with mechanical and structural incompatibility between implanted devices and host tissue.[47] A novel low permeable woven polyester fabric (Bard DeBakey) has been developed for use in outflow tract repairs, patch graft angioplasty and septal defects. The woven polyester patch was utilised to develop a new technique which promotes active exercise immediately after surgery without cast immobilisation.[48]

14.6 Conclusions

It is demonstrated in this chapter that the application of woven textiles and their products in the healthcare and hygiene sectors is highly significant. Conventional weaving technology still plays a crucial role in designing and engineering hi-tech medical devices such as prosthetic grafts and heart valves. The application of woven textiles in wound dressings, bandages, implant products and meshes has been critically appraised. Medical products for hygiene and healthcare uses have also been discussed. The contribution of woven structures for use in hospital fabrics is well recognised, although nonwovens are equally used as disposable products. Commercial names of some of the medical products and their corresponding uses have been tabulated which may serve as 'ready-reckoners'. This review contradicted the myth that medical application of woven fabrics is limited and they are used only for manufacturing cotton gauze for wound management. With the arrival of novel woven technologies such as 3D weaving, it is expected that the contribution of woven structures in medicine will be enormous.

14.7 References

1. Market Overview, *Healthcare*, India Brand Equity Foundation (IBEF), October 2007.
2. Anon. India's medical textile market expands, *Medical Textiles*, December 2007, 11.
3. Report, Centre for Exploitation of Science and Technology (CEST) Demographic Factbook, CEST, London, 2000.
4. Morrell, C.J. *et al.* Cost effectiveness of community leg ulcer clinics: randomised controlled trial. *British Medical J*, 1998, 316, 1487.
5. Rajendran, S., Anand, S.C. Developments in medical textiles, *Textile Progress*, 2002, 32, 4.
6. Chang, H., Wind, S., Morries, M.D. Moist wound healing. *Dermatology Nursing*, 1996, 8: 3, 174–176, 204.
7. Alginate dressings, *British Pharmacopoeia*. London: HMSO, 1995, 1706.
8. Edwards, J.V., Yager, D.R., Cohen, I.K., Diegelmann, R.F. *et al.* Modified cotton gauze dressings that selectively absorb neutrophil elastase activity in solution. *Wound Repair and Regeneration*, 2001; 9:1, 50–58.
9. Foster, L., Moore, P. Acute surgical wound care 3: fitting the dressing to the wound. *British J Nurs.*, 1999, 8:4, 200–210.
10. Rajendran, S., Anand, S.C. Application of textile materials and products in healthcare, *Technical Textiles Market*, 2nd quarter, No. 45, 2001, 25–49.
11. Rajendran, S., Anand, S.C. Hi-tech textiles for interactive wound therapies, in V.T. Bartels (ed.), *Handbook of Medical Textiles*, Woodhead Publishing, Cambridge, 2011, 38–75.
12. Hampton, R., Hanes, Jr., F.P., US Patent 4207885, Woven elastic compression bandage, 1980.

13. Gosh, S., Mukhopadhyay, A., Sikka, M. Compression bandages, in V.K. Kothari (ed.), *Progress in Textiles: Science & Technology*, IAFL Publications, Delhi, 2008, 280–320.
14. Teufel, L., Pipal, A, Schuster, K.C. *et al.* Material-dependant growth of human skin bacteria on textiles investigated using challenge tests and DNA genotyping, *J Applied Microbiology*, 2010, 108: 2, 450–461.
15. Voorhees, A.B., Jaretzki, A., Blackemore, A.H. *Ann Surg*, 1952, 135, 332.
16. Edwards, W.S. Clinical experience with Teflon grafts, in S.A. Wesolowski and C. Dennis (eds), *Fundamentals of Vascular Grafting*, McGraw-Hill, New York, 1970, 367.
17. Noon, G.P., DeBakey, S., Debakey Dacron prothesis and filamentous velour graft, in P. N. Sayer and M. S. Kaplitt (eds), *Vascular Grafts*, Appleton-Century Crofts, New York, 1978, 177.
18. Grigiom, M., Daniele, C., Avenio, G.B. Barbaro, V. Biomechanics and hemodynamics of grafting, in M. Rahman and M.G. Satish (eds), *Vascular Grafts Experiment and Modelling*, WIT Press, Southampton, 2003, 41.
19. Voorhees, A.B. Jr., Jaretzki, A., Blakemore, A.H. The use of tubes constructed from vinyon 'N' cloth in bridging arterial defects. *Ann Surg*, 1952, 135, 332–336.
20. Blakemore, A.H., Voorhees, A.B. Jr. The use of tubes constructed from vinyon N cloth in bridging arterial defects; experimental and clinical. *Ann Surg*, 1954, 140, 324–334.
21. Kakisis, J.D., Liapis, C.D., Breuer, C., Sumpio, B.E. Artificial blood vessel: the Holy Grail of peripheral vascular surgery. *J Vasc Surg*, 2005, 41, 349–354.
22. Grigiom, M., Daniele, C., Avenio, C.B., Barbaro, V. Biomechanics and hemodynamics of grafting, in M. Rahman and M.G. Satish (eds), *Vascular Grafts Experiment and Modelling*, WIT Press, Southampton, 2003, 41.
23. Golden, M.A., Hanson, S.R., Kirkman, T.R., *et al.* Healing of polytetrafluoroethylene arterial grafts is influenced by graft porosity. *J Vasc Surg*, 1990, 11, 838–844.
24. Leaflet, Gelweave™ Aortic Root Designs, Vascutek, Renfrewshire, Scotland, 2011.
25. Quarmby, J.W., Burnand, K.G., Lockhart, S.J.M., *et al.* Prospective randomised trial of woven verses collagen-impregnated knitted prosthetic grafts in aortoiliac surgery, *British J Surgery*, 1998, 85: 6, 775–777.
26. Tardito, E., Blondo, B., Caputo, V., *et al.* Biodegradation of Dacron vascular prostheses: Physico-chemical, histological, morphometric and ultrastructural study. *Minerva Cardioangiol*, 1993, 41, 59–80.
27. Scott, S., Berman, M.D., Glenn, C., Hunter, M.D. *et al.* Application of computed tomography for surveillance of aortic grafts, *Surgery*, 1995, 118: 1, 8–15.
28. Alimi, Y., Juhan, C., Morati, N. Dilation of woven and knitted aortic prosthetic grafts: CT scan evaluation, *Annals of Vascular Surgery*, 1994, 8: 3, 238–242.
29. Mattens, E., Engels, P., Hamerlijnck, R., Kelder, J. Gelseal® versus Gelweave® Dacron prosthetic grafts in the descending thoracic aorta: A two-year computed tomography scan follow-up study, *Cardiovascular Surgery*, 1999, 7: 4, 432–435.
30. Chlupac, J., Filova, E., Bacakova, L. Blood vessel replacement: 50 years of development and tissue engineering paradigms in vascular surgery, *Physiol. Res.*, 2009, 58 (Suppl. 2), S119–S139.
31. Yu, T.J., Chu, C.C. Bicomponent vascular grafts consisting of synthetic biodegradable fibres. Part 1. *In vitro* study. *J Biomed Mater Res.*, 1993, 27, 1329–1340.
32. Anon. *Medical Textiles,* 1999, 6, 7.

33. Sun, G., Xu, X., Bickett, J.R., Williams, J.F. Durable and regenerable antimicrobial finishing of fabrics with a new hydantoin derivative, *Industrial Engineering Chemistry Research*, 2001, 40, 1016–1021.
34. Rajendran, S., Anand, S.C. Development of a versatile antimicrobial finish for textile materials for healthcare and hygine applications, *Proceedings, Medical Textiles '99 Conference, 24–25 August, Bolton, UK*, Woodhead Publishing, Cambridge, 2001, 107–116.
35. Aibibu, D., Lehmann, B., Offermann, P. Qualitative evaluation of the barrier effect of textiles in use, in S.C. Anand, J.F. Kennedy, M. Miraftab and S. Rajendran (eds), *Medical Textiles and Biomaterials for Healthcare*, Woodhead Publishing, Cambridge, 2006, 168–176.
36. Joshi, M., Wazed Ali, S., Rajendran, S. Antibacterial finishing of polyester/cotton blend fabrics using neem (*Azadirachta Indica*): A natural bioactive agent, *J Appl Poly Sci*, 2007, 106, 793–800.
37. Chatterjee, A., Pakashi, S. *The Treatise on Indian Medicinal Plant*, 1994, 3, 76.
38. Wazed Ali, S., Rajendran, S., Joshi, M. Synthesis and characterization of chitosan and silver loaded chitosan nanoparticles for bioactive polyester, *Carbohydrate Polymers*, 2011, 83, 438–446.
39. Mariscal, A., Lopez-Gigosos, R.M., Carnero-Var., M.B., Fernandez-Crehuet, J.B. Antimicrobial effect of medical textiles containing bioactive fibres, *European Journal of Clinical Microbiology and Infectious Diseases*, 2011, 30: 2, 227–232.
40. Harpa, R., Piroi, C., Cristian, I. Study regarding the physical-mechanical properties of cotton/amicor woven fabrics for medical use, *ITC and DC: Book of Proceedings of the 4th International Textile, Clothing and Design Conference – Magic World of Textiles*, 2008, 769–774.
41. Zieba, J., Frydrysiak, M., Tesiorowski, L., Tokarska, M. Textronic clothing to ECG measurement, *MeMeA 2011 – IEEE International Symposium on Medical Measurements and Applications, Proceedings*, 2011.
42. Rice, R.D., Ayubi, F.S., Shaub, Z.J. Comparison of Surgisis, AlloDerm, and Vicryl woven mesh grafts for abdominal wall defect repair in an animal model, *Aesthetic Plast Surg.*, 2010 34:3, 290–296.
43. Minns, R.J. Development of a knitted mesh/woven composite implant for the repair of abdominal defects in the human, in S.C. Anand (ed.), *Proceedings, International Conference on Medical Textiles '96*, Woodhead Publishing, Cambridge, 1997, 102.
44. Anon. *Medical Textiles*, 1992, 8, 6.
45. Heim, F., Gupta, B.S. Textile heart valve prosthesis: The effect of fabric construction parameters on long-term durability, *Textile Res J*, 2009, 79: 11, 1001–1013.
46. Heim, F., Durand, B., Chakfé, N. A new concept of a flexibel textile heart valve prosthesis, *Transactions – 7th World Biomaterials Congress, Sydney, 17–21* May 2004, 1656.
47. Karamuk, E., Mayer, J., Raeber, G. Tissue engineered composite of a woven fabric scaffold with tendon cells, response on mechanical simulation *in vitro*. *Composites Science and Technology*, 2004, 64: 6, 885–891.
48. Ishizuki, M. The polyester patch: a new technique to promote early motion exercises in extensor tendon transfers, *Techniques in Hand and Upper Extremity Surgery*, 2002, 6: 3, 114–118.

Index